BARRY
BALLISTER'S
FRUIT AND
VEGETABLE STAND

BARRY BALLISTER'S

FRUIT
AND
VEGETABLE
STAND

A Complete Guide
to the Selection,
Preparation and Nutrition
of Fresh Produce

The Overlook Press
Woodstock, New York

TO MY FAMILY

First published in the United States by
The Overlook Press
Lewis Hollow Road,
Woodstock, New York 12498

Copyright © 1987 Barry Ballister

Data on "fresh fruits" (pages 74-80)
Data on "fresh vegetables" (pages 145-161)
from FOOD VALUES OF PORTIONS COMMONLY USED
by Jean A.T. Pennington and Helen Nichols Church
Copyright © 1980, 1985 by Helen Nichols Church
and Jean A.T. Pennington
Reprinted by permission of Harper & Row, Publishers, Inc.

Library of Congress Cataloging-in-Publication Data

Ballister, Barry.
Barry Ballister's fruit and vegetable stand.

Includes index.
1. Vegetables   2. Fruit   3. Marketing (Home economics)
I. Title: Fruit and vegetable stand
TX401.B35   1987   641.3'4   86-43061
ISBN 0-87951-272-5

BOOK AND JACKET DESIGN BY JAYE ZIMET
JACKET PHOTO BY FRANK SPINELLI

CONTENTS

INTRODUCTION

*T*here is a scene in *The Godfather* in which Marlon Brando goes shopping at a sidewalk fruit and vegetable stand on Manhattan's West Side. He carefully points out each piece of fruit he wants, and the clerk picks them up and packs them in paper bags. To most people it was the scene before Don Corleone gets machine-gunned down in the street. Actually, it's a scene that gives a lesson in the best way to buy fresh fruits and vegetables: *Select with your eyes, and allow professional hands to handle the merchandise.*

The creation of the self-service supermarket has done more to eliminate sweet, fresh, crisp produce from our stores and lives than any other factor. Most people have no experience in the art of selecting fruits and vegetables. More is not known than what is known about fresh produce, the greatest source of health, beauty, and value in the food world.

As the old-time "fruit men," as we call ourselves, die away, leaving empty stores and sons and daughters with college degrees, the techniques and knowledge of selection, purchase, and preparation of fruits and vegetables fade away as well. "Please no touch the tomatoes" was never meant to scare off trade. It was meant to keep the tomatoes perfect and to respect their delicate nature. The fruit and vegetable stands of the American immigrants, the roadside stands of the American family farm, and the wagons of the American peddler delivered healthier, prettier, sweeter fruits and vegetables than the modern supermarket can ever hope to provide.

Supermarket fruit is almost never sold ready to eat. Varieties are limited to the most commercially acceptable and most resistant to damage. "Self-service" means no service. It also means no information and no access to the buyers, who deal in numbers, not sweetness. The real old-time fruit stands are almost all gone except in the big cities. The latest influx of new immigrants from Asia, however, seems to be revitalizing the concepts of the fresh produce stand. These new Americans, with their backgrounds in agriculture and nontechnological food supplies, seem to be repeating the patterns of retail produce activity established in the early years of this century. Centuries of a cultural background that found beauty in the simple essences of life have produced a new generation of "fruit people" whose presentation of fresh produce is both beautiful and compelling to the customer.

The Farmers' Markets in Los Angeles and San Francisco, the Amish Market in Lancaster, Pennsylvania, and the Haymarket in Boston are the most outstanding examples of fresh produce outlets that stimulate our basic reaction to beautiful food. They draw huge crowds and rack up enormous sales as tourists and locals alike are overwhelmed by dazzling displays of giant red strawberries, golden peaches with rose blushes, lettuce so crisp the leaves crack or so soft they melt like ice cream. Imagine a fruit stand brimming with high pyramids of ripe, colorful fruit, wafting a honeylike aroma and a vibrancy that fairly shouts, "Take me, peel me, eat me!" Imagine one in every small town, one in every neighborhood or, in the cities, one, two, or three on every block. That's the way it was when I was a boy. There were fruits and vegetables of every variety stacked three and four feet high from the back of the store halfway out to the curbstone. And with it all the trucks, and men shouting and laughing, hoisting one-hundred-pound sacks of potatoes *all day long*. Carrots came forty-eight bunches to the crate, green tops and bright orange roots packed solid in crushed ice in heavy wood boxes all the way from California. Now *they* were heavy. Oranges, grapefruits, and lemons came in wood also, long boxes divided into two squares packed so full the top bellied up. And you had better stack 'em sideways so you never bruised a single piece, or the boss put it to you good.

I used to spend hours in the back room prepping salad greens, plating strawberries, weighing potatoes and onions, and studying California fruit-crate labels like Airship and Navajo, Athlete and Paisano. There was one called Camel that depicted a whole caravan crossing a desert, with hooded Arabs, pyramids, and an oasis all in one scene. The colors were both hard and soft, with Maxfield Parrish pinks and misty clouds. It was art like nothing before: bold, colorful, and sexy. There was one brand called Buxom. The label came on California grapefruits and melons. It had a blonde lady in a little blouse and shorts. She had long legs and lots of cleavage. I loved her for years. Later on when I was creating my own fruit stand, I recalled Lady Buxom and painted a price card for some very high-quality California ruby-red grapefruits. I called them Marilyn Monroe grapefruits. "Because they're beautiful, pink, and sweet, and they come from California," I answered my customers. We sold many, many boxes of those wonderful fruits.

The local fruit-stand operators always had a line of chatter going. In between the hawking and calling out for more of this and a crate of that, they had a way of teasing and flirting and selling, selling, selling. A handful of cherries passed around a crowded store, a sweet, ripe honeydew melon cut up in slices, or a few split peach halves can start a buying frenzy, with every customer trying to get into the act. An act it was, full of fun and promises and bags of sweet fruit that were just as good when you got home as when you were laughing back at Marty's or Tony's or Midtown Produce. The old-time clerks would drop a little Yiddish or some Polish something. Italian ladies received lots of pretty phrases and expressions that invoked God and the Virgin Mary over some tomatoes or fresh olives or a box of Zinfandel wine grapes. Of course, every store had its own ethnic flavor, depending on proprietorship and location. There were Jewish stores that specialized in items like fresh-ground horseradish, pickles, and *chav* (sorrel to the *goyim*). Cabbages were kings in the Polish neighborhoods, along with kohlrabi, kale, and potatoes. Italians loved big fruit, mushrooms, tomatoes, basil, and garlic. Every fruit man knew his trade, knew his holidays and holy days, knew how to fix up a basket for sitting shiva or how to send the right thing over to the Machetti house for after the funeral. Petty arguments over pennies would rage for weeks between customers and clerks, but business went on as usual. Serious anger was solved with a little extra on the scale or a bunch of grapes gratis. The people knew what they wanted. They knew what was good, what was fresh, and what was ripe. So did the clerks, each one an encyclopedia of knowledge and experience. They were aware of the produce seasons and when to stop buying and selling plums and start buying and selling apples. The fruit-stand business was ever-changing, spring to summer, fall through winter. New items meant new display stands and different locations for items leaving with the season and those arriving. People shopped every day. Trucks came every day. Every day was new and fresh and exciting. Maybe that's why everything tasted so good.

All that excitement, all that personal involvement, all that ever-changing display of green and gold, red and orange, purple, pink, and earth is gone, replaced by the chrome, mirror, and plastic of the supermarket produce department. Hard, wooden plums, mealy apples in plastic bags, hydro-vacuum dry lettuce wrapped in heat-sealed poly paper, dull green string beans in paper trays, and pale strawberries with white hearts suffer under the fluorescent lights as a thousand hands pinch, poke, and puncture the produce. *At least twenty per cent of all supermarket produce goes in the garbage unsold,* an incredible waste of money, work, and vital food supply.

My father wore his first apron at the age of seven, sweeping up Solly Cohen's fruit stand in downtown Newark. He also shined shoes and sold papers. "Big ship sinks in subway," he used to tell me was his cry. "Read all about it." He could make change, add a column of numbers in his head, read scales and unload trucks by the time he was ten years old. My grandfather came here from Sicily in the early 1900s. (He went back once to get a wife.) Everyone in his family was a farmer or an artisan. His backyard was a garden thick with green vegetables, huge red tomatoes, peach trees, fig trees, and flowers. His cellar was a craft shop where he fashioned furniture, made repairs, and did metal work. This thing my family had for fruits and vegetables and making beautiful displays, it was in our blood from as far back as the Etruscans, the world's great farmers and developers of so many of the fresh foods we eat today. The Phoenicians and the Moors and the Persians crossed the Mediterranean and then crossed their wares, their ideas, and their souls on that island we call Sicily. The foods of the world came to that place, things from as far away as China. Seeds. Plants. Knowledge. We call them oranges, lemons, persimmons, broccoli (which is really a cabbage), zucchini, tomatoes, and apricots. My family and so many others like them brought this heritage here. And together with the American farmer, who cultivated land as no other has ever done before or since, they created a bounty of beautiful food unparalleled in the history of the world. I have spent my whole life among these people.

By sixteen school was over for my father. He worked full-time, all the time, as a "fruit man." He unloaded barrels of apples, 110 pounds each one. Truckloads of melons, freight cars of lettuce, shiploads of bananas. He did it all and knew it all. Every size, every color, every which way you could buy and sell fruits and vegetables he knew. And he had to know because the fruit man next door knew just as much as he did. And the fruit man across the street knew more, maybe. And there was much to know: 48-size grapefruit, 120 tangerines, super-colossal giant garlic, bushels of spinach heavy with a piece of ice in the bottom that weighed more than the spinach itself. By the time he was nineteen, he had a horse and wagon. He was a straight-load peddler, with a wagonload of melons or tomatoes or string beans piled mountain-high behind a horse he rented that morning just outside the market. The peddlers moved around the neighborhoods, crying out their wares. The straight-load fruit men had their corners. There were always fights over whose corner it really was, and petty bribes to cops on the beat. My father could yell like hell and push out twenty-five bushels of beans at ten cents a pound or strawberries on a hot June day at twenty-five cents a quart all day long. And the days were long. A fruit man's day starts at 3 A.M. or earlier, hustling in the market, bargaining, buying, comparing, and getting loaded. From the market to the store with a full load, then the unloading, the setting up, and doing business until 6 P.M. Then close up, clean up, count up, and lock up. Eat. Sleep. And do it again. Saturdays were easy. No market. Fruit men love Saturdays—big business and a short day, 7 till 7.

In 1937 my father opened his first fruit stand. It was a stack of boxes in front of a meat market. He used the tops of cauliflower crates as trays strong enough to support the familiar fruit-stand pyramids. The stand was small, beautiful, and busy. He called it Marty's Market. My brothers still operate Marty's Market today. Nearly fifty years old, its clientele has changed from Polish, Italian, German, Jewish, and Irish families to blacks and Orientals and Hispanics. The inventory has shifted from potatoes to yams, from broccoli and kohlrabi and horseradish to turnip, collard, and mustard greens. But big red apples, sweet navel oranges, and summer fruits of all kinds still make the folks' mouths water, along with the new fruits like mangos, papayas, red bananas, plantains, and seedless red grapes.

In two years my father owned the whole store. By 1942 I was working. So was my little brother, Joe. I was seven, he was six. All the men had gone to war. We went to work early in the morning while my father went to the market. We'd start carrying out what we could and take care of customers. It was easy for me. Even at seven. It was in my blood. I learned to make displays, make change, add long columns of two-digit numbers in my head, read scales, and wait on trade. My brother and I had to stand on boxes to reach things, and tough old immigrant ladies would beat us around for a penny. And there was hell to pay if something less than perfect managed to "slip" into their shopping bags.

I remember one time during the war we got a load of potatoes. Potatoes were so scarce they became a black-market item. We stacked those one-hundred-pound bags like a sandbag bunker in front of the store and dumped loose potatoes in the center. My brother and I stood all day, weighing potatoes five pounds at a time for a long line of ladies that stretched around the block. We sold these potatoes five pounds for a quarter. Of course, if you wanted the potatoes, you had to buy our tomatoes too. I think that was the deal my father had to make with the railroad company. If he wanted the potatoes, he had to take the tomatoes off their hands. He just passed the deal along, as always, to the customer.

I saw it all happen; the wholesale market, the fruit auctions, the farmers' market, and the changes that came after the war that affected the fruit and vegetable business forever. But not for me. I loved it the way it was, and I hope some of us fruit men, including my own son, will always keep it that way.

I went to school but never spent my time the way the other kids did. I went to Cub Scouts and played some playground sports, went to dances and got good grades, but I worked. My brother and I worked after school, every Saturday, and all those heavy holidays like Thanksgiving, Easter, and Christmas. We worked almost every day all summer. It made us different and I think smarter. I mean, how many kids knew the difference between an artichoke and a Jerusalem artichoke? And how many cared? How many knew a seedless grapefruit seldom has more than eight seeds? I ate persimmons, pomegranates, Bosc pears. I knew Bing cherries were harder and sweeter than Lamberts and that a Kelsey plum will turn yellow if you wait long enough. And along with all those tiny details I learned, I developed a deep appreciation for the independent way of life and for relying on one's self-knowledge and skill to provide a way of life independent of the corporate, bureaucratic, or wage-earning system. It was very simple: Buy for five; sell for enough to pay the bills and leave a little extra for tomorrow, the years ahead, and dinner tonight. In the long run a fruit man answers only to nature and himself.

Sometimes my father went peddling on Sundays even though he had the store. My mother never liked that, but with eight children at home it took everything he earned, sometimes more. My father used to gamble as well, like many other fruit men. Cards and craps were his games, mostly. After work on a Saturday night he'd roll up the receipts in small wads and tuck them in my socks, my sleeves, and my underpants. If he was losing, he'd come over to me, usually asleep on a stack of potatoes, and take a wad out of my socks or sleeves. But no matter how bad it went, he'd always leave one to take home to my mother. I guess it was the Sundays after a bad night that we'd go peddling. After Mass we'd go down to the store and load the truck with whatever and go somewhere and start shouting and causing excitement, and soon we'd have something going. I was always amazed at how all of a sudden we'd be empty. Sold out with fat pockets. I'd fall asleep in the cab, exhausted. I always looked forward to school on Mondays.

By the age of sixteen I was a main man in my father's main store. My brother ran a store in the black section of Newark. He was smaller than me and quieter, and it was a tough neighborhood, but he managed very well as part of the growing family business. My uncle ran a fancy store in the suburbs. I was fortunate to be able to see the different styles of fruit-stand retailing and fascinated by the cultural and socioeconomic differences that caused certain people to identify with very specific fruits and vegetables and fruit and vegetable prices. My father spent most of his working time in the markets. One day he brought in a load of honeydew melons. Sixes they were, in big, wide wooden boxes. At that time a 6-size honeydew was a very large melon. These were perfect, golden-ripe, with a sticky skin that oozed the stuff we call honeydew. My father was going to peddle them. I asked if I could take the load. We agreed, and he drove me to a corner with a phone booth and left me. I shouted. I yelled. I cut pieces. The price was fifty cents a melon. My father had paid twenty-five cents each. Around noon I called in to report that I had sold ten cases. I had forty cases left. Within minutes my father was on the scene and created such a commotion that people on the top floors of the tenements were shouting down for melons. I was carrying honeydews up neverending flights of stairs. I remember a black woman shouting down for two melons. I ran them up to her floor and knocked on her door. The door opened, and this huge woman opened her housedress and showed me everything. "How 'bout a trade fo' those melons, honey?" I don't remember what scared me more, the woman or going back without the dollar. I told her I could come in, but I needed the dollar. She gave me a big laugh and a dollar. I never sell a honeydew melon to this day without recalling that lady. My father hustled melons to the luncheonette on one corner, two crates to a deli around the corner, and some more to a neighborhood diner. By three o'clock we were sold out. As we drove home, he laughed about how he sold forty crates in a couple of hours and I sold ten all morning. He knew his stuff and was showing me everything.

By twenty-one I thought I knew all I wanted to know about the fresh fruit and vegetable business and put it aside for a career in advertising. I managed good grades in marketing and advertising along with history, English, and philosophy, at Villanova University. I accepted a job in the mailroom of Grey Advertising for forty dollars a week. My father freaked out. "Four years of college and you go to work for forty dollars a week? You can make forty dollars a day in your own fruit business." He was right, but nevertheless I thought I'd try my hand at something a little more glamorous. The first day I walked into the offices of Grey Advertising on Park Avenue, I took a look around at my new colleagues and contemporaries and said to myself, "Barry, after the fruit business this is going to be a piece of cake." And it was. All those years. All those hard-working immigrant ladies with large families and small pockets. All those nickel-and-dime transactions, the hot days in the sun when the tomatoes had better be sold and the flecks on the bananas were turning into blotches. They were my constant reminders of what the real people need and want. I seemed to have a sixth sense for creating the precise slogan or presentation that made people respond. I never forgot those old ladies who carried their shopping money in a tiny knotted handkerchief and fed a family of five or six for a dollar.

I spent fourteen years on Madison Avenue. At thirty-one I was elected Senior Vice-President in charge of the Creative Department of Ted Bates & Company, then the largest advertising agency in the world. I managed 250 personnel with an annual budget of four million dollars. I traveled around the world, teaching the Ted Bates advertising techniques to our international companies in Australia, Japan, India, Thailand, Greece, France, Spain, and England. I had personally written highly successful television campaigns for Wilkinson Sword razor blades, Viceroy and Kool cigarettes in the United States, Players and Peter Jackson in Canada, Wonder Bread, Bic Pen, Colgate, Marx Toys, and many others. I had three secretaries and a limousine, and I was bored, angry, and empty at thirty-four. I was missing something and didn't know what, and I was tired of selling people poor-quality things they didn't need at prices they couldn't afford.

A lot of people have the idea I achieved some sort of Everyman's dream: Walk away from the daily grind of corporate life; escape from the noise, dirt, and pollution of the city to live in the mountains, breathing clean air and ticking to my own clock. Looking back, I did it for all the right reasons, but in terms of a work trade-off I came out on the tough end. I traded forty hours of top executive life with all its perks for a ninety-six-hour week of intense physical labor, combined with constant concentration on a thousand details every day. I built my new life from the ground up. I did everything from the cinder-block foundation to the roof. I even designed the aprons we wear in the store. My life changed from lunch at "21," dinner at La Grenouille, followed by ringside at the Garden and weekends in Puerto Rico or sailing on Long Island to bellowing "buy orders" at the market at 4 A.M., loading an old truck with a bad radiator, unloading it with a bad back and a storeful of inventory for which I had only two options: sell it or toss it. My family thought I was crazy. Maybe so. But there are no rules that say a man must keep to just one task in a lifetime. Before I pass on and come back, I expect to experience many roles and enjoy the pursuit of challenge and the glowing joy that comes from achievement. As I said, I did it for the right reasons, but I also did it with no plan. Life presents opportunity to all of us all the time. The fruit business was one of those opportunities, and it came to be quite by chance.

In December of 1968 I visited Woodstock, New York. Something in me moved. I returned the following spring and saw the land come alive. I was excited by the rolling landscapes full of fruit trees in blossom, the endless fields of newly planted corn. I met farmers and orchard men. I saw gardens that took me back to my grandfather's garden. By October of 1969 I had resigned from my job and began a new life doing nothing. I wrote a few articles and waited out my first winter in the country. In the spring I planted a huge garden and began restoring an eight-acre apple orchard. It was peaceful and healing but not enough. That winter I left for Europe, and it was there I began to feel a new spirit. As I traveled, without realizing why, I prowled the produce markets from Morocco to Amsterdam. "Don't touch the fruit" was the cry in each one. The early-morning hustle, the beautiful displays and incredibly perfect fruit overwhelmed me. I related. I loved the merchants, and they knew I knew. They filled my basket with extras. I saw the seagoing fruit men of the Greek Islands tie up their small boats and build displays in them to do business with the customers on the little wharves. The Moroccan *souks* featured more oranges than I had ever seen in my life. The French country markets were breathtaking in their variety and beauty. I traveled to Australia, where the fruit displays were made in convex patterns that burst with color. One shop had all the walls, ceilings, and entrance covered with pineapples. I saw fruit stands in Japan that featured flat walls of fruits in spirals and pinwheels. And wherever the produce was most beautiful, most perfect, most delicious, the policy was, "Don't touch the fruit."

I came home to early spring the next year with no idea of what to do next. It was late June or early July when I drove past a small, derelict, shuttered roadside stand on the edge of town. It was surrounded by two acres of overgrown weeds, junk cars, refrigerators, and display cases. There was a crude hand-lettered sign on the door: "Out of Business." Something in my blood stirred. How could this be? A fruit stand in the summer, in a high-activity tourist town like Woodstock, closed? I walked around the eerie scene, looked inside, and realized the owner had shut down with all the produce still in the cases and on the stands, weeks and weeks ago. I drove to a funky one-room shack in the back of the property, where I met a strange man who told me he had inherited the place from his old Italian uncle who had died a year earlier. More Italian blood.

Without thinking, I blurted out, "I want to rent the fruit stand."

"No way," he said. "You'll only fail. Woodstock hates outsiders."

"I think I can do it," I answered. "I'm the world's best fruit man." (I have no idea what compelled me to make that statement, even though I probably believe it to be true.)

"Oh yeah? How much do you want to pay?"

"How much is it worth?" I asked.

"Six hundred a month," he said, grinning.

"I'll give you six hundred dollars for three months total, but I'll pay in advance." We agreed. It was that easy. I was back in the fruit and vegetable business. I gutted the building, and recalling every detail from the past that I could apply to the present, I recreated an old-fashioned fruit stand complete right down to the hand-painted red, white, and blue signs. It was beautiful, clean, and charming. I called it Sunfrost and opened on August 3, 1972. It was quite by accident or by subconscious design my father's birthday. The name "Sunfrost" came from an old farmer I had met in an apple orchard one May day. All the trees were in bloom. As we stood there talking in the sunshine, it started to snow. I had never seen anything like it before. Sunshine, apple blossoms, blue sky, and snow. "Yep," the old guy said, "we call it sunfrost." Opposite perfections, existing together to create a unique concept. So began Sunfrost. Our slogan was "Beautiful Fruits & Vegetables"; our policy was "Please don't touch the fruit."

Most people wouldn't notice, but almost all fruit and vegetable stores or departments are laid out the same way. Salads are always together. So is citrus. Summer fruits like peaches, plums, cherries, and nectarines are in a group. So are potatoes, onions, yams, and sweet potatoes. Garden vegetables like string beans, peas, broccoli, cauliflower, and zucchini make up another section. During the summer months the space given to potatoes and onions dwindles, while that of the fruits and melons expands and dominates. As the sweet fruits finish up in the fall, apples dominate; then squash, potatoes, pumpkins, and onions take over. By winter the citrus items like big oranges and shiny grapefruits get more space and display. This is more than just a method of merchandising. It is actually determined by nature as it provides us with what seem to be the foods most welcome for that time of year. During the winter our bodies seem to crave citrus and vitamin C. We also like hot soups made with carrots and potatoes and leeks, which are so plentiful from October through February. By early spring we're fed up with starches and acid fruits, and we welcome those first sweet and tender salad greens. We begin to eat watercress and fresh soft Boston lettuce, red and white radishes, sweet green onions, and those sugary Honeybelle Murcotts up from Florida to tide us over the last cold days. By May we start to feel warm breezes, and the early peaches come from southern California and Georgia. Real tomatoes arrive from Mexico, melons from Texas, and green beans from the Carolinas. Out west the seasonal changes are a little less dramatic, but even there our bodies respond to the new things we see in our fruit stores, and there is a natural want and desire. Give in to it. Let your eyes look seriously at your produce source, and they will tell you what's best and when! *The freshest, most plentiful, healthiest fruits or vegetables always look fresh, bountiful, and healthy, and that's the time to buy them.*

Sunfrost Farms never sold apples out of season or tomatoes out of the refrigerator. We never sold cello-wrapped cauliflower or lettuce. Fresh does not just mean when the truck delivered it. It means when it was picked. An apple picked over six weeks ago is not a fresh apple. It is a storage apple. Tomatoes that are picked green, stored in a dark, cold room, then sorted for size and ripened with ethylene gas are not fresh tomatoes. They are manipulated tomatoes. All supermarket tomatoes are refrigerated. Refrigeration retards redness, increases water content, and decreases flavor. A tomato will actually last longer outside a refrigerator, sitting quietly on a shelf, than if it were left to ripen after having been chilled. A tomato that is not squeezed, that is not punctured or bruised, will last in its red stage at least a week and still be hard. Kiwis will last for six weeks before they pass acceptable ripeness. The American colonists kept apples from October to June without refrigeration and without rot. The secret is in the careful handling of these precious things called fruits and vegetables.

When customers come to Sunfrost, they are overwhelmed by the colors, the freshness and vitality of the fruits and vegetables they see. They see tiny pink potatoes next to tiny white potatoes. We carry six kinds of onions, and itty-bitty zucchini with the blossoms still on them. Tomatoes are our specialty, and most of the time we feature five, six, or seven different varieties, from giant lumpy Beefmasters to tiny teardrop plum tomatoes. People want it all. I remember one lady customer who kept buying and buying. Finally she ran outside to the parking lot and shouted to her husband in their car, "Harry, come in here. I can't stop myself."

I try to know everything I can about every item, for example whether the string beans are hand-picked or machine-gathered. It makes a difference. Machine-gathered beans have more vines in the bushel, which reduces saleable weight and adds to the cost. There are more cut beans and a wider range between the fattest, older beans and the smallest, immature ones. Hand-picked beans have no waste and are almost always uniform, without the overgrown or undergrown pods at either extreme. I know the specific qualities of every apple, and our displays carry little cards that indicate the hardness of each variety, its sweetness or tartness, its best use, and how well the apple keeps. We can tell our customers that 3 pounds of apples make a flat 8-inch pie and 5 pounds make a high 9-inch pie. We sell certain apples for baking, others for sauce, and those which should only be eaten fresh, like Macouns.

There are peaches that arrive in late June that have a greenish-white skin with a pink blush and a red stone; they are so juicy they must be eaten with a napkin. Those who doubt now have a peach-decorated shirt or blouse in their wardrobes. The first year I found them at a local orchard, it was a slow sell. White peaches? Peaches should be yellow, shouldn't they? Not true. Yellow peaches should be yellow, red peaches should be red, and white peaches white. After lots of juicy samples and lots of juicy chins, our white peach sales have outstripped our white peach supply. Today customers who know call up weeks in advance to order them by the bushel.

People used to say to me how much they loved my store but that they couldn't afford it. Too expensive. Yet I knew that item by item everything was either competitively priced or priced lower than at either of the two supermarkets in town or any of the three local grocery stores. What was happening was that the customers were used to going to the store for a head of Iceberg lettuce, two tomatoes, four onions, some potatoes and a vegetable, bananas, and another fruit maybe. At Sunfrost there was such a variety, such freshness and assurance of sweetness and quality, that they purchased two or three kinds of lettuce, watercress or arugula, red and white radishes, and our specially selected unwaxed cucumbers. Then they'd try some of our giant vine-ripened Beefsteak tomatoes, our fiery-red Mexican cherry tomatoes or carefully ripened round California salad tomatoes. They'd select onions in three colors, baking potatoes (all exactly the same size) as well as tiny pink boiling potatoes, and parsley for the Parsley Potato recipe we suggested. We turned our customers on to mangos and papayas. Then we'd teach them how to cut a pineapple. Our melons were always ripe, and they'd want one of those. As their eyes and imaginations feasted, their shopping bags got bigger. It was true. They were spending more money on fruits and vegetables than ever before. But they were enjoying and benefiting from them more than ever before. The tiny Sunfrost fruit stand, only twenty-by-twenty feet, changed the eating and shopping habits of an entire community.

One day I bought over a hundred boxes of sweet, ripe parrot-colored mangos in all sizes. They were imported from Mexico, they were perfect, and the price was right. I made a grand display in the middle of the store. They died. People knew what they were and said they were beautiful, but still no sale. After two days it dawned on me that nobody knew how to eat a mango. I was trying to sell a common Mexican street fruit to a Birdseye generation. I quickly did some crude drawings of how to peel and eat a mango, wrote some short instructions and some mango history, made five hundred copies, and handed them to customers. Today Sunfrost sells enormous numbers of mangos to a steady stream of mango aficionados, many of whom got their first taste in our store. As a boy I never sold or even saw a mango. Imported fruit was rare, and the immigrant families of Newark didn't include mangos anywhere in their diets. The ability to present the mango or the white peach comes from way back in the beginning. It's in the blood.

As the business grew and the trickle became a flood, we encountered customer-relations problems with our policy, "Please don't touch the fruit." There was a beautiful poster in the store, with mountains and cornstalks and a statement that read, "Please Let Us Serve You," and our clerks wore Sunfrost T-shirts that said, "Please Let Me Help You." Yet people persisted in trying to serve themselves. It was totally understandable. Almost every store and supermarket has self-service. Customers are allowed to poke, pinch, and pick out anything they want. I've seen supermarket customers pick stems off cherries while eating one for every one that goes in the bag. I've seen them break ends off beans and stems off broccoli. I've seen them rummage through grapes like they were stones. The supermarkets don't care. What they lose in merchandise they save on labor costs. But in the long run it is the customers who lose out in quality, ripeness, freshness, and variety.

It took almost three years to overcome resistance (and often resentment) to the Sunfrost policy. People took it personally, and perhaps we were too inflexible. Over time we learned to soften our approach and offer more and more information. We hired bright young college men and women with personality and good manners. Gradually people recognized Sunfrost as the fruit stand where you couldn't touch the fruit but everything you buy is beautiful and perfect. And did you taste those giant strawberries? And their tomatoes are always real! Our business grew, while our total waste volume remained less than ten pounds a week.

A young mother came into the store one day with her little boy. We were featuring some especially wonderful Sunkist juice oranges. They were thin-skinned and seedless, with deep orange flesh, and gave a sweet, thick juice. "Should we get some oranges for orange juice?" the mother asked her son.

"No," he answered. "Let's get the regular kind in the can." Bye-bye, OJ. We are past the beginning of the end for real orange juice. You cannot buy a glass of *fresh-squeezed* orange juice in almost any restaurant, even in Florida and California, where juice oranges grow. You can get *fresh-thawed* but not fresh-squeezed, except at a juice bar or specialty restaurant. Supermarkets have almost eliminated the juice orange from their inventory. When was the last time you bought a dozen oranges, took them home, and squeezed a glass of this deeply refreshing, healthful orange juice? Too much trouble? Too much time? Really, now. Is it more trouble than shaving? Does it take more time than putting on makeup?

At Sunfrost we always sell two kinds of juice oranges, Florida Valencias and California Valencias. A Valencia orange tree is one that grows fruit while it continually blossoms. Both the California and Florida varieties produce for about eight or nine months a year. Each has a different dormant period, and both states have highly organized Grower's Associations that control distribution and maintain storage facilities in order to keep a constant year-round supply. Florida oranges reach their low point in late fall, California ones in late spring. But there are always fresh juice oranges available. Before frozen concentrates existed, every store had at least two kinds of juice oranges, each in two or more sizes. It is the food processors and marketers who are narrowing the selection of fresh food. Years ago television commercials proclaimed that Minute Maid orange juice was *better for your health than fresh-squeezed.* It is a fact that this claim was based on the evidence of some samples of fresh home-squeezed juice that contained bacteria from the hands of the person who squeezed the juice, hands that weren't washed before making breakfast. The "better for your health" claim had nothing to do with the juice itself, but it was those commercials that created the first successful orange-juice-concentrate company and began the end of fresh orange juice in the home. Entire orange harvests are purchased by the processors in advance of picking. This eliminates sorting, grading, packing, storage, and shipping of heavy, bulky, and perishable fresh fruit. Money is made faster for fewer people in greater amounts. But the consumer loses because the food supply manipulators have made convenience and profit more important than freshness, deliciousness, and healthfulness.

With or without government price supports the American farmers and growers of table food are subject to the demands of the food processors and marketers. Processors buy many times more fresh food than fresh produce wholesalers and distributors. Processors need uniformity, mass quantities, and low prices. What grows best, processes best, and costs less is what will eventually reach the consumer. Your right to America's fresh food abundance is restricted as fresh food varieties dwindle in favor of the most efficiently marketed. It is also important to realize that processed produce is not fresh produce. It is subject to adulteration with flavorings, preservatives, and other additives and to increased costs of packaging, marketing, and storage.

As really fresh, varied, peak-of-perfection foods are made less available, those very special items that are to be truly appreciated as "the best" will appear more and more as specialty items at much higher prices. They are emerging now in the marketplace as "designer fruits" and "designer vegetables." They are packaged in small quantities, labeled, and sold by brand names. Many are imported. The Granny Smith apple is one of those fruits. Granny Smiths are imported from New Zealand, South Africa, Australia, and Chile. These countries are south of the Equator and produce apple crops in seasons opposite of our own. Their apple harvest time is our spring blossom time. Our domestic apple supply is at its low point, and the very hard, tart, and juicy Granny Smith is far superior to the mealy storage apples being sold in the American off-season. Imported Granny Smiths sell for almost a dollar a pound, while domestic apples in season sell for three or four pounds to a dollar. Holland now exports tons of hydroponically grown tomatoes, each packed in its own cell within a single-layer wooden box. Each tomato retains its stem and leaves, and each carries a bright-yellow sticker reading "Holland Hydroponically Grown." Their cost? Anywhere from three to four dollars a pound. Very superior American farm tomatoes, which are available at real fruit and vegetable stands but never supermarkets, are priced at two or three pounds to a dollar. Tiny zucchinis and baby teardrop plum tomatoes never get to the supermarket. It takes too long to pick them and requires too many trips into the field every season. They are now becoming available as specialty items in select stores at three to four times the price of the ordinary club-sized zucchini.

In addition to designer fruits and vegetables there are the new inventions. These are man-created but nature-grown varieties that are part of government-sponsored food marketing programs. New Zealand, of course, was very successful in the creation and marketing of the giant Chinese gooseberry renamed the kiwi fruit. Israel succeeded in marketing the Jaffa orange, a seedless, easily peeled, juicy version of the California navel orange. The star fruit is another interesting development that has yet to reach enough people to make it a popular item but is well on its way. Seedless red grapes are a successful hybrid fruit developed by the California grape-growers and their South American counterparts in Chile, Argentina, and Mexico. Israel is currently trying to market a new fruit called the Sharon, a persimmonish-looking fruit with tiny white seeds and deep orange flesh. It has an inverted heart shape and sits up on its petalish bracts.

Whether the new varieties are designer versions of the old standbys like Holland tomatoes, or the all-new never-seen-before inventions of man and science, the endless bounty of earth food is ever increasing and changing. And as technology shrinks the world, we will encounter exotic foods of the world like the pink-skinned white perfumed tangerines of Thailand, the rose-colored peaches the size of cantaloupes that grow in Japan, paw-paw and guava from the tropics, and all the other yet-to-be-discovered wonders. There is no reason why our tables and kitchens should suffer from the same dull over-and-over-again potatoes, string beans, Iceberg lettuce, and white-hearted tomatoes. This unimaginative routine has created a generation that identifies with more varieties of pizza pie than of fresh fruits and vegetables.

Recipes always helped a customer to accept a new idea. I had a connection with a local farmer who was willing to pick his peas for me every two days. Pea fields are picked all at once, maybe twice. That means that the peas that are the latest to mature are picked at the same time as those that matured earlier. Result: bushels of peas with mixed size, tenderness, and sweetness. This man's peas were uniform, perfectly tiny and sweet, but they sold slowly. "Too much trouble to shell. Too much waste with the pod." I heard this too often until I worked out a recipe to cook peas in their pods that was quick, easy, fun, and delicious. After giving my customers the recipe, I would always say, "Then you eat the pods like an artichoke leaf, and the peas pop in your mouth." I began to sell a lot more peas. Busy people want their kitchen time to be easy. And the supermarket people know it. They made convenience the main indication of food value in the consumer's mind. And it's easy for the store manager, easier to sell canned peas than give out recipes and ideas. All too often, however, what is gained by convenience is lost in freshness, flavor, and nutrition.

If you develop a rapport with your local fruit and vegetable guy, he'll recognize your individual preferences and cater to your tastes. He'll look for things in the market just for you. He knows if he turns you on to something good, your whole family will enjoy it. They in turn will become fruit-stand fans and good customers. I loved to hear a customer tell me she was planning a dinner party or a reception or some affair. I always helped her buy something unusual or particularly beautiful. I recall the time that the wife of the local Chevrolet dealer was giving an important party and wanted me to make her a spectacular fruit bowl. I had just gotten in some Arizona seedless grapes grown in limestone soil. They had never been sprayed or chemically fertilized. I had fifty boxes and a high price. I suggested she fill two huge silver bowls with nothing but these grapes cut up into small clusters and dotted with a few strawberries. She was slightly reluctant but agreed. The grapes were a huge success, and the following days were a grape sales festival. I bought thirty more boxes and sold out again. Years later those grapes are still remembered as one of our best items. If you have a rapport with your fruit man, he will introduce you to the best items at their best. Every fruit man I know loves to wait on appreciative customers. When customers place their trust in me to serve them food for their homes, I accept that as an honor. They recognize my experience and my professionalism. They trust my abilities to select to their taste and needs. I recognize their appreciation of quality, value, and beauty. The experience is uplifting for both of us. When customers come back and tell me how good it was, it's a big kick, like the feeling a painter gets when he passes a painting he finished and loves.

In the hustle of a busy fruit stand on a Saturday afternoon, "Please don't touch the fruit" sometimes comes out gruff. Sometimes people are offended. Sometimes they leave and don't come back. You try to be gentle and you try to explain, but I've learned that for all the people who know a little about fruits and vegetables, there are many, many more who know nothing about them and even less about how to be waited on. The only experience most people have with being waited on is in a restaurant, and there the waiter or waitress is a neutral zone between the customer and the chef, who is really responsible for pleasing the patrons. A produce clerk performs a very personal, sophisticated service: the suggestion, selection, and packaging of food for another's taste and standards. People who take the time and make the effort can learn to be waited on. Don't rush your thoughts; think about your real needs. Try to determine when you plan to serve or prepare certain items. If your fruit selection is to be eaten over a period of several days, tell your clerk. Peaches, pears, and plums that are ready to eat on Monday will be overripe by Wednesday or Thursday. If you are buying tomatoes to last all week, be specific; tell your produce man you want six tomatoes, some ripe for now and some for later in the week. If you have children, ask for small fruit. If you're making dinner, tell the clerk how many you're serving. As side dishes garden vegetables like string beans, broccoli, zucchini, and carrots serve three to a pound. Three pounds of potatoes make mashed for six, one medium-size eggplant will serve three or four, and a large baked butternut squash will feed four or more. Tell your produce clerk exactly what you want and how you need it to be. Some folks like their seedless grapes greenish-yellow, some prefer a touch of deep yellow with a fleck of brown. It's the clerk's job to give you exactly what you prefer in the best condition possible. Shop with confidence, and realize the clerk enjoys pleasing you. If he doesn't, he should be tightening bolts in Detroit.

Different fruit men enjoy different aspects of the business best. Some love the quiet pre-dawn ride to market that explodes into noise and hustle and action as soon as they enter the gates. I remember the drive past the misty low fields as the sun rose in early pink and yellow, lighting up the gray fields with lavender and soft green. Moments later I'd be assaulted by farmers hawking, "Hey, Sunfrost, where you buying beans? Hey, Sunfrost, I got raspberries today if you can pay the price." And the old Italian and Jewish guys conning everybody all the time. "Okay, Sunfrost, let's get hot! It's your kind. Your kind. Whaddya gotta look for? It's the candy kind. Two-fifty. Two and a quarter. Gaw ahead, clean 'em up for two dollars. Nunzio, put them berries on Sunfrost's truck 'fore I sell 'em to someone else." Some fruit men like that part of the game best. It's the kind of action where a dime or twenty-five cents can make the difference between an item that *makes money* or one that just cuts a little profit. In two or three hours a fruit and vegetable buyer makes a thousand decisions. While he's buying, he's rearranging the store and mentally counting his inventory at home and figuring today's load against tomorrow's load, and everything's building to the weekend. By 9 A.M. the fruit man's worked six hours, and he's got miles to go before he sleeps, miles to go before he sleeps.

Some of us like setting up our stores best. We know what has to be sold first, what items make the most money, what is bountiful and plentiful and in demand, like corn in season, peaches at their best, crystalline sweet watermelon in July and apples in October. Never put items of the same color next to each other. Never put corn near strawberries or other delicate items. The corn silk never comes off and spoils beautiful displays. Tomatoes can't stand sun, and bananas turn gray in a cold draft. Fruit-stand people know these things and manipulate their displays to accommodate the facts while they create appetizing fantasies that will turn on your tastes and make you an eager customer. I loved setting up the store, painting descriptive signs ("Easy-Peeling Giant California Navels, 3 For $1.25"; "California Leatherback Baking Potatoes, 30 minutes at 350°, 3 pounds $.89"), and waiting on customers. I watched people come in and look around. I knew what they would buy. Their eyes locked on the strawberries or glazed over at the giant white cauliflowers and the tiny baby Italian eggplants. The store was my game. I would set it up to move certain items that would, in turn, sell other items. Big beautiful Beefsteak tomatoes will stimulate the purchase of fresh basil and oregano, Italian red onions, garlic, and a bottle of imported extra-virgin olive oil. We sell heavy cream to go with the fresh-picked strawberries, and limes to squeeze on the melons, papayas, and watermelon. I loved to watch it work, and it was fun for me and my customers. Every fruit and vegetable operator has his favorite part of the trade, but one thing's for sure: We all like counting up, and nobody likes paying out.

No matter how hard you try, the retailer cannot please everyone. One day a lady shopper bought a nice order. She was skeptical and questioned everything. Okay, that's her right, and I managed to work through it pleasantly enough. She bought a pint of blueberries for ninety-nine cents.

The next day she came back to a crowded store, waving the empty basket and shouting, "These blueberries are no good. You have some nerve. Look at all these bad ones."

I took the basket from her and looked inside at three little green berries rolling around on the bottom. "Where's the bad ones?" I asked.

"There they are. On the bottom, where you hide all the junk." The original package was full of prime-quality fat dusty-blue berries from New Jersey under the Bluck Buck label. The best! I set the little basket on the counter and gave the lady a big pink-and-white peach. "This cannot even be discussed," I said as I walked her to the door. She has never returned.

I consider myself the world's best clerk. I taught our clerks as much as I could and as much as they could absorb. My son was like a sponge, with an additional creative energy of his own. We run rehearsals that include exercises like "How much is a pound, five ounces of cherries at one eighty-nine a pound?" They're taught to read dial scales, even though we have digital scales, so they will understand the mechanics of price and weight. All orders are added up on bags by hand. They are taught recipes, varietal differences, fruit preparation, and vegetable recipes and are constantly briefed on new items. They are taught places of origin, taste qualities, seasonality, nutrition, and a wide spectrum of tips that will help their customers select and enjoy our fruits and vegetables. About eight years ago a young Oriental boy came to work for us, doing the sweeping up. He had never eaten a tomato, never drunk fresh orange juice, and never eaten a fresh vegetable (his modern mom was a supermarket shopper). He began to try everything, and it was soon evident that he had a natural perception of the beauty in fresh produce. Our customers trusted him, and he developed into one of the most valuable employees of Sunfrost. He went on to college, where part of his work in commercial art was a series of fresh fruit and vegetable drawings that could only come from a mind and a talent that truly "knew."

My son spent his early years working and growing up around the fruit stand. Like myself he found other interesting things to do with his life. And like myself he found himself drawn back to the fruit business and the kind of life it demands. We worked together every day; he learned and imagined. He began to grow things and soon built his own modern greenhouse. He began experimenting with hydro-cultivation and developed different varieties of annual flowers and herbs. One greenhouse became two greenhouses. He began to go to market and started to express his own ideas and implement his own innovations. I became less tied to the store, and we built a wonderful fresh juice and salad bar. Sunfrost began to evolve.

I had always liked the *liquado* stands of Mexico and Central America. I drank icy, fresh papaya and banana with orange juice; mango, lime, and orange juice; or cantaloupe and pineapple. The combinations were exciting and endless. They were also delicious and healthful. I studied raw vegetable juice books, and together with the fruit shake recipes I created, we presented a full menu of liquid delights. Our carrot juice extractor was handmade in California and separates the juice from the root without centrifugally forcing it through its own pulp. Our carrot juice cannot be matched for sweetness or smoothness by that made by any other process. We also use only California carrots. They are sweeter, juicier, and more alkaline than any other carrots. Our vegetable combination drinks are designed for specific purposes: weight loss, reduction of body temperature, vitality, cleansing, hangover relief, or just plain deliciousness. The juice bar offers all the tropical fruits in fresh salad and liquid forms. The peach, berry, melon, grape, and banana drinks are the best tastes in the world. To me the world of fresh fruits and vegetables is ever-expanding. There is so much more available to us than what is being channeled through the mass food outlets. Canned fruit drinks are not fruit juices. They are water, sugar, and some kind of fruit—whatever. A home juicer and a blender will open up incredible variety in your daily diet. To us, our Sunfrost Fruit Stand and Juice Bar, greenhouses, and garden business is not something we do to earn our living. It is an extension of our way of life that also sustains our economic needs.

As I learned from my father, I expanded on that knowledge. As my son absorbed the original information and what I added on, he began to expand with his own ideas and visions. Under his direction Sunfrost is brimming over with homemade fruit pies, tomato sauce, Mexican salsa, fresh-baked breads, jams, jellies, honeys, and glistening jarred fruits. His young energy is expanding the world of fruits and vegetables even further. It is the effort and knowledge of three generations keeping the world of fresh fruits and vegetables alive and thriving in the face of a society headed in the opposite direction.

Two of the fastest-disappearing factors in our society are beauty and value. A properly run fruit and vegetable stand is one of those rare things of beauty and value. You will be rewarded if you take the time to look deeper into the fresh food you prepare for yourself and family. That's what this book is mostly about. It's about simple, beautiful real food you can have every day. It's about a fruit bowl on the kitchen table that is always filled with sweet, juicy, delicious food more health-giving and life-respecting than Snickers or Captain Crunch or gooey, sticky stuff that cannot sustain life. If you'll trade your afternoon cookies for a strawberry-banana fruit shake you can make in two minutes, if you'll trade some of that canned tuna and mayonnaise for an avocado and tomato, if you'll trade your french fries (fast-food or frozen) for a great California russet baked potato, you'll do yourself and your family an immense favor. Sure, you'll have some trouble breaking old patterns and habits. Sure, the kids will demand their sugar. But the benefits of health, beauty, and value will soon become apparent. You'll spend less money. You'll look and feel better, and your food-shopping trip will become a personal service where the person you give your money to is a friend who wants to please you and serve you again and again.

The word "doctor" is Greek in origin and means a teacher, a knower of secrets, one who shows the way. Your fruit and vegetable man is your food doctor. He knows what is best, when and how it is best enjoyed. This book will help make you your own food doctor. I will try to show you how things came to be the way they are. I will point out the patterns in the fresh food supply so you will know what to expect when you make your selection. You will be able to recognize sweet, juicy grapefruits by sight. You will never be disappointed by dry, mealy peaches, nectarines, and apricots again because you'll be able to tell which ones are great and which are not before you buy them. There is broccoli and cauliflower that does not smell noxious when cooked, and you will be able to cook these vegetables so that they are crisp and sweet instead of mushy and tasteless. If you have spotted or bruised fruit in your home, you will learn to use your blender. Most melons make excellent juice. A piece of cantaloupe, a spotted banana, and a bruised peach with an ice cube or two turn into a delightfully refreshing and seriously nutritious drink that costs less money and contains more benefits than any bottled soft drink. The selection, purchase, and proper use of fresh fruits and vegetables is the most fun you can have with food. All processed, packaged food items are always the same, day after day, month after month. Fresh produce is always changing. Carolina string beans are different from Kentucky string beans in size, taste, and texture. Little white boiling potatoes are really different from large red ones and are cooked differently. Mustard greens, turnip greens, broccoli rabe, chard, and kale are all called greens, but each offers its own taste and flavor and variety to your table. It is no wonder children hate vegetables and grow up indifferent to them in their diet. Most families suffer a dull, carelessly prepared routine of the same two, three, or four vegetable side dishes. Try halved acorn squash baked with butter and maple syrup, or whole baby white eggplants stuffed with herbs, sautéed onions, bread crumbs, and raisins. Serve carrot wheels glazed in butter or honey, whole string beans steamed with almond halves and white vinegar, baked potatoes topped with chopped parsley and sour cream, whole red peppers stuffed with rice and herbs, or boiled pearl onions in white sauce. These simple ideas are just a tiny part of the enormous and colorful variety that is available to the fruit-stand shopper.

One of the major components of a healthful and fresh diet is fresh green salad. A fresh green salad is not a combination of some pathetic pieces of white Iceberg lettuce, a couple of slices of cold pink tomatoes, and some cucumber slices, smothered with a bottled creamed dressing. That's just roughage and spice. A fresh green salad should have several kinds of lettuce, endive, or greens. Green salads should have color from carrots, beets, radishes, red onions, and red cabbage. Green salads should have texture from crisp onions, white turnips, Kirby cukes, radishes, and apples. And green salads should have contrasts: sweet with bitter, soft with crisp, sharp with mild, and small with large. These combinations are what make salads interesting and delicious. The dressing is for counterpoint; it is not the main event! Learn to use all the lovely little leafy things like Belgian endive, watercress, chives, arugula, basil, dill, parsley, sorrel, New Zealand spinach, and nasturtiums. The salad-maker who uses these leafy plants will never have the waste of leftover salad. In addition to the taste experience it gives, a well-made salad stimulates digestion and activates the taste sensors to make the meal that follows more delicious. This book will show you what lettuce is really all about and how to avoid high prices for market-manipulated varieties. You'll learn which salad greens have a natural spicy taste, which ones are soft and buttery, and which ones to eat when.

The person who shops his or her local fruit and vegetable market will very often be able to avoid high prices that accompany short supply. The supermarket shopper who shops for produce by habit and buys that weekly head of Iceberg lettuce (because it's the only kind her family likes) will pay a dollar or more per head at least three or four times a year, during the predictable short-supply periods. The same kind of inordinately high price is charged for fresh string beans, seedless grapes, broccoli, cauliflower, cucumbers, and several other constantly purchased varieties of produce. If you notice, these items are the ones most people buy all the time, unlike peaches, watermelon, pineapples, and yams. When you shop the local produce market, the personal service you receive will include suggestions on how to replace those short-supply high-cost items with more plentiful, less expensive, and fresher alternatives. When Iceberg lettuce reaches its top dollar, romaine lettuce and Boston lettuce are usually cheaper and better. When broccoli becomes scarce and expensive in August and March, your local fruit-stand man will suggest turnip greens, broccoli rabe, or Swiss chard. When string beans are scarce in late spring, beet tops are plentiful, cheap, and nutritional dynamite.

Properly and easily prepared, they taste sweet and tender and are an excellent stimulant for cleansing winter starch and fat residues. These items may be new to your family table, but they are highly beneficial and easy to prepare. They add variety to your diet, and they cost a lot less. Mother Nature knows. So does your fruit and vegetable man.

Of course, there are some things that have no substitutes. Nothing can replace a tomato. If the big red tomatoes we love are unavailable or priced sky-high, look for cherry tomatoes or small Italian plum tomatoes. By late winter and early spring there are several excellent Mexican varieties. But if none exists in your store or if the ones you see are pale and watery, *do without them*. No tomatoes are better than poor tomatoes. When you see expensive, highly waxed, soft-tipped cucumbers, *pass*. Cucumbers' only value is their taste and texture, never their nutrition. There is also no substitute for peppers in any color. When peppers become high-priced, ask your fruit man to select large, firm, flat-bottomed ones. Serve them stuffed with sautéed onions, garlic, raisins, and rice or bread crumbs as the main part of the meal. You'll get all your money's worth and a happy family table. The more you learn about fruits and vegetables, the more you will be aware of the changes in availability, quality, and price. Knowing these things will help you take advantage of the endless variety and bounty available almost everywhere every day. Don't go shopping with a "must have" attitude. Select what is most beautiful for the best value, and use your imagination in its preparation and presentation. You'll be surprised at what you can do. Your family will love it, and it will make you feel good.

Don't expect to learn everything all at once. After forty years of fruits and vegetables, I learn all the time. A little old man in a Mexican market told me my pineapples were rotting after I got them home because I didn't keep them upside down. It makes sense. After the pineapple is cut from the plant, the juice wants to run down the stem out the cut end. The sweet, sugary juice builds up at the bottom of the fruit, which is the first part to ripen, and ferments. Turning the pineapple leaves down makes the juice run to the top and ripens the top half equally with the bottom. How simple.

I was shopping in a Los Angeles market and came across small, shiny, green, flattish scalloped things like tiny U.F.O.s. The sign in the supermarket said, "Squash 19¢ lb." They looked beautiful and tender whatever they were, and at nineteen cents a pound I had to buy them. Never say no to beauty and value. I asked the aisle clerk what they were. He said, "Squash." Obviously he could read as well as I. I bought a bagful, took them home, and steamed them whole with onions, garlic, pepper, parsley, and a little butter. They were great, and we ate them all. Several years later while in the market shopping for Sunfrost, I saw the same item on a farmer's truck.

"They're early patty-pan squash," he told me. Patty-pan squash is almost always sold in the early fall as a large, flattish, semihard, white-skinned squash, often the size of a dinner plate. I have now learned and tried both the early green version and the mature white version. Without a doubt the babies are always sweeter and juicier, with more flavor, and easier to prepare. Like zucchini, baby patty pans are harder to pick when small. It takes more effort to fill a basket and requires daily trips to the field to maintain a constant crop of small sizes. The cost is higher, but so is everything else: freshness, flavor, nutrition, and ease of preparation. The retailer who looks for new information and presents this kind of merchandise is able to offer yet another item unavailable in most major food outlets and continually broadens the choices available to you. Once you begin to explore the possibilities that exist in your fresh fruit and vegetable market, you will be able to select fresh fruits, even melons, that will finally satisfy your taste for sweet, juicy pleasure. Your blender will create fruit shakes and juice drinks that your family will love. Tomatoes will no longer be just a summertime thing. Potatoes will go far beyond baked, fried, and boiled. Papayas, pineapples, and mangos will lose their mystery, and you'll include these tropical delights in your regular household food supply.

There is no substitute for the real thing, nothing like the taste of fresh orange juice, a strawberry-banana fruit shake, or a properly prepared green garden or tropical fruit salad. With this book and a new familiarity with the secrets of fresh fruits and vegetables you will become an expert in the selection of fruits and vegetables. You will know when to buy and when to pass. Your ability to choose the best and to prepare hundreds of simple recipe ideas from this book will add a new dimension to your meals and menus and to your life.

In addition to fruit-and-vegetable-stand know-how this book offers important practical benefits to the modern household. Fresh fruits and vegetables cost less than canned, frozen, or dried; they require little preparation and taste fresher and sweeter than any processed varieties. They are immeasurably more beneficial to your overall health, weight, and vital functions. Fresh fruits and vegetables as part of your daily diet will cleanse, regenerate, and stimulate the body's processes. Salads aid digestion, grapefruit reduces cholesterol, green vegetables increase oxygenation, and potatoes are easily digested and extremely high in usable protein. String beans, Jerusalem artichokes, and dandelion greens help metabolize calcium and increase adrenal activity. Sweet fruits reduce body acids. Greens and fruits can help eliminate menstrual cramps, iron loss, and water retention. Papaya relieves indigestion and helps the digestive process. Fresh fruits and vegetables can be eaten raw, juiced, simply cooked, or prepared in elaborate and elegant combinations. *Your local fruit stand is the best source of taste, health, beauty, and value available in the world today.* In addition you receive the added benefits of personal service and attention. This book will provide you with the information and enthusiasm you need to turn the "fruits of the earth" into an endless variety of beautiful, delicious, healthful additions to your daily life.

—BARRY BALLISTER

SPEAKING THE SAME LANGUAGE

*F*ruit men have a language of their own. In the wholesale market they talk flats, straps, bins, loads, Red Balls, all kinds of numbers, and a range of quality descriptions like commercial, candy, fancy, and a "real money-maker." A "real money-maker" is a cheap item of low to downright poor quality that retailers sort, trim, clean, and offer at sale or special prices. Berries come twelve pints to a flat, twenty-four to a strap, or sixteen quarts to a crate. Watermelons come in bins from thirty to forty-five, depending on the size of the fruit. So do apples and pears. A bin is twenty bushels of unsized but even-quality fruit. Usually a bin of cider fruit (or drops) is the lowest grade, while firm, unmarked, tree-picked fruit is called orchard-run. Only sized, perfect fruit of any variety is called fancy.

In the citrus business Red Ball refers to second-level California oranges, lemons, and grapefruit. It is a grade, not a label or brand. This grade has only the slightest skin mark or irregularity in shape. It represents excellent quality but is always sold at significantly lower prices than the top of the line, which is graded Sunkist and bears the stamped name. Sunkist is always perfect in every way. Once again Sunkist is a grade, not a label or brand. For instance, a box of perfect California navel oranges will have a label that reads "Mountain Top Sunkist Navel Oranges." A box of less-than-perfect California navel oranges will read "Two Amigos California Navel Oranges." The Florida equivalent of Sunkist is Florigold.

At the retail level the common terms refer to individual pieces or sizes. A head can take two different forms. A head as in broccoli or cauliflower is a group of flowers that grow tightly packed together. As in Iceberg lettuce a head is leaves growing protectively and closely together, surrounding the flowering part of the plant.

"Fresh" is an indiscriminate word. In its largest sense fresh is applied to any fruit or vegetable that is not processed. I say fresh means crisp, bright, flavorful, and glimmering with the bloom of life.

Ripe is a word that has lost its meaning. Ripe should only mean ready-to-eat. A speckled banana is ripe. A yellow banana is almost ripe.

A cantaloupe with a stem is not ripe and never will be. A pink tomato is not ripe. Ripe only happens after maturity, which can only take place on the plant. If allowed to mature on the plant, bush, or tree, all fruits and vegetables will ripen un-refrigerated. The trick is to find mature fruit. As you read this book, you will learn the unmistakable signs that will help you determine the difference between ripe, ready-to-eat produce and immature, overripe, or downright poor produce. When you follow the recipes, keep in mind that all ingredients are always fresh, ripe, and mature unless otherwise indicated, as in Fried Green Tomatoes.

Green is a color, of course. It also refers both to unripe and to immature produce. The discerning eye of the informed shopper, however, can distinguish a green that will never ripen from a green that will. Fruit men call a green that will never ripen dead green. When you see well-formed, unbruised, quality produce whose green is a live green, take it home and carefully ripen it to perfection.

Waxed produce is one of the side effects of the year-round long-distance fresh food marketing system. Rutabaga turnips are waxed. Winter and spring peppers, cucumbers, tomatoes, and eggplants are waxed. It's a drag, but it's the only way to satisfy the American consumer's demand for summer-fall produce during winter-spring months. The wax is a vegetable derivative, nutritionally useless and tasteless too. Years ago packers used melted beeswax. Today so much produce is shipped so far that there just aren't enough bees. Waxing prevents shriveling, frostbite, and nutrition and flavor loss; it is essential to the maintenance of high-quality year-round availability. Even though the wax can be removed *almost* completely with vinegar and warm water and the residue is harmless, I recommend peeling waxed produce whenever possible. It just makes me feel better.

Produce is subject to endless variables. Nutritional values are among the most variable areas of fresh fruits and vegetables. Primarily it all begins with the seed and the variety. Soil affects nutritional content. Weather affects nutrition. So do fertilization, cultivation, picking, packing, and shipping. Refrigeration affects nutrition. So do cooking and degree of ripeness. Above all, the degree of freshness can alter the nutritional values of fresh fruit and vegetables. Optimum nutrition can only be found in produce that glows with vitality. The nutritional values listed in this book for fresh fruits and vegetables have been obtained from inventories commonly found in stores and markets. Without question, the fresher you get it, the better nutrition you get from it.

I hear customers use the word "gassed" in reference to artificially ripened fruit. The term conjures up eerie scenes of masked people in space suits spraying foggy toxins over innocent, helpless apples and bananas. It is nothing like that. Mature but unripe fruit is stacked in insulated rooms that are sealed and environmentally manipulated. Nitrogen is removed and replaced with a controlled amount of ethylene gas, and the temperature and humidity are set to an optimum level for each specific fruit. Mangos, bananas, and papayas have a different set of numbers than tomatoes, avocados, and lemons. The whole process activates the natural pectin in the fruit and causes an increase in the sugar content. There is no spray or chemical on the fruit itself. There is no toxic residue. The ethylene process simply activates the ripening process. It is harmless, and without it we would be unable to enjoy many of the beautiful, healthful foods that fill our stores and bellies.

All fruit have two ends: the stem end and the blossom end. The stem end, obviously, is where the fruit was attached to the plant, bush, or tree. It can indicate freshness or age. The blossom end is directly opposite and in almost all cases ripens first. It is also the part that most people squeeze. Please don't. Look with your eyes for the signs. Cradle the whole fruit in your hand. Ripe, ready-to-eat fruit is obvious when you know what to look for.

Fruit and vegetables are sold by piece, by pound, and by package. Everyone knows twelve pieces are a dozen and sixteen ounces are a pound. It's the pints, pecks, and bushels that often confuse the shopper. Two of the most misleading packaged items are prewashed spinach and mushrooms. They are not pound packages. They are almost always either eight, ten, or twelve ounces. The price may be attractive, but be aware that it represents something other than one pound. Raspberries come in half-pint trays and one-pint baskets. Blueberries and strawberries come in pints and quarts. A quart is equal to two pints. Apples, pears, potatoes, onions, and peaches are often sold in different-sized baskets and bushels. The smallest size is one-half peck, which is four quarts and weighs about five to six pounds. A full peck is about ten to eleven pounds. Four pecks make a bushel. Full of apples or peaches, a bushel weighs about forty pounds. A bushel of onions weighs fifty pounds, of potatoes fifty-five pounds, and a bushel of pears almost sixty pounds. During the harvest season farm stands and roadside markets sell pecks and bushels at excellent prices for canning, winter storage, or fall feasting.

There are other terms slightly more esoteric and less common, but they nevertheless have meaning and expand your understanding of the fruit and vegetable bounty. A bract is a modified leaf either supporting the fruit or flower, as in the persimmon, or surrounding the unopened bud, as in the artichoke. A drupe is any fruit with a hard kernel or stone encasing a fleshy seed. Apricots, cherries, peaches, and plums are drupes. Coriaceous fruit has a leathery skin, like pomegranates and kiwis. Rhizomes are creeping underground stems that send up leafy shoots or flowers. Jerusalem artichokes are rhizomes. Tubers are thick underground branches that produce above-ground leaves with below-ground stems, as in the potato. Taproots are usually long, but sometimes round, single vertical roots with small lateral hairy roots. Radishes, turnips, beets, parsnips, and carrots are taproots. They all produce abundant above-ground greens, some of which are edible and some of which are not.

Of all the terms and phrases used in the world of fruits and vegetables, none causes more discussion than the essential difference between a fruit and a vegetable. There are many definitions, from the very scientific to the very wrong. The one I subscribe to is simple and meaningful. If you pick the food and the plant dies, the food is a vegetable. If you pick the food and the plant lives, the food is a fruit. It is part of the karmic classification of all living things and raises fruit to the highest spiritual level of all earthly edibles. Think of that the next time you see someone battering a pile of plums or squeezing the life out of the cantaloupes. Thank you.

CANNING AND PICKLING

PREPARING JARS

All jars must be cleaned and sterilized according to the manufacturer's instructions. Follow package directions exactly and completely to avoid bacterial growth, spoilage, and disappointment. Use only new metal lids or rubber rings.

PICKLING PROCESS

Immerse filled, capped jars in enough boiling water to reach one inch or more above the tops of the jars. Boil steadily for 5 minutes for pint jars, 10 minutes for quarts. Corn relish requires 15 minutes for pints and 30 minutes for quarts. Pickled fruit must be boiled 20 minutes for any size jar.

CANNING PROCESS

Although fruit and tomatoes may be canned with their skins intact, I seriously recommend that most fruit and all tomatoes be peeled, as well as cored or pitted, before processing. Pack fruit or tomatoes tightly into jars and cover with liquid (see "Simple Syrup," below), leaving ½-inch head space. Remove any air bubbles with a flat knife blade. Begin by heating water in a standard canner with a rack. A 7-pint canner requires 4 inches of water, a 7-quart canner requires 4½ inches of water, and a 9-quart canner requires 5 inches of water. Place the packed jars in hot water, but begin counting the boiling time only when the water begins to boil. If more water is needed, add boiling water only.

BLANCHING

Blanching is the simple process of immersing fruit or vegetables into boiling water for 15 to 30 seconds. This causes skins or peels to slip off easily. Whole, sliced, or cut-up vegetables may be blanched for several minutes to achieve a softness that prepares the stems for other culinary steps, e.g., french- or home-fried potatoes, broccoli and cauliflower salads, or breaded and fried vegetable recipes.

SIMPLE SYRUP

Most fruit is processed in its own juice or in a simple sugar-water syrup or in a combination of both. Fruit that is ripe, juicy, and sweet in its natural state will require less sugar in its processed state. Prepare your sugar syrup to taste after tasting the fresh fruit first. 1 cup sugar with 4 cups water or juice makes 4½ cups light syrup, 2 cups sugar with 4 cups water or juice makes 5 cups medium syrup, 4 cups sugar with 4 cups water or juice makes 6 cups heavy syrup. Mix the sugar with the water and/or juice. Heat until the sugar is dissolved. Pour the boiling-hot syrup over the fruit in the jars. Be sure that the packed fruit is completely covered.

PREPARATION AND TIMETABLES

Tomatoes. Use only perfect, ripe tomatoes. Core and blanch them. Drain, saving the juice. Peel. Pack whole into jars, adding ½ teaspoon salt and a full stem of basil along the side of each jar. Press down to fill any spaces, and fill to the top with the saved juice if necessary. Seal. Boil for 45 minutes.

Apples. Peel. Cut large pieces from around the core. Pack into jars and fill to top with sugar syrup (use light syrup for sweet apples, heavy for tart), adding ½ teaspoon cinnamon or a cinnamon stick to each jar. Seal. Boil for 20 minutes.

Apple Sauce. Pack apple sauce into jars to ½ inch from the top. Seal. Boil for 30 minutes.

Apricots. Wash. Blanch. Remove skins and stones. Cut in halves, or pack whole in jars. Cover with boiling-hot syrup. Seal. Boil for 25 minutes.

Berries (except strawberries). Wash and drain until dry. Pour 1 cup boiling simple syrup into each quart jar. Fill with berries and shake to pack tightly. Add boiling syrup to fill the jar to ½ inch from the top. Seal. Boil for 20 minutes.

Strawberries. Gently mix ½ cup sugar with each quart of hulled strawberries in a large pot. Let stand 5 hours in a cool place. Cover and heat until the sugar dissolves and the berries are hot. Pack into jars. Seal. Boil for 15 minutes.

Cherries. Wash, drain, stem, and pit. Cover with boiling syrup, leaving ½ inch of head space. Seal. Boil for 25 minutes.

Mangos. Peel and remove seed leaving pieces as large as possible. Pack tightly into jars and cover with boiling medium syrup to ½ inch from the jar tops. Seal. Boil for 30 minutes.

Papayas. Peel, remove seeds, and cut in long wedges. Pack tightly into jars and cover with boiling light syrup to ½ inch from the jar tops. Seal. Boil for 25 minutes.

Pears. Peel. Cut in halves and core. Pack tightly into jars and cover with boiling syrup to ½ inch from the jar tops. Seal. Boil for 30 minutes.

Pineapple. Peel. Remove core and cut fruit into chunks. Pack tightly into jars and cover with boiling syrup to ½ inch from the jar tops. Seal. Boil for 30 minutes.

Plums. Use only blue-skinned prune plums or purple freestone damson plums. Wash. Prick the blossom ends. Pack tightly into jars and cover with boiling syrup to ½ inch from the jar tops. Seal. Boil for 25 minutes.

COOLING AND SECURING SEALS

Remove the jars from the rack after the boiling time. Tighten screw rings. Let stand at room temperature, jars not touching, for 12 hours. Metal lids will pop during cooling. Test them by pushing down the centers. If the lids don't give, the jars are sealed.

STORAGE

Wipe the jars and label with the canning date. Store in a cool, dark place. Refrigerate after opening.

GARDEN VEGETABLES

1. *SWEET PEPPERS*
GREEN, YELLOW, RED, ITALIAN

2. *HOT PEPPERS*

3. *STRING BEANS*

4. *PEAS*

5. *LIMA BEANS*

6. *ZUCCHINI*
YELLOW, GREEN

7. *BROCCOLI*

8. *CAULIFLOWER*

9. *EGGPLANT*

10. *CORN*

11. *PATTY-PAN SQUASH*

12. *OKRA*

13. *SQUASH BLOSSOMS*

Garden vegetables can make or break a produce business. They can turn you on with their bright green crispness or turn you away when they look wilted, broken, or yellowed. Nothing is less appealing than broccoli with yellow buds, rusty cauliflower, or wrinkled gray string beans.

The garden vegetables are the most common, the least consistent in price and quality, and the most-consumed category of fruits and vegetables. There is an immeasurable difference between "garden fresh" vegetables that are truly garden fresh and what happens to the poor things after they've been picked, packed, shipped, stored, refrigerated, trucked, displayed, handled by hundreds of hands, sold, and finally cooked to their boiling death. Awwrrrgggh!

Every fruit man or supermarket produce manager knows his store looks best when his garden vegetables are bright, perky, and sparkling in endless verdant shades. When you see those kinds of "live" vegetables, buy them!

Garden vegetables require that you *shop for them without a list*. If your list says broccoli and string beans and you've planned a meal or menu around them, chances are you'll be disappointed. Broccoli and string beans may be "off items" that day—wilted, picked over, overpriced, and over-age. Meanwhile the peas may be excellent, the zucchini small and tender, the eggplants big and black, and the red peppers bigger than coffee mugs. Stuff an eggplant, or bread some whole baby zucchini instead. Your meal will be exciting, you'll have a different vegetable experience, and you'll have purchased the best instead of settling for yesterday's news. Garden vegetables require very little cooking time, and aside from washing and trimming, they need little preparation (maybe more than a "boil-in-a-bag" but certainly a minimum effort for the reward).

Garden vegetables are the ones all home gardeners cultivate and enjoy. To come anywhere near the taste and vitality of homegrown, *you must buy fresh*. Insist on freshness wherever you shop, and you'll get it. Stores respond to customer demand. Some of our town's citizens put enough pressure on the local chain store to make it install a fresh salad bar and begin to carry loose eggs. Now people don't have to buy a dozen eggs if they only want two.

Garden vegetables are highly seasonal. During the gardening months these items are at their lowest price and best quality. They are always more expensive during the winter. Nevertheless garden vegetables are vitally important to your winter health and table variety. Their bulk, along with their mineral and vitamin content, is an important counterbalance to the increase in starch and carbohydrate foods we tend to consume in cold weather. Shop carefully and eyefully. Look for the leaves, buds, and pods that glisten and wink at you. Sometimes I wink back. Try it yourself.

SWEET PEPPERS

What we are used to calling peppers are actually capsicums. The true pepper is an East Asian spice that is dried, powdered, and used as seasoning.

As late as the nineteenth century peppercorns were used as legal tender. In earlier times a serf could buy his freedom with pepper; a king could buy an army or navy with pounds of peppercorns worth more than pounds of gold. Our common, less costly capsicums grow in a variety of colors, shapes, and sizes. The most common is the green bell pepper with thick green flesh, a square shape, a minimum of seeds, and a heavy outside skin that some folks find hard to digest. Red peppers are actually green peppers that have been allowed to ripen on the plant. Red peppers are softer than green peppers and sweeter, since they are riper. They also tend to spoil faster and cost more money. You will also find peppers with mixed green and red skins called Suntans. Suntans are sweeter than green peppers and sell for less. There are also white, deep-purple, and bright-yellow capsicums. These colorful capsicums taste much like the red varieties but tend to be a little crisper. The yellow variety is extremely sweet-tasting and a delightful vegetable both as a raw salad ingredient or cooked as a stuffed or roasted pepper. There are long pale-green peppers known as Italian frying peppers. They are sweet and tender and should be cooked whole, stem and all, as the seeds impart a gentle tang to the sweet pepper taste. After cooking, the stem and seeds are easily and completely removed.

Peppers are available all year round, with two periods of extremely high price, in midwinter and late spring. Winter-season green peppers are imported from Mexico. Colored peppers are usually imported varieties from Holland, which has made a national industry of large, high-quality, select capsicums. The American varieties of red and colored capsicums come from California and Mexico in May, the Florida and Southern harvest in June, and the many Northern local harvests in early fall.

Peppers are always at their best at peak harvest from local sources, but high-quality, reasonably priced selections are available almost always. All domestic peppers except local crops are waxed for shipping and keeping. No matter your color choice, select crisp, firm, smooth peppers with bright stems and wrinkle-free skins. Peppers to be used for frying or sautéing need not be as perfect, but those to be stuffed or eaten raw should be of the highest quality.

Capsicums offer very high amounts of vitamins A, C, and E, high fiber, and low calorie content.

NUTRITIONAL DATA

Weight (g)	Calories	Protein (g)	Carbohydrates (g)	Crude Fiber (g)	Water (g)	Fat (g)	Cholesterol (mg)
100	22	1.2	4.8	1.4	93.4	0.2	—

Vitamin A (IU)	Vitamin C (mg)	Thiamine B-1 (mg)	Riboflavin B-2 (mg)	Niacin (mg)	Vitamin B-6 (mg)	Vitamin B-12 (mcg)	Folic Acid (mcg)	Pantothenic Acid (mg)
420	128	0.08	0.08	0.5	—	—	—	—

Sodium (mg)	Potassium (mg)	Calcium (mg)	Phosphorus (mg)	Magnesium (mg)	Iron (mg)	Zinc (mg)	Copper (mg)	Manganese (mg)
13	213	9	22	18	0.7	—	—	—

Raw—1 large

Roast Whole Peppers

Per medium green or red pepper:

1 shallot
1 clove garlic
2 tablespoons olive oil or vegetable oil
Dash dried oregano
Dash onion powder
Dash freshly ground black pepper

Preheat oven to 400° F. Finely chop garlic and shallot. Place peppers in shallow metal or glass ovenproof dish. Pour half the oil into dish. Place peppers on their sides. Pour remaining oil over each pepper. Cover with garlic and shallot. Dust peppers with light coat of oregano, onion powder, and black pepper. Cover with aluminum foil and place in oven for 30 minutes. Remove foil. Turn oven setting to broil. Replace peppers in oven, turning them at 3 to 5 minute intervals until skins blister and begin to blacken. Remove. Let cool. Peel blistered skins with fork and fingers. Reheat at 350° F. for several minutes. *Serve whole individually.*

Barbecued Peppers

Per medium green or red pepper:

2 shallots
2 cloves garlic
Dash dried oregano
Dash onion powder
Dash freshly ground black pepper
¼ cup olive oil

Finely chop garlic and shallots. Mix all ingredients except peppers thoroughly in deep bowl. Dip peppers before setting them on grill over hot coals after flames have died down. Baste and turn peppers until skins blister and begin to blacken. Remove from grill and peel. Cool before serving. *Serve individually.*

STUFFED PEPPERS

Per large green or red pepper:

1 tablespoon chopped onion
1 clove garlic
1 tablespoon chopped parsley
Butter or margarine
1 anchovy, chopped (optional)
1 tablespoon pignola nuts
¼–½ cup bread crumbs
Light vegetable oil

Preheat oven to 350° F. Chop onion, garlic, and parsley together and sauté until soft but not brown in butter or margarine. Mix sautéed ingredients with anchovy, nuts, and bread crumbs. Cut pepper around stem and remove stem with seed pod. Break off seeds and discard. Fill each pepper with stuffing and replace stem cap. Set peppers upright in heavy roasting pan or deep pot with a touch of water and vegetable oil just covering the bottom. Sprinkle extra stuffing over peppers. Cover and place in oven for 30 to 40 minutes. Remove cover. Raise oven temperature to 425° F. for 5 minutes. *Serve individually.*

STEAMED PEPPERS

Per medium green or red pepper:

Dash onion powder
Dash garlic powder
Dash dried oregano, rosemary, marjoram, or coriander

Cut peppers in half. Remove seeds. Place in shallow saucepan cut side down. Dust with onion powder and garlic powder. Sprinkle with oregano. Fill pan with ¼ inch water. Cover tightly. Cook over low heat for 20 to 30 minutes.

Pepper halves can be served plain or filled with Spanish rice, couscous, or ground beef simmered in tomato sauce.

STIR-FRIED PEPPERS

The trick to cooking stir-fried peppers evenly along with softer vegetables like mushrooms and peas is blanching them first. Wash and de-stem peppers and cut into wide strips. Drop quickly into boiling water for 30 to 45 seconds. Remove. Drain. Add to heated stir-fry mixture, using margarine, vegetable oil, or olive oil base. Season to taste. All spicier stir-fry mixtures utilize more garlic, onion, black pepper, curry powder, and tamari sauce. *1 to 2 peppers per serving.*

PEPPERS AND ONIONS IN EGGS

½ red or green pepper
½ small yellow onion
2 sprigs parsley
2 tablespoons margarine
Dash freshly ground black pepper
2 eggs
1 teaspoon butter
Dash freshly ground black pepper

Wash, de-stem, and seed red pepper. Cut into thin slices or chop. Chop onion and parsley. Sauté red pepper in black pepper and margarine. As pepper softens, add onion. Sauté until soft but not brown.

Beat eggs well, adding butter, black pepper, and parsley. Remove pepper and onion from sauté pan and place in dish. Scramble eggs in sauté pan over medium heat. As eggs begin to cook, add pepper and onion gradually. Serve hot, or allow to cool and serve as sandwich filling.

SAUSAGE, PEPPERS, AND ONIONS

6–8 inches Italian sausage
1 medium green or red pepper
1 small yellow onion
1 clove garlic
Light vegetable oil
Dash salt
Dash freshly ground black pepper
Dash dried oregano
Bun or roll, split

Cut sausage into 2 pieces crosswise. Fry thoroughly, draining off and reserving fat, or broil until crisp. Remove and keep hot in oven or over low heat.

Wash, de-seed, and slice pepper into wide strips. Slice onion into medium rings. Coarsely chop garlic. Using minimum amounts of sausage drippings or oil, fry pepper over medium heat. Add garlic a little at a time. Turn frequently with wide-tined fork. As pepper begins to soften, add onion slices. Cook separately in same pan, turning often. Sprinkle lightly with salt, pepper, and oregano. Slice cooked sausage lengthwise. Place in bun or roll. Top with pepper and onion. Salt entire sandwich to taste.

MEXICAN BREAKFAST

This is an excellent brunch dish.

⅓ cup diced potato
½ green or red pepper, chopped
½ small yellow onion, chopped
4 tablespoons butter or margarine
Dash salt
Dash freshly ground black pepper
¼ cup tomato sauce
2 eggs
½ teaspoon chopped fresh or dried parsley, oregano,
** or coriander**
Freshly grated Colby, Monterey jack, or cheddar cheese

Blanch diced potato. Sauté pepper and onion in butter or margarine. Season with salt and pepper. Add potato when pepper softens and sauté for 2 to 3 minutes. Add tomato sauce. Cover. Simmer over low heat. Whisk eggs. Add parsley. Pour into greased, heated pan over medium heat. Cover. When eggs are half cooked, add vegetables and tomato sauce. Cover. Lower heat and cook for 3 to 5 minutes. Serve in slices, with grated cheese. *Makes 1 serving.*

HOT PEPPERS

One day Columbus dined on a spicy native South American dish and mistakenly called it peppers. It was a dish of chiles. Chiles are an American original, and nobody handles them better than Mexicans and South Americans.

All chiles are members of the capsicum family, but they have their own definite personalities. They start out green; some are left on the plant to ripen and turn red. Green hot peppers are hotter than almost all red hot peppers. But some red hot peppers are hotter than anything you can imagine. Chiles are rated in the trade for their "heat" on a scale from 1 to 120. The familiar lip-searing jalapeño is rated 15. Anyone ready to tear into a genuine 120?

There are thousands of varieties of chiles, but the most common are the cayenne, jalapeño, cherry, and *cuerno del diablo* ("horns of the devil"), along with a general collection of many hues and shapes.

The jalapeño is short, green, and shiny. Cherry chiles are either red or green and round like cherry tomatoes, and the cayenne is very short and conical, either green or red.

Hot peppers can be pickled whole, chopped in salads and sauces, or cooked along with any recipe that calls for a little added excitement, like a guacamole or memorable salsa. Spanish, Chinese, Indian, South American, and Mexican recipes demand the heat as part of the pleasure. If you plan to dry your fresh hot chiles, make sure they are free from rot, cracks, and wet spots. String them together through the stems with a needle and heavy thread. Hang the bunch in a warm, dry area with good air circulation; avoid direct sunlight and dampness. Wash your hands thoroughly before touching your eyes or lips. Like other capsicums chiles should be firm, bright, and fresh-stemmed. Chiles are available year-round, but their most plentiful season is late summer through the fall.

Red hot peppers are extremely high in vitamin A, with huge amounts of niacin, potassium, and calcium. Unfortunately, they cannot be eaten by North Americans in an amount significant enough to nourish anything. The cayenne variety, however, is an excellent stomach and liver cleanser, and the powdered form may be used regularly in vegetable juices and salad dressings.

NUTRITIONAL DATA

Weight (g)	Calories	Protein (g)	Carbohydrates (g)	Crude Fiber (g)	Water (g)	Fat (g)	Cholesterol (mg)
100	93	3.7	18.1	9.	—	2.3	—
50	160	6.4	29.9	3.1	—	4.5	—

Vitamin A (IU)	Vitamin C (mg)	Thiamine B-1 (mg)	Riboflavin B-2 (mg)	Niacin (mg)	Vitamin B-6 (mg)	Vitamin B-12 (mcg)	Folic Acid (mcg)	Pantothenic Acid (mg)
21600	369	0.22	0.36	4.4	—	—	—	—
38500	6	0.11	0.66	5.2	—	—	—	—

Sodium (mg)	Potassium (mg)	Calcium (mg)	Phosphorus (mg)	Magnesium (mg)	Iron (mg)	Zinc (mg)	Copper (mg)	Manganese (mg)
9	420	29	78	27	1.2	—	—	—
186	600	65	120	—	3.9	—	—	—

Red, raw—3½ ounces
Red, dried—1¾ ounces

STRING BEANS

*B*eans are beans. Dried beans, shell beans, pod beans, green and yellow string beans, are all the same. Dried beans are the last stage in the development of the many different cultivars of the general category known as beans. Many beans, such as kidney, cranberry, Scotch, cannellino, and long beans, are sold in the pod or shell, then removed and dried and, later, soaked and cooked. Harvesting at an earlier stage of development produces the green string bean and yellow wax bean. These are the most popular of all the fresh beans and deserve the most attention. Fresh string beans have a wonderful, bright taste and are an excellent source of vegetable nutrition.

There are several varieties of green string beans, including the larger flat-pod string bean called Kentucky Wonder. When fresh and velvety, Kentucky Wonders have a marvelous sweet taste and an excellent crunchy texture. Eight inches long and bright green, they are grown on poles. Try them when you see them.

String beans are available all year long, but with short supplies and high prices in midwinter, midsummer and midfall. I prefer the early crop of spring beans. They are smallish and round, with a very noticeable velvet down and tender sweetness. No matter the season, the very best beans come from local sources and are in the markets within one or two days of picking. At any time of the year look for string beans with that fresh velvety coat. The official U.S.D.A. designation for string beans is "snap beans," so if they don't, don't buy them. The pods should be wrinkle-free, with only subtle definition of the beans within. Pods with engorged, heavy beans will taste tough and bitter. Children will hate them even more.

There is a definite difference between hand-picked and machine-picked beans. Machine-picked beans are picked indiscriminately all at one time—immature beans, overgrown beans, and peak beans. The machine also accidentally cuts and crushes some of the beans. It also takes along leaves, vines, and stems. Machine-picked beans sell for less at the wholesale level, but after the storekeeper sorts out the culls and the plant material, he must raise the price to compensate for the weight loss. Hand-picked beans, usually superior in every way, sell for the same price or less. Although a slight mix of smaller and larger beans makes for a good texture when cooked, the best, sweetest taste comes from evenly picked, evenly matured beans. The only person who seems to have gained from the use of the bean-picker is the guy who sold the machine!

String beans are very high in mineral content, with significant amounts of vitamins A, B-1, and B-2. Raw string beans help metabolize calcium, support adrenal production, and make excellent high-energy vegetable drinks when blended with carrots, celery, and spinach.

NUTRITIONAL DATA

Weight (g)	Calories	Protein (g)	Carbohydrates (g)	Crude Fiber (g)	Water (g)	Fat (g)	Cholesterol (mg)
125	31	2	6.8	1.2	115.5	0.2	—

Vitamin A (IU)	Vitamin C (mg)	Thiamine B-1 (mg)	Riboflavin B-2 (mg)	Niacin (mg)	Vitamin B-6 (mg)	Vitamin B-12 (mcg)	Folic Acid (mcg)	Pantothenic Acid (mg)
675	15	0.09	0.11	0.6	—	—	—	—

Sodium (mg)	Potassium (mg)	Calcium (mg)	Phosphorus (mg)	Magnesium (mg)	Iron (mg)	Zinc (mg)	Copper (mg)	Manganese (mg)
5	189	62	46	—	0.8	—	—	—

Cooked—1 cup

String Beans in Tomato Sauce

1½ pounds string beans
3 cloves garlic, sliced
Olive oil
2 cups tomato sauce

Snap off string bean ends and break into 1-inch pieces. Wash and keep in cold water. Sauté garlic in olive oil to coat bottom of saucepan until just before it browns. Add wet string beans. Cover. Simmer over medium heat. When string beans are hot, add tomato sauce. Cover. Simmer over low heat until beans soften. *Serves 4.*

String Beans Vinaigrette

1½ pounds string beans
3 large shallots or garlic cloves
6–8 sprigs curly parsley, chopped
2–3 sprigs fresh thyme, marjoram, or winter savory, chopped
½ cup apple cider vinegar
¼ cup salad oil
¼ cup lemon juice

Snap off string bean ends, leaving beans whole. Wash and steam in 1 inch water until firmness gives way but beans are still bright green. Crush shallots with herbs. Mix vinegar, salad oil, and lemon juice with herb mixture. Stir. Drain beans and place in medium deep bowl. Pour vinaigrette over beans and sprinkle with parsley. Cover and chill. Serve cold, with dollop of mayonnaise, sour cream, or cottage cheese. *Serves 4.*

Bright Green Beans

1½ pounds string beans
3–4 sprigs dill, chopped
3–4 sprigs summer savory, chopped
2 tablespoons butter
Salt and freshly ground black pepper

This recipe is especially designed for garden-fresh, locally produced, tender velvet-skin beans.

Add ¼ inch water to shallow pan. Snap off string bean ends. Place whole beans neatly in pan no more than 2 or 3 deep. Sprinkle with herbs. Top with butter. Cover. Cook over medium heat for 6 to 8 minutes. Lift beans from pan. Spoon liquid over beans. Season to taste. *Serves 4.*

BUTTER BEANS

1½ pounds butter beans (yellow beans, wax beans)
¼ cup melted butter or margarine
Juice of ½ lemon
2 tablespoons maple syrup

Snap off butter bean ends. Break into 1-inch pieces. Place beans in saucepan with ½ inch water. Cover. Cook for 8 to 12 minutes over medium heat. Drain, leaving some water. Add melted butter, lemon juice, and maple syrup. Heat over low heat. Mix. *Serves 4.*

STRING BEAN SALAD

1 pound string beans
2 medium Jerusalem artichokes
3–4 radishes
4 sprigs summer savory, dill, or tarragon
4 Boston lettuce leaves
4 Romaine lettuce leaves
3–4 scallions
Oil and vinegar dressing
1 small clove garlic, chopped
1 small piece ginger, grated

Snap off string bean ends. Break into 1-inch pieces. Drop into boiling water and cook for 1 to 2 minutes. Drain and chill. Scrub and slice artichokes and radishes. Chop summer savory. Wash and tear lettuce. Chop scallions. Mix all ingredients with beans. Serve with oil and vinegar dressing flavored with garlic and ginger. *Serves 4.*

RED, WHITE, AND GREEN GREEN BEANS

1½ pounds string beans
8 small white onions
1 red pepper
2 tablespoons butter or margarine
3–4 sprigs summer savory, tarragon, or dill, chopped

Snap off string bean ends. Break into 1-inch pieces. Peel and blanch onions. De-seed and dice red pepper. Cook beans in covered pot in ¼ inch water for 8 to 10 minutes. Sauté red pepper in butter along with summer savory. Drain beans until almost dry. Add onions and beans to herb sauté. Mix. Cover. Simmer over low heat for 3 to 4 minutes. *Serves 4.*

STRING BEANS WITH BUTTERED ALMONDS

> **1½ pounds string beans**
> **4 tablespoons butter or margarine**
> **1 tablespoon brown sugar**
> **½ cup blanched almonds, sliced**

Snap off string bean ends. Cook whole beans in covered pot in ½ inch water over medium heat for 8 to 10 minutes. Melt butter in shallow pan. Add brown sugar. Add almonds when butter is hot. Stir for 1 to 2 minutes over low heat. Drain beans. Place over almond mixture. Cover. Heat for 3 to 4 minutes. *Serves 4.*

STRING BEAN CRISP

> **1½ pounds string beans**
> **4 tablespoons butter or margarine**
> **3 shallots, chopped**
> **1 cup bread crumbs**

Snap off string bean ends. Cook whole beans in shallow pan in ¼ inch water for 8 to 10 minutes. Drain. Melt butter in sauté pan. Add chopped shallots. Sauté until shallots begin to brown. Add beans mixed with bread crumbs. Over medium heat turn beans until coated and thoroughly heated. *Serves 4.*

FRENCH-FRIED STRING BEANS

> **1½ pounds string beans**
> **1 cup vegetable oil**
> **3–4 cloves garlic, sliced**
> **Dash freshly ground black pepper**
> **½ cup flour**

Snap off string bean ends. Blanch in boiling water for 1 minute. Remove. Place on paper towel. Heat oil and brown garlic. Remove garlic. Add pepper. Roll beans in flour. Place in hot oil, turning until crisp and slightly brown and blistery. Remove. Drain on paper towels. *Serves 4.*

PEAS

Peas are one of my favorite vegetables. They are the first vegetable out of the garden in the spring after lettuce and beet thinnings. I love to see peas in our store. It is a sure harbinger of the fresh garden produce that fills our land every summer. Fresh peas have the sweetest, nicest taste of all the vegetables, and I even enjoy the mindless work of shelling peas. I sit right on the porch, shelling and enjoying the late afternoon and watching the bowl fill up with bright green baby pearls.

Most peas are available either canned or frozen. Canned peas have a dull, grayish look, and their canning liquid is inedible. Frozen peas collapse after cooking. I don't mind a few wrinkles, but a dish of chemically green, totally wrinkled peas is more than I want to deal with. Fresh green peas are bright, round, and firm. They can be eaten raw or can be cooked to perfection quickly and easily. A covered saucepan with a touch of sautéed onion and garlic and freshly picked and washed peas will produce perfect peas in less than five minutes.

Fresh peas, if available at all, begin to appear in late April and are available until mid- to late June. Look for pods that are firm and unwrinkled, with fresh leaves at the stem end. Fat, overgrown peas will taste bitter and tough. Inside the pod the peas should be bright, uniform, unsprouted, and firm.

In addition to the spring peas there are Chinese snow pod peas (which are also known as Mennonite peas and are grown heavily around Lancaster, Pennsylvania). These are eaten whole and are available year-round. They make excellent side dishes, mix well with other vegetables, and add life and crunch to salads. Many local stores do not carry this variety, and they are never inexpensive. A half pound of snow peas, however, is a lot.

A new variety of peas called Sugar Snap is beginning to find its way to the stores. This variety looks like the common fresh pea pod but can be eaten whole, pod and pea, and has a delicious, sweet flavor.

Peas seem to have been with us since around 10,000 B.C. It was the Italians who developed the fine art of preparing fresh peas and weaned Europe from the dried version. Later the French under Louis XIV made peas a palace rage, and the tiny Italian *piselli novelli* became *petits pois* the world over forever. Christopher Columbus planted peas in the New World, and Thomas Jefferson carried on with thirty varieties in his Monticello garden.

Peas of all kinds go lots of ways. They mix well with all vegetables, not just carrots. I add them to salads, *ceviche,* sauces, gravies, soups, and stir-fry combinations and especially like them barely blanched in a cold tuna or salmon salad. If shelling peas is just too much for you, I offer a simple way of quickly cooking the whole pea, pod and all, in a very hot sauté. You then eat the peas, still in the pod, like an artichoke leaf. Adults love the taste. Children love the idea.

Although fresh peas lose a little in cooking, they are extremely nutritious, with a broad range of healthful food elements. They are almost seven per cent protein (dried peas are twenty per cent), with very high amounts of potassium, phosphorus, and iron. Peas are also an excellent source of many B vitamins, A, and C.

NUTRITIONAL DATA

Weight (g)	Calories	Protein (g)	Carbohydrates (g)	Crude Fiber (g)	Water (g)	Fat (g)	Cholesterol (mg)
100	84	6.3	14.4	2	78	0.4	—
100	71	5.4	12.1	2	81.5	0.4	—

Vitamin A (IU)	Vitamin C (mg)	Thiamine B-1 (mg)	Riboflavin B-2 (mg)	Niacin (mg)	Vitamin B-6 (mg)	Vitamin B-12 (mcg)	Folic Acid (mcg)	Pantothenic Acid (mg)
640	27	0.35	0.14	2.9	—	—	—	—
540	20	0.28	0.11	2.3	—	—	—	—

Sodium (mg)	Potassium (mg)	Calcium (mg)	Phosphorus (mg)	Magnesium (mg)	Iron (mg)	Zinc (mg)	Copper (mg)	Manganese (mg)
2	316	26	116	35	1.9	—	—	—
1	196	23	99	—	1.8	—	—	—

Raw—¾ cup
Cooked—⅔ cup

PEAS AND ONIONS

1 pound peas
1 medium Spanish onion
Light vegetable oil

Shell peas and place in cold water. Chop onion and sauté in oil until lightly browned. Place in saucepan while still hot. Drain peas. Place in saucepan. Cover. Simmer over low heat for 3 to 5 minutes. Stir. *Serves 2.*

PEAS SAUTÉED WHOLE IN THEIR PODS

1 pound peas
1 small yellow onion
1 shallot
1 clove garlic
4 tablespoons butter
Dash tamari or soy sauce

Wash unshelled peas. Drain. Finely chop onion, shallot, and garlic. Melt butter in sauté pan. Add onion, shallot, and garlic. Sauté until soft. Add unshelled peas. Stir. Sprinkle with tamari. Cover. Simmer for 3 to 5 minutes.

Peas are eaten whole like artichoke leaves. The outer flesh of the pod has a delicious nutty flavor, and the peas will pop out in your mouth. *Serves 2 to 3.*

FRESH PEA SALAD

3–4 radishes
2–3 scallions
1 shallot
1 small red onion
2 sprigs parsley
1 pound peas
2 tablespoons apple-cider vinegar
Juice of 1 lemon

Finely chop first five ingredients. Shell peas and blanch for about 30 seconds. Mix peas with chopped ingredients. Add vinegar and lemon juice. Mix. *Serves 2 to 3.*

SAUTÉED PEAS

1 pound peas
2–3 sprigs thyme, dill, or marjoram, chopped
2 shallots, chopped
1 small yellow or white onion, chopped
2 tablespoons butter or margarine

Shell peas. Melt butter and sauté thyme, shallots, and onion. Add peas. Stir. Sauté for 5 to 7 minutes. *Serves 2.*

LIMA BEANS

Lima beans are fast disappearing from our fresh food stores. Blame it on convenience. They are not easily opened and are not inexpensive to buy, but should you find them, the taste and flavor of properly prepared fresh lima beans will reward you many times over for the effort involved. The delicate texture and the aromatic flavor of fresh lima beans have nothing to do with the common taste of lima beans that have been frozen, canned, or poorly prepared. Search out the fresh lima. It is an outstanding garden vegetable that is easily cooked once depodded, and offers outstanding nutritional value along with its tender, melt-in-your-mouth flavor.

I wish I could say more about fresh lima beans, but unfortunately they are almost unavailable. In the spirit of "half a loaf is better than none," though, try using frozen Fordhook baby lima beans, just barely cooked, with a touch of butter and pepper.

Lima beans are almost eight per cent protein. They provide carbohydrate energy, lots of B vitamins, and some of the highest mineral contents of any vegetable.

NUTRITIONAL DATA

Weight (g)	Calories	Protein (g)	Carbohydrates (g)	Crude Fiber (g)	Water (g)	Fat (g)	Cholesterol (mg)
100	111	7.6	19.8	1.8	71.1	0.5	—

Vitamin A (IU)	Vitamin C (mg)	Thiamine B-1 (mg)	Riboflavin B-2 (mg)	Niacin (mg)	Vitamin B-6 (mg)	Vitamin B-12 (mcg)	Folic Acid (mcg)	Pantothenic Acid (mg)
280	17	0.18	0.1	1.3	—	—	—	—

Sodium (mg)	Potassium (mg)	Calcium (mg)	Phosphorus (mg)	Magnesium (mg)	Iron (mg)	Zinc (mg)	Copper (mg)	Manganese (mg)
1	422	47	121	—	2.5	—	—	—

Cooked—⅝ cup

BUTTERED LIMAS

1 pound lima beans
1 tablespoon honey
3–4 tablespoons butter or margarine

Place ¼ inch water in saucepan. Add honey and heat until it melts. Add lima beans. Cook over medium heat for 10 to 12 minutes. Add butter just before serving. *Serves 2.*

SUCCOTASH

1 pound lima beans
2 ears white or yellow corn
¼ teaspoon freshly ground black pepper

In large pot place 1 inch water. Cover and bring to boil. Uncover and add husked corn. Boil for 3 to 5 minutes. Place ½ inch water and shelled lima beans in saucepan. Cover and cook over medium heat for 5 to 7 minutes. Drain lima beans and remove corn. Let corn cool. Remove all kernels from cobs and mix with cooked lima beans. Add pepper. Mix. Place in saucepan in ¼ inch water. Cover and cook for 3 to 5 minutes over low heat, stirring frequently. *Serves 2.*

ZUCCHINI

Zucchini is the most bountiful and the most common of all garden vegetables. It is easily grown and easily shipped, and it keeps fairly well when properly handled. Zucchini, however, does not like very cold temperatures and will lose its flavor if over-refrigerated. Zucchini is originally an Italian summer squash. Once pale with dark-green stripes, the most common variety now is a dark-green club-shaped vegetable, grown universally.

The best zucchini are the smallest zucchini. Sure, larger zucchini make excellent ratatouille, parmigiana, bake boats, and stir-fries, all of which are delicious, but smallest is still best. Some farm stands and knowledge-able vegetable stores sell very small finger-size zucchini with the blossom still attached. Without a doubt this is the most innocent vegetable you will ever eat.

Zucchini also represents a general class of summer squashes that are sold and displayed together in almost every store. These include the yellow zucchini (which looks exactly like its green brother), yellow squash (which is rounder and bumpier), the striated zucchini, and the round zucchini. All summer squashes, if allowed to remain on the vine, will grow to enormous size and develop a thick, hard skin that helps, with proper care, keep the vegetable for many, many weeks. Unfortunately, much of the sweet, tender summer taste is lost in the winter state. Small summer squashes, however, are now available year-round from Mexico, the Southern states, and California. Sometimes, when supplies are short in early spring and late fall, prices get out of hand, but mostly the summer squashes are reasonably priced and of excellent quality. Look for bright colors, fresh stem end cuts, and firmness. Really fresh summer squash has a light coat of fine prickly hairs I call the bloom of life. Soft, black-spotted, balloony zucchini is worthless.

To grow zucchini is to know zucchini. Once it is planted, only a hard frost will stop its incessant production. Centuries of abundance have created endless variations on the zucchini theme. Zucchini can be eaten raw, steamed, fried, stuffed, baked, sautéed, or added to salads. Zucchini can be pickled, baked in bread, or cooked at its peak and frozen for a winter meal. During the gardening months even the blossoms are eaten.

Zucchini is an excellent vegetable as part of a regular diet. It is very low in calories, is easily digested, and has high amounts of vitamin A, potassium, and iron. Zucchini is a recommended vegetable for people taking medication for high blood pressure, as it replaces vital minerals lost to antihypertension medicines.

NUTRITIONAL DATA

Weight (g)	Calories	Protein (g)	Carbohydrates (g)	Crude Fiber (g)	Water (g)	Fat (g)	Cholesterol (mg)
100	19	1.1	4.2	0.6	94	0.1	—
100	14	0.9	3.1	0.6	95.5	0.1	—

Vitamin A (IU)	Vitamin C (mg)	Thiamine B-1 (mg)	Riboflavin B-2 (mg)	Niacin (mg)	Vitamin B-6 (mg)	Vitamin B-12 (mcg)	Folic Acid (mcg)	Pantothenic Acid (mg)
410	22	0.05	0.09	1.0	—	—	—	—
390	10	0.05	0.08	0.8	—	—	—	—

Sodium (mg)	Potassium (mg)	Calcium (mg)	Phosphorus (mg)	Magnesium (mg)	Iron (mg)	Zinc (mg)	Copper (mg)	Manganese (mg)
1	202	28	29	16	0.4	—	—	—
1	141	25	25	—	0.4	—	—	—

Raw—½ cup
  Boiled—½ cup

Zucchini Salad

Juice of 1 lemon
¼ cup vegetable oil
1 tablespoon apple-cider vinegar
¼ teaspoon freshly ground black pepper
1 clove garlic
½ small red onion
2 stalks celery
2 sprigs parsley
2–3 zucchini
Dash dried basil

Blend lemon juice, oil, vinegar, and pepper. Finely chop garlic, onion, celery, and parsley. Slice unpeeled zucchini. Toss with chopped ingredients. Add basil and lemon juice mixture. Chill. *Serves 4.*

Breaded Zucchini Slices

1 small yellow onion
3–4 cloves garlic
6–8 sprigs flat-leaf parsley
2 cups bread crumbs
Dash freshly ground black pepper
2 eggs, beaten
3 medium or 6 small zucchini
Light vegetable oil

Chop onion, garlic, and parsley. Add bread crumbs and pepper. Mix. Beat eggs and place in shallow bowl. Slice unpeeled zucchini into thin slices lengthwise. Cover bottom of frying pan with thin layer of oil. Heat until hot. Dip zucchini slices in egg, then bread crumbs, coating both sides thickly. Place in frying pan and brown. Turn and brown other side. Remove. Drain on paper towels. Serve on flat plate as appetizer or hors d'oeuvre. *Serves 4.*

Zucchini Sticks

1 medium zucchini
¼ small red onion, chopped
1 clove garlic, chopped

6–8 sprigs parsley, chopped
3–4 sprigs oregano, chopped, or ½ teaspoon dried oregano
Juice of 1 lemon
2 tablespoons light vegetable oil
2 tablespoons apple-cider vinegar

*P*eel zucchini and slice lengthwise. Place each slice flat on cutting board and cut lengthwise into sticks ¼ to ½ inch wide, discarding seed strips. Cut sticks into 3- to 4-inch lengths.

Chop onion, garlic, and herbs. Puree in blender with remaining ingredients. Pour into small bowl, and place zucchini sticks upright in sauce. *Serves 2.*

ZUCCHINI CHUNKS

Light vegetable oil
1 tomato, peeled and cut up
Pinch chili powder
Freshly ground black pepper
2 cloves garlic, chopped
1 small yellow onion, chopped
1 large or 2 medium zucchini (1 pound)
4 ounces mozzarella or cheddar cheese, grated

*H*eat thin layer of oil in shallow pan. Add tomato, chili powder, and pepper. Cover. Simmer over low heat for 10 minutes. Add garlic and onion. Stir. Simmer for 5 minutes.

Slice zucchini into quarters lengthwise. Cut into chunks. Add zucchini to tomato sauce mixture. Cover. Simmer for 5 minutes. Uncover. Add cheese. Do not stir. Cover. Simmer until cheese melts. *Serves 3 to 4.*

ZUCCHINI PARMIGIANA

Light vegetable oil or virgin olive oil
2 cloves garlic, chopped
1 small yellow onion, chopped
1 cup tomato, peeled and cut up
6–8 leaves basil, chopped
¼ teaspoon freshly ground black pepper
10 sprigs oregano, chopped, or 2 tablespoons dried oregano
1 large or 2 medium zucchini (1 pound)

½ cup ricotta cheese
4 ounces mozzarella or cheddar cheese, grated
2 ounces Parmesan cheese, grated

*H*eat thin layer of oil in shallow pan. Add garlic and onion. Brown lightly. Add tomato, basil, pepper, and half the oregano. Cover and simmer for 10 to 15 minutes.

Peel and slice zucchini lengthwise. Place in shallow baking pan. Sprinkle with oil and remaining oregano. Broil until zucchini slices begin to brown. Remove. Cool. Add ricotta cheese to tomato sauce. Stir. Simmer for 5 minutes. In shallow baking dish place layer of tomato sauce, followed by layer of zucchini slices topped with layer of sauce, then grated mozzarella cheese, then layer of zucchini slices. Repeat layers. Finish with dusting of grated Parmesan cheese. Bake in preheated 350° F. oven for 15 to 20 minutes. *Serves 2 to 3.*

CREAMY GARLIC AND ZUCCHINI SALAD

4–6 cloves garlic
6–8 sprigs flat-leaf parsley
¼ cup light vegetable oil
2 tablespoons white-wine vinegar
Dash freshly ground black pepper
1 large or 2 medium zucchini (1 pound)
½ red pepper

*C*oarsely chop garlic and parsley. Blend all ingredients together in blender except zucchini and red pepper. Peel and cut zucchini into small chunks. Finely chop red pepper. Place zucchini in bowl. Pour dressing over, and sprinkle red pepper on top. Chill. *Serves 2 to 3.*

STUFFED ZUCCHINI BOATS

1 very large hard-skinned zucchini
1 small eggplant
Light vegetable oil
1 small yellow onion, chopped
2 cloves garlic, chopped
¼ teaspoon freshly ground black pepper
½ teaspoon dried oregano

2 tablespoons pignola nuts
1 tablespoon seedless raisins
8–10 sprigs flat-leaf parsley, chopped
2–3 ounces freshly grated Parmesan cheese

Cut zucchini in half lengthwise. Remove center, leaving 1 inch zucchini meat intact all around. Cut zucchini and eggplant meat into medium chunks. In heated oil sauté zucchini, eggplant, onion, garlic, pepper, and oregano. When soft, remove from pan and place in bowl. Mix with nuts, raisins, and parsley.

Place zucchini shell face down in shallow baking dish in ¼ inch water. Bake in preheated 350° F. oven until shell can be pierced with fork. Do not overcook shell. Spoon in sautéed ingredients. Sprinkle with grated Parmesan cheese. Place back in oven for 10 minutes. Then broil quickly for 1 to 2 minutes until top is slightly crisp. *Serves 4.*

ZUCCHINI-VEGETABLE MIX

Light vegetable oil
1 small yellow onion, chopped
1 clove garlic, chopped
2 shallots, sliced
½ pound broccoli
½ pound string beans
1 small eggplant
1 medium or 2 small zucchini

Heat thin layer of oil in sauté pan. Brown onion and garlic. Slice shallots. Blanch in boiling water until slightly soft. Drain. Cut up broccoli florets and slice stalk. Blanch stalk until soft. Drain. Snap off string bean ends. Blanch until soft. Drain. Peel eggplant and zucchini and cut up into medium chunks. Place eggplant chunks in garlic sauté. Cover pan and simmer until eggplant softens. Add zucchini and blanched vegetables. Cover and heat over low heat for 3 to 5 minutes. *Serves 4.*

PICKLED ZUCCHINI

10 pounds zucchini
3 cloves garlic, sliced
8 cups sugar
4 cups apple-cider vinegar

2 tablespoons pickling spice
5 teaspoons pickling salt

*P*lace zucchini in large container and cover with boiling water. Leave overnight, draining next day. Repeat each day for four days. Cut zucchini into 2-inch chunks and return to container. Combine remaining ingredients in saucepan. Heat to boiling and pour over zucchini chunks. After 24 hours drain off liquid and heat to boiling. Pour over zucchini. Repeat again after 24 hours. Pack zucchini into sterilized jars and fill with liquid, leaving ½-inch space at top of jar. Seal (see "Canning process" entry in Glossary, page 000).

BROCCOLI

*B*roccoli is another gift from the Italians. Prized by the Romans, broccoli was developed from cabbage ancestry by the Etruscans, wizards of edible horticulture. It is one of the few Western-world vegetables that have been adopted by the Chinese, and so far only the Chinese and Italians seem to have mastered the art of cooking it.

Despite centuries of careful cultivation, broccoli was introduced commercially in the United States only around 1930. Maybe that's why so much broccoli is carried home to a gray, noxious, mushy death. I shudder to think of all the dinner-table conflicts gray broccoli has caused for youngsters. I have a picture in my head of thousands of wilted, sad broccoli portions being scraped from the dinner plates of American broccoli-haters. Broccoli deserves better. Properly prepared so it reaches the table crisp and green and sparkling, broccoli is a delicious, tender vegetable. Actually, broccoli is eaten as an unripe plant. If allowed to continue life in the garden, the tiny, tiny, blue-green buds of the broccoli head will burst open and shoot forth hundreds of yellow flowers.

Most broccoli sold in the United States is grown in California. These California varieties are usually heavily stalked, with large floret heads. Obviously California broccoli is shipped great distances, and one should choose very carefully, with an eye for brightness and a nose for freshness. Look for tight blue-green buds with many small branches forming the head. Choose the smaller stalk bunches when possible, and avoid dry, woody stems, yellowish or wilting florets, and any hint of strong aroma. In late spring and early fall many roadside stands and vegetable markets carry broccoli from local farms. These varieties are always thin-stalked, blue-budded, and tender-textured. Local fresh-picked broccoli cooks with no odor.

Excellent broccoli is available year-round except for the hottest days of midsummer. Like all produce, broccoli has its short-supply season when prices become uncomfortably high. If you see broccoli with a high price and less than great quality, pass.

The best way to cook broccoli is as little as possible. Some purists say never let the flower head touch the water. In quick cooking it doesn't matter. I try to cut the stalks and the heads lengthwise into spears of equal thickness. After sautéing a bit of garlic and onion, I add the washed, wet broccoli with two to three ounces of water, cover, and cook over a medium heat for five to seven minutes. No more. Sometimes I also sprinkle on grated cheese, or add a little lemon-butter sauce, or sprinkle finely chopped nuts on the broccoli before cooking. Broccoli should be cooked and served in large pieces, and served and eaten bright-green and crisp. An absolute broccoli rule is that the smaller the pieces, the less time and water should be used. Raw broccoli is also an excellent vegetable served with cheese dips as an appetizer; with lemon, pepper, and celery chunks as a salad; or blanched and chilled, with lemon or a vinaigrette.

Broccoli is a highly nutritious vegetable. It is low in calories and carbohydrates, with massive amounts of vitamins A and C, calcium, niacin, potassium, and iron. Broccoli loses twenty to thirty per cent of its vital nutrients in cooking, so once again the less you do to broccoli, the more it will do for you.

NUTRITIONAL DATA

Weight (g)	Calories	Protein (g)	Carbohydrates (g)	Crude Fiber (g)	Water (g)	Fat (g)	Cholesterol (mg)
100	32	3.6	5.9	1.5	89.1	0.3	—
100	26	3.1	4.5	1.5	91.3	0.3	—

Vitamin A (IU)	Vitamin C (mg)	Thiamine B-1 (mg)	Riboflavin B-2 (mg)	Niacin (mg)	Vitamin B-6 (mg)	Vitamin B-12 (mcg)	Folic Acid (mcg)	Pantothenic Acid (mg)
2500	113	0.1	0.23	0.9	—	—	—	—
2500	90	0.09	0.2	0.8	—	—	—	—

Sodium (mg)	Potassium (mg)	Calcium (mg)	Phosphorus (mg)	Magnesium (mg)	Iron (mg)	Zinc (mg)	Copper (mg)	Manganese (mg)
15	382	103	78	24	1.1	—	—	—
10	267	88	62	—	0.8	—	—	—

Raw—1 stalk
  Cooked—²/₃ cup or 1 large stalk

Broccoli Hollandaise

  1 bunch broccoli
  4 tablespoons lemon juice
  4 egg yolks
  ½ teaspoon salt
  ¼ teaspoon cayenne pepper or freshly ground black pepper
  1 cup butter or margarine

Trim tough bottoms off broccoli stalks. Slice heavy stalks lengthwise. Leave smaller stalks whole. Place in saucepan with 1 inch water with half the lemon juice. Cover. Cook until soft but still bright green. Drain liquid. Cover. Keep warm.

  In blender combine egg yolks, salt, and pepper. Blend slightly. Heat butter until melted. Do not burn. Slowly pour butter into blender while running on low speed. Blend until just thickened. Blend in remaining lemon juice. Place broccoli in shallow bowl. Cover with sauce. Serve immediately. *Serves 4 to 5.*

Steamed Whole Broccoli With Lemon

  1 bunch broccoli
  Juice of 2 lemons
  Grated peel of 1 lemon

Trim tough bottoms off broccoli stalks. Place in large saucepan in 1 inch water. Pour lemon juice over broccoli. Top with grated peel. Cook over medium heat until broccoli softens but is still bright green. *Serves 4.*

Baked Broccoli in Cheese

  1 large bunch broccoli
  Butter
  3–4 sprigs dill, thyme, or tarragon, finely chopped
  ½ pound Monterey jack, Swiss, or white
  cheddar cheese, grated

Trim tough bottoms off broccoli stalks. Slice broccoli lengthwise through centers of stalks. Drop into boiling water and blanch until slightly softened. Butter shallow baking dish. Sprinkle with dill and pepper. Place broccoli in pan cut side down.

Cover with grated cheese. Cover pan with aluminum foil and place in preheated 350° F. oven for 15 minutes. Remove foil. Raise oven temperature to 450° F. and bake for 5 minutes.   *Serves 4.*

CREAMED BROCCOLI SOUP

1 bunch broccoli
1 small onion, chopped
1 stalk celery, chopped
½ teaspoon butter or margarine
3 tablespoons flour
3 cups milk
1 teaspoon salt
Dash freshly ground black pepper
2 sprigs savory, chopped

*R*emove heavy stalks from broccoli and break or cut florets into small pieces. Sauté onion and celery in butter until tender. Stir in flour. Gradually stir in milk and add broccoli pieces. Heat until mixture thickens and boils, stirring constantly. Add salt, pepper, and savory.   *Serves 4.*

CRISPED BROCCOLI

1 bunch broccoli
4 eggs
1 cup flour
1 small yellow onion, chopped
1 clove garlic, chopped
Dash freshly ground black pepper
Light vegetable oil

*T*rim tough bottoms off broccoli stalks. Break florets into medium pieces and slice stalks. Blanch stalks in boiling water until soft. Drain and dry. Separately blanch florets quickly. Drain.

Beat eggs. Place flour in flattish bowl and mix with half of the onion, garlic, and pepper. Heat ⅛ inch oil in frying pan. Dip broccoli pieces in egg, roll in flour, and place in heated oil along with remaining onion, garlic, and pepper. Once browned and crisp, remove and drain on paper towels.   *Serves 3 to 4.*

CAULIFLOWER

Cauliflower is the queen of the garden vegetables. I am always startled and impressed when I unwrap the leaves of a fresh-picked cauliflower and see that brilliant white "flower" glistening in its own dew. Wow! Farm-fresh cauliflower is picked, sprayed with cold water, packed with all its leaves, and chilled. It is usually sold the next day or so to local wholesalers, farm stands, and fruit and vegetable markets. This beautiful fresh vegetable is hardly related to the soft, pale, cellophane-wrapped variety. Shipping, temperature, and handling take their toll on cauliflower. So does cooking. Cauliflower suffers badly from overcooking, which produces a mushy, bland, bad-smelling vegetable that frightens children and ruins dinnertime. Cauliflower is actually thousands of tiny white flower buds closely packed into even larger buds, forming florets that in turn form the head or "flower."

Like all flowers, cauliflower must be treated carefully in all stages. While the plant is growing, the cauliflower leaves are raised up and tied around the head. The white head grows inside the cool, shady leaf cave. Fresh raw cauliflower is sweet and tender and makes an excellent appetizer with dips or in salads and vegetable mixes. I prefer to cook the whole head in one to two inches of water for about ten minutes. I also love my breaded cauliflower recipe, but the best is the whole head sprinkled with pepper and melted butter, or cheese and parsley, or whatever taste delights come to mind.

Cauliflower greens are delicious and edible. Perhaps our tastes shy away from a whole passel of cauliflower leaves, but some can be cooked with the head, stir-fried, braised, or added to soup like cabbage.

Look for big, hard, dewy-fresh heads with bright-green leaves and plenty of them. Cauliflower that is shipped and stored has all the leaves removed to save weight. Fresh-picked cauliflower keeps its leaves to keep it fresh. Sometimes the supermarket or vegetable stand will offer really fresh cellophane-wrapped cauliflower; fruit men call it cello-wrapped. It is not fresh-picked but will serve you well and should be enjoyed whenever local produce is not available.

There is also a purple cauliflower variety, an excellent seasonal vegetable that combines the taste of broccoli with the cauliflower form and texture. I like it.

The best season for cauliflower is fall and winter, but good-quality cello-wrapped Canadian cauliflower is available most of the year. Cauliflower is relatively high in protein and mineral content, with a low calorie and carbohydrate count.

NUTRITIONAL DATA

Weight (g)	Calories	Protein (g)	Carbohydrates (g)	Crude Fiber (g)	Water (g)	Fat (g)	Cholesterol (mg)
100	27	2.7	5.2	1	91	0.2	—
100	22	2.3	4.1	1	92.8	0.2	—

Vitamin A (IU)	Vitamin C (mg)	Thiamine B-1 (mg)	Riboflavin B-2 (mg)	Niacin (mg)	Vitamin B-6 (mg)	Vitamin B-12 (mcg)	Folic Acid (mcg)	Pantothenic Acid (mg)
60	78	0.11	0.1	0.7	—	—	—	—
60	55	0.09	0.08	0.6	—	—	—	—

Sodium (mg)	Potassium (mg)	Calcium (mg)	Phosphorus (mg)	Magnesium (mg)	Iron (mg)	Zinc (mg)	Copper (mg)	Manganese (mg)
13	295	25	56	24	1.1	—	—	—
9	206	21	42	—	0.7	—	—	—

Raw—1 cup pieces
Cooked—⅞ cup

Steamed Whole Cauliflower with Pepper, Butter and Lemon Sauce

1 head cauliflower
Juice of 3 lemons
4–6 tablespoons melted butter or margarine
Freshly ground black pepper

*R*emove all leaves from cauliflower. Mix lemon juice with melted butter. Pour half the mixture over cauliflower in pot. Top with generous amount of pepper. Cover and cook over medium heat for 6 to 8 minutes. When cauliflower is tender but not mushy, turn off heat. Add remaining lemon-butter sauce. Cover and allow to stand for several minutes. Remove cauliflower from pot. Serve whole head cut in wedges. *Serves 4.*

Cauliflower Cheese Melt

1 head cauliflower
¼ pound Gruyère, Parmesan, or
  cheddar cheese, grated
3–4 sprigs tarragon, savory, or marjoram, chopped
Freshly ground black pepper

*R*emove all leaves from cauliflower. Place in pot in ½ inch water. Sprinkle grated cheese, tarragon, and black pepper over cauliflower. Cover and cook for 6 to 8 minutes over medium-high heat. Let stand covered for several minutes. When tender, remove and serve whole. *Serves 4.*

Baked Cauliflower in Cheese Sauce

1 head cauliflower
4 tablespoons butter
¼ cup flour
¼ teaspoon salt
Freshly ground black pepper
2 cups milk

8 ounces Monterey jack, Parmesan, or
   cheddar cheese, grated
2 teaspoons Worcestershire sauce
Dash chili powder
Dash Tabasco sauce
Butter

*R*emove all leaves from cauliflower. Break head into medium florets. Place in pot in 1 inch water. Cover and cook over medium heat until slightly soft but not mushy. Drain.

Melt butter in saucepan. Blend in flour, salt, and pepper. Cook until smooth. Gradually stir in milk. Stir until mixture boils and thickens. Add cheese, Worcestershire sauce, chili powder, and Tabasco sauce. Butter baking dish. Place florets in dish and pour cheese sauce over. Bake in preheated 350° F. oven for 5 to 8 minutes. *Serves 4.*

BROILED CAULIFLOWER IN GARLIC BUTTER

1 large head cauliflower
¼ pound butter, softened
2–3 large cloves garlic, finely chopped
6–8 sprigs flat-leaf parsley, finely chopped
Freshly ground black pepper
½ cup bread crumbs
4 ounces Parmesan cheese, freshly grated

*R*emove all leaves from cauliflower. Break into large florets. Blanch in boiling water until slightly soft. Drain. Mix butter with garlic, parsley, and pepper. Place blanched cauliflower florets in baking dish. Spread butter mixture over florets. Sprinkle bread crumbs and cheese over entire surface. Broil for 3 to 5 minutes until top becomes crisp. *Serves 4.*

CRISPED CAULIFLOWER

1 head cauliflower
4 eggs
1 cup flour
1 small yellow onion, chopped

1 clove garlic, chopped
Freshly ground black pepper
Light vegetable oil

Remove all leaves from cauliflower. Break into medium florets. Beat eggs. Place flour in flattish bowl and mix with half the onion, garlic, and pepper. Heat ⅛ inch oil in frying pan. Dip florets in egg, roll in flour, and place in hot oil along with remaining onion, garlic, and pepper. Once brown and crisp, remove and drain on paper towels. *Serves 4.*

EGGPLANT

The idea of a large purple-black shiny mass with spiny green leaves does not suggest the egg at all. However, the original American eggplant, a member of the potato family, was the shape of a large white egg and the same color. The purple variety was developed because it showed bruise marks less and grew to a larger size. White, black, or purple, the eggplant is a large berry that contains a small number of seeds. Look for firm, heavy, shiny fruit with fresh leaves and sharp spines. An oval mark at the blossom end indicates fewer seeds and firmer flesh; a round mark indicates the opposite.

Eggplants have a wide variety of menu applications and can be prepared as blandly or spicily as you like. They can be stuffed, fried, sautéed, stir-fried, roasted whole, layered, baked, pickled, and added to a spaghetti sauce in a traditional Sicilian recipe. They cannot, however, be eaten raw.

Eggplants are available year-round, and although most stores carry only the common roundish purple eggplant, the long varieties, the small Italian eggplants, white eggplants, and Chinese eggplants each have subtle differences in taste and cooking applications. Eggplants are an excellent addition to your kitchen fare and can be prepared literally thousands of ways in simple recipes from China, India, the Arab world, and Sicily.

Eggplants are not high in nutritional content, but they are excellent cooking companions for meat, cheese, nuts, raisins, couscous, rice, and other high-powered foods.

NUTRITIONAL DATA

Weight (g)	Calories	Protein (g)	Carbohydrates (g)	Crude Fiber (g)	Water (g)	Fat (g)	Cholesterol (mg)
100	25	1.2	5.8	0.9	92.4	0.2	—
100	19	1	4.1	0.9	94.3	0.2	—

Vitamin A (IU)	Vitamin C (mg)	Thiamine B-1 (mg)	Riboflavin B-2 (mg)	Niacin (mg)	Vitamin B-6 (mg)	Vitamin B-12 (mcg)	Folic Acid (mcg)	Pantothenic Acid (mg)
10	5	0.05	0.05	0.6	—	—	—	—
10	3	0.05	0.04	0.5	—	—	—	—

Sodium (mg)	Potassium (mg)	Calcium (mg)	Phosphorus (mg)	Magnesium (mg)	Iron (mg)	Zinc (mg)	Copper (mg)	Manganese (mg)
2	214	12	26	16	0.7	—	—	—
1	150	11	21	—	0.6	—	—	—

Raw—½ cup diced
   Cooked—½ cup diced

BREADED EGGPLANT

1 large eggplant
1 small yellow onion, chopped
2 cloves garlic, chopped
2 cups bread crumbs
8–10 sprigs flat-leaf parsley, chopped
2 ounces Parmesan cheese, freshly grated
¼ teaspoon freshly ground black pepper
4 eggs
Extra-virgin olive oil

*P*eel and slice eggplant, or cut into squares or sticks. Mix onion, garlic, bread crumbs, parsley, cheese, and pepper. Beat eggs. Dip each slice eggplant into egg, then bread crumb mixture. Place in hot oil in pan and cook until brown and crisp. Drain on paper towels. *Serves 3 to 4.*

STUFFED MIDEAST EGGPLANT HALVES

1 large eggplant
1 green pepper
Olive oil
4 cloves garlic, chopped
1 medium onion, chopped
4 scallions, chopped
2 tablespoons pignola nuts
¼ cup raisins
Optional: ½ pound ground lamb, beef, or sausage
¼ cup bread crumbs
1 small piece Monterey jack or Parmesan cheese, grated

*C*ut eggplant in half. Remove meat in large pieces, leaving a shell ½ inch thick. Cut green pepper into small squares. Heat a thin layer of olive oil in pan. Sauté peppers quickly. Then add garlic, onion, scallions, and eggplant. After a minute or two add nuts and raisins. If using meat, brown separately with garlic and pepper before adding. After cooking, add bread crumbs and mix. Stuff eggplant shells with cooked mixture. Place in baking dish with shallow layer of water and bake in a preheated 350° F. oven for 20 minutes. Remove and top with grated cheese. *Serves four; serves six with meat.*

CURRIED FRIED EGGPLANT CHUNKS

1 pound lamb cubes
¼ teaspoon curry powder
½ cup raisins
1 large eggplant
Light vegetable oil
¼ cup mixed nuts, finely chopped
¼ cup shredded coconut
1 cup pineapple chunks
8–10 tiny white onions

Sauté lamb cubes until browned. Add half the curry powder. Boil ½ cup water and add raisins. Stir over medium heat until mixture thickens and raisins soften. Peel eggplant and cut into medium chunks. Add remaining curry powder and fry in thin layer of oil until soft but not mushy. Arrange lamb cubes and eggplant chunks on platter. Serve with individual bowls of nuts, coconut, pineapple, onions, and raisin sauce. *Serves 6.*

EGGPLANT PARMIGIANA

Light vegetable oil or virgin olive oil
2 cloves garlic, chopped
1 small yellow onion, chopped
1 tomato, peeled and cut up
6–8 leaves basil, chopped
¼ teaspoon freshly ground black pepper
10 sprigs oregano, chopped, or 2 tablespoons
  dried oregano
1 large eggplant
½ cup ricotta cheese
4 ounces mozzarella or cheddar cheese, grated
2 ounces Parmesan cheese, grated

Heat thin layer of oil in shallow pan. Add garlic and onion. Brown lightly. Add tomatoes, basil, pepper, and half the oregano. Cover and simmer for 10 to 15 minutes.

Peel and slice eggplant. Place in shallow baking pan. Sprinkle with oil and remaining oregano. Broil until eggplant slices begin to brown. Let cool. Add ricotta cheese to tomato sauce. Stir. Let simmer for 5 minutes. In shallow baking dish place layer of tomato sauce, followed by layer of eggplant slices topped with layer of sauce, then grated mozzarella cheese, then layer of eggplant slices. Repeat layers and finish with grated Parmesan cheese. Bake in preheated 350° F. oven for 15 to 20 minutes. *Serves 4.*

TOMATO-EGGPLANT PASTA SAUCE SICILIANA

2 cloves garlic
1 small yellow onion
Olive oil
1 teaspoon dried oregano
1 tablespoon dried basil
4 tomatoes, peeled and cut up
1 6½-ounce can tomato paste
1 medium eggplant

Chop garlic and onion. Brown in olive oil along with oregano and basil. Mix tomatoes and tomato paste in saucepan. Cover and cook over medium heat. Peel eggplant and slice into strips the full length of the eggplant. Add all ingredients to tomato mixture. Cover. Cook over low heat, stirring frequently, for 4 hours. Remove cover and simmer for 30 minutes. *Serves ½ pound pasta.*

STUFFED WHOLE WHITE EGGPLANT

4 small mushrooms
1 small yellow onion
2 cloves garlic
3 sprigs savory, tarragon, or marjoram
Olive oil
1 cup bread crumbs
4 white eggplants

Chop mushrooms, onion, garlic, and savory and sauté in olive oil. Mix thoroughly with bread crumbs. Cut a small pocket in each eggplant along its side and remove a small section of the seeds with a spoon. Pack pockets with bread crumb mixture. Place part of removed section back as a cap. Place eggplants in shallow baking dish with ¼ inch water. Cover with aluminum foil. Bake in preheated 350° F. oven for 20 to 25 minutes. Remove foil and bake at 425° F. for 10 minutes. *Serves 4.*

Barbecued Whole Eggplant

1 small yellow onion
2 cloves garlic
¼ cup olive oil
2 tablespoons Worcestershire sauce
1 teaspoon dried oregano
Dash dried red pepper flakes
1 eggplant

Chop onion and garlic and mix with olive oil, Worcestershire sauce, oregano, and red pepper. Wash eggplant and remove stem. Cut 8 to 12 1-inch slits in skin of eggplant. Roll eggplant in onion mixture. When coals are white hot, place whole eggplant on grill at medium height. Turn frequently while cooking, brushing the remainder of the mixture into slits. When eggplant is completely soft, remove. Split in half. *Serves 2 to 3.*

Corn

No vegetable in the world is more a part of a culture than is corn in America. It appears in some form at almost every meal on every table. It is used in breads and cake flour, cooking oil, baking needs, condiments, margarine, and prepared foods. Corn is the principal feed of all livestock raised for food, including meat, milk, butter, cheese, and eggs. Fresh sweet corn is one of the most popular vegetables ever "invented." Only rice is grown in greater quantity than corn. The United States produces more corn than any country in the world. Every state grows corn. Corn is being grown somewhere in the United States all the time.

Corn is an American phenomenon. Before Columbus, before Cortes, before Montezuma, the Toltecs, the Aztecs, the Incas, and the Mayas there was corn. There was corn growing between the rocks of a civilization that stretched from Canada to Peru. The people knew sun, rain, god, and corn. Corn was here when they came from other planets and carved their names in the Mayan temples high as the pyramids of Egypt. And it grew nowhere else in the world.

Indians call corn maize. "Corn" is an old European word meaning kernel. To the Indian civilizations of America the kernel's inner part, the milk, was the important part, the soul. Even today, when sweet, tender eating corn could be grown in South America, the corn that is grown is field-dried hard and ground into flour. This flour makes the meal for tortillas. Machine-stamped or hand-patted corn tortillas are the basic food of millions and millions of Latin Americans.

United States Americans like to get their corn direct from the cob. We like our corn hot, sweet, juicy, and tender, even if it has to come from a thousand miles away. Americans love summer-sweet corn. No other people in the world eat corn on the cob. Europeans think of cob corn as cow food. It's just that they don't know. Only Americans know how good corn on the cob can be—the kind with kernels that pop soon as you touch 'em with your teeth. The kernels come flying off that butter-sucking tender little cob, and before you get one done, you're reaching for another to go with the two you've already finished. Nobody knows that except Americans. The trouble is that Americans want fresh-picked summertime corn all year round. It's just not going to happen.

Thousands of tons of sweet corn are shipped all over the United States in field-filled, field-chilled trucks, trying to get it from farm to family within hours. Somewhere between the truck and the family dinner table something happens. The wholesaler has it a day or two. The retailer keeps it a day or two. When the supermarket produce manager gets it, he husks it, clips it, puts two stumpy cobs in a poly-wrapped tray, puts it on display, and calls it fresh sweet corn. That is not fresh, sweet corn. That kind of corn is a "chewee" for humans. Real sweet corn is only available in season from local sources, when it can be picked, bought, and sold in one day, if not a few hours. For most of America real homegrown sweet corn is available for three to five months each year. That's plenty of time to eat fresh, sweet, powerfully nutritious corn, plenty of time to husk and freeze a couple of bushels of fresh corn. How good it'll be in January!

There are four kinds of corn, with hundreds of varieties, hybrids, colors, and sizes. The first is field corn, or feed corn, which we won't discuss. The second kind is sweet corn, which we will.

Sweet corn is the kind we crave. Most sweet corn is yellow corn. One of the most popular varieties is Golden Bantam, a small-kerneled, light-gold, tender ear about eight inches long and not too thick. America used to like heavier, deeper-yellow, chewier corn, but tastes are changing quickly. The new bicolored varieties like Gold and Silver or Butter and Sugar are in great demand because of their novelty and downright good taste. The yellow kernels have a chewier texture than the white kernels and different degrees of sweetness. I'm a white-corn fan. Years ago there was a long, elegant ear of corn with small, very white kernels, called Country Gentleman. It was very sweet and seriously delicious but not very popular. People wanted heavy yellow corn. Nowadays, however, there is a very strong shift to white corn like Silver Queen, Silver Chief, and Bantam Queen. They are not white-white like Country Gentleman but an ivory-white. They are very sweet, with that tender skin that bursts at the bite. Today's white corn, especially Silver Queen, can be eaten raw when picked and sold fresh. It is that good.

Do not boil corn. After husking and desilking it, place the ears in one inch of boiling water, cover the pot, and let it steam for three minutes. Or pull back the husks without removing them, desilk the corn, replace the husks, soak the corn for ten minutes, place it in one inch of boiling water, cover, and let it steam for ten minutes. Or desilk, leaving the husks, soak the ears, and place them in a 350° F. oven for twenty minutes. Or desilk and soak the corn, then place it on a barbecue grill for fifteen minutes, turning frequently. Ceramic corn dishes are great for really enjoying corn on the cob. Place butter, salt, and pepper in the corn dish, and roll the hot cob in the buttery seasoning.

When buying corn there is absolutely no substitute for freshness. Any corn bought out of season is "*caveat emptor.*" All supermarket corn is questionably fresh, but sometimes what they got is all you're going to get. In season shop the roadside markets, the farm stands, and the best fruit and vegetable markets you can find. Look for bright-green, dewy husks, soft silks with dry tips, and plump, even kernels. Avoid enlarged, fat kernels or anything that is even slightly shriveled past the very tip. Worms are no longer a corn problem; the pesticides are. Actually, if I see a worm or two in a batch of fresh corn I just bought, I like it. It means fewer pesticides or none at all. I'd rather have a caterpillar or two I can see and pick off than some deadly chemical I can't.

Don't strip corn when you buy it, please. There is more to corn than ripping an ear open, jamming a thumbnail into a fat kernel, and tossing it aside if it doesn't squirt. First of all, corn has two sides, the sun side and the stalk side. The kernels on the sun side form a convex curve. They are usually broader and fuller. Kernels on the stalk side form a concave curve and are somewhat smaller and seem less developed. A customer might strip an ear of corn, look at one side, and reject it, when she might have liked the other side and made the purchase. Small, thin ears are almost always sweeter than big, fat ears. Choose corn from one or two samples, and make your choice without opening the ears. Or ask professional clerks, if you can find any, to help you. If you like what you see, feel the husks for fullness and look for the signs of freshness. If the ears feel full and fresh, and they look fresh and tender, chances are they'll taste the same way. Maybe every ear won't be perfect, but ninety-nine per cent of the time you'll be more than satisfied if the corn is fresh and you know your store. Remember, God grows 'em all, big and small.

Sweet corn is a nutritional powerhouse. It has a very high protein content. It is high in vitamins A, B-1, and B-2 and delivers extraordinary amounts of vital minerals. There is little wonder that this single food sustained incredibly advanced civilizations for thousands of years.

A third variety of corn is Indian corn, which appears in stores and markets as part of the autumn scene every year. Indian corn is left on the stalks to dry, then the husks are pulled back to reveal jewel-like kernels in endless color combinations of blue, yellow, red, purple, and white. A bundle of Indian corn on the front door invokes prosperity through the winter. The kernels can be loosened and fed to winter birds and other animals, or strung as a ceremonial necklace, or ground and made into the famous "blue" corn cakes of the American Indians of the Southwest.

The fourth variety of corn also appears in the fall as an ornamental variety. These ears, small and cone-shaped with tiny, hard kernels, are the strawberry, white, and confetti popcorns. Like Indian corn, popcorn is dried on the stalk. A survival food, popcorn could be stored for very long periods. These little ears of pretty-colored kernels can be used for fall decoration, then husked and popped or hung outside your windows to feed small winter birds.

Corn is so good to us in so many ways. Buy it carefully. Buy it at its best, and cook it right. It is one of the great foods of history.

NUTRITIONAL DATA

Weight (g)	Calories	Protein (g)	Carbohydrates (g)	Crude Fiber (g)	Water (g)	Fat (g)	Cholesterol (mg)
100	100	3.3	21	0.7	74.1	1	—

Vitamin A (IU)	Vitamin C (mg)	Thiamine B-1 (mg)	Riboflavin B-2 (mg)	Niacin (mg)	Vitamin B-6 (mg)	Vitamin B-12 (mcg)	Folic Acid (mcg)	Pantothenic Acid (mg)
400	9	0.12	0.1	1.4	—	—	—	—

Sodium (mg)	Potassium (mg)	Calcium (mg)	Phosphorus (mg)	Magnesium (mg)	Iron (mg)	Zinc (mg)	Copper (mg)	Manganese (mg)
trace	196	3	89	—	0.6	—	—	—

Yellow, cooked—on cob, 4-inch ear

PATTY-PAN SQUASH

Along with all the zucchini family and its yellow squash cousins, summertime produces a flattened, scallop-edged white squash with a firm, delicious flesh called scallop or patty pan. Patty pans are best when picked very young, about three inches in diameter or less, and still slightly greenish. They are excellent cut into pieces and sautéed, or steamed whole with a sautéed sauce of onions, garlic, salt, pepper, dill, and parsley. The larger white patty-pan squash develops a harder skin and requires more cooking. It likes being served with cheese, butter, salt, and pepper, or stir-fried with other summer squashes.

Patty-pan squash is the first summer squash to appear in the markets after zucchini. In its green stage it feels tender, and its skin has a fine coat of tiny white hairs. The larger, firmer white patty pans should be free from rust or dark spots. Look for fresh-cut stems in both the green and white stages. If your local markets don't carry patty-pan squash, ask for it. Chances are you'll get it.

Patty-pan squash offers the same kind of nutrition as the green and yellow zucchini. It is an easily digested low-fiber food that provides important minerals, especially potassium.

NUTRITIONAL DATA

Weight (g)	Calories	Protein (g)	Carbohydrates (g)	Crude Fiber (g)	Water (g)	Fat (g)	Cholesterol (mg)
100	19	1.1	4.2	0.6	94	0.1	—
100	14	0.9	3.1	0.6	95.5	0.1	—

Vitamin A (IU)	Vitamin C (mg)	Thiamine B-1 (mg)	Riboflavin B-2 (mg)	Niacin (mg)	Vitamin B-6 (mg)	Vitamin B-12 (mcg)	Folic Acid (mcg)	Pantothenic Acid (mg)
410	22	0.05	0.09	1	—	—	—	—
390	10	0.05	0.08	0.8	—	—	—	—

Sodium (mg)	Potassium (mg)	Calcium (mg)	Phosphorus (mg)	Magnesium (mg)	Iron (mg)	Zinc (mg)	Copper (mg)	Manganese (mg)
1	202	28	29	16	0.4	—	—	—
1	141	25	25	—	0.4	—	—	—

Raw—½ cup
   Boiled—½ cup

SMALL GREEN PATTY-PAN SQUASH

2 shallots
1 small yellow onion
2 cloves garlic
6–8 sprigs flat-leaf parsley
Light vegetable oil
6 small green early-season patty-pan squash

Chop shallots, onion, garlic, and parsley and sauté in oil. Wash squash and add while still wet to sauté. Cover and cook over medium heat for 5 to 8 minutes. *Serves 2 to 3.*

LARGE WHITE PATTY-PAN SQUASH

2–3 small mushrooms
1 small yellow onion
2 cloves garlic
4 tablespoons butter or margarine
½ cup bread crumbs
¼ cup chopped walnuts or cashews
1 large white late-season patty-pan squash

Chop mushrooms, onion, and garlic and sauté in butter. Mix with bread crumbs and nuts. Remove top of squash. Scoop out seed pocket. Fill with bread crumb mixture and place part of removed top back as cap. Place in covered baking dish in ¼ inch water. Bake in preheated 350° F. oven for 20 minutes. *Serves 2 to 3.*

OKRA

Okra is an African vegetable that found its way to North America during a time when the United States supported a black slave society. The okra seed was one of the very few details of the black African culture to take hold in America. Okra is a staple item in Southern cooking. It cooks well with other vegetables, especially tomatoes, corn, peppers, and onions. Okra is an excellent rice companion, and it is the essential ingredient in Creole gumbo, contributing a thickening quality that is part of the soul of that famous Louisiana recipe. This unusual but versatile vegetable can be stewed, steamed, batter-dipped, breaded and fried, or cooked with curry and spicy meat dishes. Along with tiny baby corn and imported Italian pimientos, pickled okra is my favorite jarred appetizer.

Okra is available almost all year round. It is very expensive in the winter and early spring. All of the okra consumed in the United States is grown here, with almost no imports during the off-season. When purchased, okra should be bright-green, with fresh stem ends, firm blossom ends, and a fine coat of sticky white hairs. Avoid limp, blackish-gray okra. Try some fresh okra in soups or vegetable mixes, or added to rice and peas. Or be really brave and cook up a whole mess with hot sauce and biscuits.

Okra is a moderately nutritious vegetable, with relatively high amounts of niacin and vitamins B-1 and B-2.

NUTRITIONAL DATA

Weight (g)	Calories	Protein (g)	Carbohydrates (g)	Crude Fiber (g)	Water (g)	Fat (g)	Cholesterol (mg)
100	29	2	6	1	91.1	0.3	—

Vitamin A (IU)	Vitamin C (mg)	Thiamine B-1 (mg)	Riboflavin B-2 (mg)	Niacin (mg)	Vitamin B-6 (mg)	Vitamin B-12 (mcg)	Folic Acid (mcg)	Pantothenic Acid (mg)
490	20	0.13	0.18	0.9	—	—	—	—

Sodium (mg)	Potassium (mg)	Calcium (mg)	Phosphorus (mg)	Magnesium (mg)	Iron (mg)	Zinc (mg)	Copper (mg)	Manganese (mg)
2	174	92	41	—	0.5	—	—	—

Cooked—8–9 pods

SQUASH BLOSSOMS

*T*he flowers of squash plants, especially zucchini, can be picked, washed, chilled, dipped in a flour batter, and lightly fried. I remember my grandmother making up a batch two or three nights a week during zucchini season. The blossoms are delicious and colorful, and they become the focal point of any meal. Fried squash blossoms can be served as an hors d'oeuvre, appetizer, vegetable side dish, or snack. The blossoms should be picked before they unfurl but well past their bud stage. Look for the orange-colored petals just beginning to open. The batter can be an ordinary light pancake batter that is spiced with oregano, basil, salt, pepper, chopped garlic, onion, and parsley. The batter should be chilled before using. Dip each blossom into the batter and fry over medium heat in light olive oil, about ¼-inch deep, until the batter is golden-brown and crisp. What you get is a crunchy, spicy crust with a soft, buttery, sweet inside. Mmmmmmm!

Although I have no indication of the nutritional value of squash blossoms, the fried batter must be a carbohydrate festival. I feel something good and vital is to be gained by eating the flowers of a healthful plant. They certainly taste wonderful.

Since the blossoms are fragile and commercially scarce, you must gather your flowers from your own garden or deal with friends for their zucchini blossoms. The effort, I assure you, is worth it.

ZUCCHINI BLOSSOMS

12 slightly unfurled zucchini blossoms
1 cup pancake mix
1 cup milk
1 egg
1 clove garlic, chopped
1 small red onion, chopped
6–8 sprigs flat-leaf parsley, chopped

2–3 ounces stale beer
¼ teaspoon freshly ground black pepper
Light vegetable oil

Wash and dry zucchini blossoms thoroughly. Place in refrigerator. Mix pancake flour with milk and egg. Beat until smooth. Add garlic, onion, parsley, beer, and pepper. Chill in refrigerator.

Heat ⅛ inch oil in pan. Do not allow oil to smoke. Dip each zucchini blossom into batter. Allow excess to drip off. Brown each side in hot oil. Drain on paper towels. Serve as appetizer or hors d'oeuvre.   *Serves 3 to 4.*

LEAFY GREEN VEGETABLES

1. BRUSSELS SPROUTS
2. CHINESE CABBAGE
3. SPINACH
4. GREEN CABBAGE
5. RED CABBAGE
6. BROCCOLI RABE
7. MUSTARD GREENS
8. COLLARD GREENS
9. TURNIP GREENS
10. SWISS CHARD
11. RHUBARB
12. KALE
13. KOHLRABI

When was the last time you prepared Swiss chard? Have you ever braised Chinese cabbage? Is rhubarb a fight at the ball game? The leafy green vegetables are not so popular these days. They are losing ground in the world of fast, easy, and sweet. But if you shop the local fruit and vegetable stand or visit different ethnic markets, you'll see bright-green sparkling leaves and stalks that look more exciting than you imagined.

In the Korean and Chinese markets around our cities there are piles and piles of different leafy greens. Every leaf is brighter than the next, crisp, and alive. When I worked in the Italian and Polish neighborhoods, we sold truck loads of broccoli rabe, cabbage, and Brussels sprouts. Brussels sprouts used to come packed in crushed ice in little wooden barrels. Digging your hands into that ice in late fall and winter was not the best part of the fruit business. Italians love broccoli rabe simmered with garlic, olive oil, and salt. It tastes incredibly alive, with a natural spice and a zing that tells you you're absorbing high vegetable protein, B vitamins, magnesium, calcium, and potassium. Stores in black neighborhoods put their greens out first and out front. Collard greens, mustard greens, turnip greens, and kale are the basic ingredient of Southern and black cooking. Throughout history the peasants, the poor, and the disadvantaged have developed a basic, cheap, powerful food for survival. The Italian peasantry invented tomato sauce. The Irish lived and died on potatoes. Black Americans have made greens a tremendous fresh food item.

Leafy green vegetables are inexpensive, always available, and, properly prepared, delicious. Brussels sprouts, spinach, Swiss chard, and cabbage can taste as good as any vegetable ever. With a touch of tamari, a garlic-onion sauté, a cream sauce or a cold vinaigrette, the leafy green vegetables can be a bright part of every meal instead of the lifeless things we've made them be.

Look for fresh, crisp, bright leaves. Ask your clerk to identify the variety, and begin to experiment. You'll get excellent value, you'll actually spend less, you'll get high-quality nutrition, and you'll add more style and variety to every meal.

BRUSSELS SPROUTS

Brussels sprouts don't have a popular image as a table vegetable. To most people they're cute little cabbages that cook up with a terrible odor and a mushy texture. The fault is not with the little sprouts. The blame lies with the cook, who usually boils them to death.

I love Brussels sprouts. I like to eat the tiny ones raw. Even large ones need very little cooking time. Their popularity in England, which has a reputation for dull, mushy food, has not helped the Brussels sprout's image. Neither does the fact that it is a relative of and looks like the cabbage. I think they're great. They have no waste, require little preparation, cook in minutes, and taste sweet and delicious.

After cleaning away any wilted or yellow outside leaves and scoring the stem end, steam or simmer Brussels sprouts just beyond the point of rawness. They will retain their crisp green outsides and delicious white centers. They will also retain their nutrients.

Although they grow in a spiral around a full, thick stalk, most Brussels sprouts are sold loose in cello-wrapped paper cups or pint baskets. Select fresh, green, young sprouts free from wilt, spots, or rot. They are at their best in late fall and throughout the winter.

Brussels sprouts are low in fat and high in vegetable protein, vitamin C, and minerals. They are also excellent digestive stimulants.

NUTRITIONAL DATA

Weight (g)	Calories	Protein (g)	Carbohydrates (g)	Crude Fiber (g)	Water (g)	Fat (g)	Cholesterol (mg)
100	36	4.2	6.4	1.6	88.2	0.4	—

Vitamin A (IU)	Vitamin C (mg)	Thiamine B-1 (mg)	Riboflavin B-2 (mg)	Niacin (mg)	Vitamin B-6 (mg)	Vitamin B-12 (mcg)	Folic Acid (mcg)	Pantothenic Acid (mg)
520	87	0.08	0.14	0.8	—	—	—	—

Sodium (mg)	Potassium (mg)	Calcium (mg)	Phosphorus (mg)	Magnesium (mg)	Iron (mg)	Zinc (mg)	Copper (mg)	Manganese (mg)
10	273	32	72	—	1.1	—	—	—

Cooked—6–8 medium

BRUSSELS SPROUTS VINAIGRETTE

1 pound Brussels sprouts
Juice of 2 lemons
¼ cup apple-cider or white-wine vinegar
1 tablespoon Worcestershire sauce
Dash salt
¼ teaspoon freshly ground black pepper

Clean sprouts and score stem ends. Steam in 1 inch water until barely tender. Mix lemon juice with vinegar, Worcestershire sauce, salt, and pepper. Place sprouts in shallow bowl. Pour vinaigrette over sprouts. Chill for 3 to 4 hours.   *Serves 2 to 3.*

BRUSSELS SPROUTS IN BUTTER-CREAM SAUCE

1 pound Brussels sprouts
¼ pound butter
½ cup half-and-half
½ cup milk
¼ cup instant-blending flour

Clean sprouts and score stem ends. Steam in 1 inch water until barely tender. Melt butter and slowly add half-and-half and milk. Lower heat. Add flour 1 teaspoon at a time, stirring constantly. Cook, stirring, until sauce thickens. Place sprouts in shallow bowl. Cover with butter-cream sauce.   *Serves 2 to 3.*

CHINESE CABBAGE

Although many varieties of Chinese cabbage are grown for Chinese kitchens and restaurants, the long, white-stemmed, green-leafed pak choi and the roundish, wide-leafed, white *wong bok* are the varieties usually found in stores and markets. The stem end of Chinese cabbage is usually cut up in one- to two-inch pieces and stir-fried by itself or with other vegetables. The leafy section is excellent for soup or added quickly at the end of a stir-fry, and the whole plant can be cut up and added raw to salads.

Chinese cabbage makes an excellent side dish braised or steamed with tamari sauce because it goes so well with all meats, chicken, and seafood. It loves ginger, hot spices, and Chinese condiments or is good simply served with salt and pepper. Chinese cabbage is easily grown, inexpensive, and available year-round almost everywhere. Add Chinese cabbage to your kitchen, and you will improve your salads, soups, vegetable mixes, and mealtime variety. Don't let the name "cabbage" fool you. Pak choi and *wong bok* are semispicy leafy vegetables with a great deal of character.

Chinese cabbage is an excellent source of minerals, including calcium, and niacin. It has almost no calories and very little fiber. It is easily digested and adds a spicy sparkle to the taste of a vegetable-juice combination.

NUTRITIONAL DATA

Weight (g)	Calories	Protein (g)	Carbohydrates (g)	Crude Fiber (g)	Water (g)	Fat (g)	Cholesterol (mg)
100	14	1.2	3	0.6	95	0.1	—
82	8	1.2	1.3	—	—	—	—

Vitamin A (IU)	Vitamin C (mg)	Thiamine B-1 (mg)	Riboflavin B-2 (mg)	Niacin (mg)	Vitamin B-6 (mg)	Vitamin B-12 (mcg)	Folic Acid (mcg)	Pantothenic Acid (mg)
150	25	0.05	0.04	0.6	—	—	—	—
40	26	0.07	0.07	0.8	—	—	—	—

Sodium (mg)	Potassium (mg)	Calcium (mg)	Phosphorus (mg)	Magnesium (mg)	Iron (mg)	Zinc (mg)	Copper (mg)	Manganese (mg)
23	253	43	40	14	0.6	—	—	—
—	—	26	32	—	0.2	—	—	—

Raw—2¼ cups shredded
Cooked—½ cup

CHINESE CABBAGE SALAD

1 small head Chinese cabbage
2 tablespoons soy sauce
2 tablespoons apple-cider vinegar
Dash garlic powder
Dash onion powder
¼ teaspoon freshly ground black pepper

Remove leaves from cabbage head. Cut off coarse portion. Break into small pieces. Mix soy sauce, vinegar, garlic powder, onion powder, and pepper in small bowl. Sprinkle over cabbage while tossing until completely coated.   *Serves 2 to 3.*

BRAISED CHINESE CABBAGE

2 tablespoons sesame oil
2 tablespoons tamari sauce
1 head Chinese cabbage
Dash garlic powder
Dash curry powder

Heat sesame oil and tamari sauce in frying pan. Remove leaves from cabbage head. Wash and place wet leaves whole in pan. Season with garlic powder and curry powder. Cover and simmer over low heat for 3 to 5 minutes. Uncover. Cook, turning leaves until they begin to brown.   *Serves 3 to 4.*

SPINACH

More than a thousand years ago spinach was introduced to Europe by the Moors. Since then its cultivation has increasingly spread and its uses multiplied until it is now found in almost every part of a meal. It's a great hot vegetable and also makes wonderful broth, cream soup, ravioli, puddings, omelettes, tarts, croquettes, and mixes with fresh soft cheese like ricotta or cottage cheese. Spinach is best eaten raw or barely cooked. It loves tamari sauce, oil, garlic, and shallots.

The best way to clean spinach is first to remove the purple root stem and break away the toughest part of the leaf stem. Then allow the spinach to soak in very cold water for twenty minutes. After it's been soaked, carefully wash the leaves one or two at a time under running water. Soaking causes much of the sand and grit to fall to the bottom and loosens the remaining amount. After soaking and washing the leaves, place them in a colander and gently toss them as water runs through them.

Spinach varieties are either very crinkly or almost flat. The very crinkly variety is slightly tougher, less sweet, and harder to clean. Both, however, are available year-round in fresh-picked bunches or partially washed and packed in plastic bags. Fresh-picked spinach from local sources or a home garden is a sweet, refreshing vegetable, but I have no objection to the prewashed variety if it is bright, fresh, and free from bruising and breakdown. Look for unbroken stems and leaves without wilt or dark edges. Spinach can be washed, towel-dried, and stored in a plastic bag in your refrigerator for several days.

Some stores now offer a triangular-shaped, flat green-leafed spinach with a short green stem, called New Zealand spinach. It is not spinach. It is a once-wild, now cultivated, plant closely resembling true spinach. It is delicious as a raw vegetable or salad but has neither the unique taste and texture nor the very high nutritional content of Popeye's favorite.

Spinach was originally grown in Italy as a Lenten food planted in autumn, wintered over, and harvested just in time for the annual self-denial season. Perhaps that's when it got its reputation as something you had to eat even though you hated it. Properly prepared, however, spinach is a thoroughly enjoyable vegetable with outstanding nutritional value.

Spinach contains large amounts of vitamins A, B, C, E, and K. The iron content is as high as that of fish and eggs, although it is difficult for the body to absorb spinach's form of iron. Spinach also contains a large amount of oxalic acid, which increases when cooked and is harmful to those with kidney or liver problems.

NUTRITIONAL DATA

Weight (g)	Calories	Protein (g)	Carbohydrates (g)	Crude Fiber (g)	Water (g)	Fat (g)	Cholesterol (mg)
100	26	3.2	4.3	0.6	90.7	0.3	—
90	21	2.7	3.2	0.3	82.8	0.5	—

Vitamin A (IU)	Vitamin C (mg)	Thiamine B-1 (mg)	Riboflavin B-2 (mg)	Niacin (mg)	Vitamin B-6 (mg)	Vitamin B-12 (mcg)	Folic Acid (mcg)	Pantothenic Acid (mg)
8100	51	0.1	0.2	0.6	—	—	—	—
7300	25	0.06	0.13	0.5	—	—	—	—

Sodium (mg)	Potassium (mg)	Calcium (mg)	Phosphorus (mg)	Magnesium (mg)	Iron (mg)	Zinc (mg)	Copper (mg)	Manganese (mg)
71	470	93	51	88	3.1	—	—	—
45	291	83	33	—	2	—	—	—

Raw—3½ ounces
  Cooked—½ cup

SPINACH

KING SALAD

1 pound spinach
6 radishes, coarsely chopped
4 mushrooms, sliced
1 small Belgian endive, sliced
2 shallots, chopped
1 small red onion, chopped
2 cloves garlic, chopped
12–14 sprigs watercress, chopped
Juice of 1 lemon
¼ cup olive oil
¼ teaspoon freshly ground black pepper

Remove spinach stems and wash leaves thoroughly. Mix all ingredients together. Chill. *Serves 4.*

CHOPPED CREAMED SPINACH

1½ pounds spinach
¼ pound butter
½ cup half-and-half
½ cup milk
¼ cup instant-mixing flour

Remove spinach stems, wash leaves thoroughly, and chop coarsely. Steam in ¼ inch water until soft. Drain. Melt butter and slowly add half-and-half and milk. Lower heat. Add flour 1 teaspoon at a time, stirring constantly. Cook, stirring, until mixture thickens. Pour cream sauce over spinach. *Serves 4.*

CREAMED SPINACH SOUP

1 pound spinach
1 small yellow onion, chopped
1 stalk celery, chopped
4 tablespoons butter or margarine
3 tablespoons flour
3 cups milk
3–4 sprigs savory, chopped
1 teaspoon salt
¼ teaspoon freshly ground black pepper

*R*emove spinach stems, wash leaves thoroughly, and chop. Sauté chopped onion and celery in butter until tender. Stir in flour. Gradually stir in milk. Add spinach. Heat until mixture thickens and boils, stirring constantly. Add savory, salt, and pepper. *Serves 3.*

GARLIC SPINACH

1½ pounds spinach
2–3 cloves garlic, coarsely chopped
¼ cup olive oil

*R*emove spinach stems and wash leaves thoroughly. Sauté garlic in olive oil in deep pan. Add wet spinach. Cover. Cook for 3 to 5 minutes over low heat. Mix. *Serves 3.*

SPINACH-CHEESE SALAD

¾ pound spinach
6 radishes, chopped
4 scallions (including greens), chopped
16 ounces cottage cheese

*R*emove spinach stems and wash leaves thoroughly. Mix chopped radishes, scallions, and spinach. Stir in cottage cheese. *Serves 4.*

SPINACH AND RICE

1½ pounds spinach
1 cup cooked rice
4 ounces butter or margarine

*R*emove spinach stems and wash leaves thoroughly. Chop and place wet spinach in pan. Cover and cook until soft. Drain. Mix with rice. Add melted butter. Mix. Place in baking dish in preheated 350° F. oven for 10 minutes. *Serves 3 to 4.*

SPINACH BROTH

1 pound spinach
4 cloves garlic, chopped
Light vegetable oil
4 cups chicken stock

Remove spinach stems and wash leaves thoroughly. Brown garlic in oil. Add spinach to hot pan and quickly mix off heat. Add spinach mixture to heated broth and simmer for 10 minutes. *Serves 4.*

CABBAGE

Cabbages are the least expensive, most consistently available, and least understood of all the common vegetables. Most cabbage is served steamed or boiled and almost always overcooked. It is hardly recognizable in the delicatessen as cole slaw, drowned in watery mayonnaise and vinegar. Actually, cabbage is a delicately flavored vegetable that cooks quickly and easily. New cabbage, with dark, crisp outer leaves, is sweet both raw and cooked. The outer leaves make excellent meat and rice wrappings, soup, and a vegetable side dish. The inner core, shredded or chopped, will add crunch, flavor, and nutrients to salads. Old cabbage tends to be a little less sweet, but its whiteness and firm texture makes it the perfect companion to simmered meats such as corned beef, smoked pork, pork knuckles, and thick German sausage.

Red cabbage or savoy cabbage, which is the cabbage with wrinkled leaves, makes an outstanding vegetable dish when it's simmered in a vinegar sauce, allowed to cool overnight, and served with cold meats. Red cabbage is also an outstanding salad ingredient. Sliced in thin ribbons or shredded, it adds a crunchy, spicy, sweet taste to a fresh garden salad. The bright color adds to the appetite appeal as well.

Select hard, round cabbages. Green cabbage should be free from rust or yellowing, savoy cabbage crisp and bright. Red cabbage should be very hard and the leaves free from black edges or rot.

Quarter cabbage, shred it, chop it, wrap it, pickle it, or steam it whole. Treat cabbage gently on the stove and generously in your salads, with apples, raisins, nuts, carrots, and tropical fruits. It will add a delicious, inexpensive dimension to your kitchen.

Red and green cabbage are close in nutritional content, with the red variety delivering slightly more of everything except vitamin A. All cabbages are excellent digestive aids and intestinal cleansers.

NUTRITIONAL DATA (Green)

Weight (g)	Calories	Protein (g)	Carbohydrates (g)	Crude Fiber (g)	Water (g)	Fat (g)	Cholesterol (mg)
100	24	1.3	5.4	0.8	92.4	0.2	—
100	20	1.1	4.3	0.8	93.9	0.2	—

Vitamin A (IU)	Vitamin C (mg)	Thiamine B-1 (mg)	Riboflavin B-2 (mg)	Niacin (mg)	Vitamin B-6 (mg)	Vitamin B-12 (mcg)	Folic Acid (mcg)	Pantothenic Acid (mg)
130	47	0.05	0.05	0.3	—	—	—	—
130	33	0.04	0.04	0.3	—	—	—	—

Sodium (mg)	Potassium (mg)	Calcium (mg)	Phosphorus (mg)	Magnesium (mg)	Iron (mg)	Zinc (mg)	Copper (mg)	Manganese (mg)
20	233	49	29	13	0.4	—	—	—
14	163	44	20		0.3	—	—	—

Green, raw—1 cup shredded
  Cooked—3/5 cup

NUTRITIONAL DATA (Red & Savoy)

Weight (g)	Calories	Protein (g)	Carbohydrates (g)	Crude Fiber (g)	Water (g)	Fat (g)	Cholesterol (mg)
100	31	2	6.9	1	90.2	0.2	—
50	12	1.2	2.3	0.4	46.4	0.1	—

Vitamin A (IU)	Vitamin C (mg)	Thiamine B-1 (mg)	Riboflavin B-2 (mg)	Niacin (mg)	Vitamin B-6 (mg)	Vitamin B-12 (mcg)	Folic Acid (mcg)	Pantothenic Acid (mg)
40	61	0.09	0.06	0.4	—	—	—	—
100	27	0.02	0.04	0.2	—	—	—	—

Sodium (mg)	Potassium (mg)	Calcium (mg)	Phosphorus (mg)	Magnesium (mg)	Iron (mg)	Zinc (mg)	Copper (mg)	Manganese (mg)
26	268	42	35	—	0.8	—	—	—
11	134	33	27	—	0.45	—	—	—

Red, raw—1 cup shredded
  Savoy, raw—1 cup shredded

SIMPLY SWEET CABBAGE

1 head green cabbage
2 tablespoons butter
½ cup milk
¼ teaspoon freshly ground black pepper

Quarter cabbage. Heat milk and butter in saucepan. Add cabbage and ¼ cup water. Sprinkle with pepper. Cover. Steam for 15 minutes. *Serves 4.*

COLORFUL CABBAGES SALAD

¼ head new green cabbage
¼ head new red cabbage
6 sprigs curly parsley, chopped
6 sprigs savory, chopped
¼ teaspoon freshly ground black pepper
Juice of 1 lemon
¼ cup mayonnaise

Wash, drain, shred, and toss cabbages. Add parsley, savory, pepper, lemon juice, and mayonnaise. Mix. Chill. *Serves 4.*

DOWN-HOME SOUTHERN CABBAGE WITH PORK

8–12 pieces pork or salt pork, cubed
3–4 small pork knuckles or 8–12 small pork ribs
2–3 cloves garlic, chopped
3–4 sage leaves, chopped
¼ teaspoon freshly ground black pepper
1 head green cabbage
¼ cup red-wine vinegar

Brown pork pieces and knuckles in own fat. Add garlic, sage, and pepper. Remove core from cabbage head and fill with water. Drain. Slice cabbage across head into 1-inch slices. Place in pan with pork. Add vinegar and ¼ cup water. Cover. Cook over low heat for 15 to 20 minutes. *Serves 4.*

Hearty Green Cabbage and Potatoes

4 medium white or red potatoes
½ head green cabbage
¼ teaspoon salt
¼ teaspoon freshly ground black pepper
4 tablespoons butter

*P*eel potatoes and cut into medium pieces. Cover with water. Boil until barely soft. Cut cabbage into 4 equal wedges. Place in saucepan. Barely cover with potato water. Add salt and pepper. Place potatoes on top of cabbage. Add butter. Cover. Simmer over low heat for 10 minutes. *Serves 4.*

Beef in a Blanket

3 cloves garlic
1 medium yellow onion
10–12 sprigs flat-leaf parsley
1 pound ground beef
1 head green cabbage with large leaves
¼ teaspoon salt
¼ teaspoon freshly ground black pepper
Light vegetable oil
4 cups cooked peeled tomatoes
1 6-ounce can tomato paste
1 cup rice, cooked

*C*hop garlic, onion, and parsley. Add half to frying pan with ground beef and brown. Remove core from cabbage. Steam entire head until soft enough for fork to begin to penetrate. Remove cabbage and cool. Sauté remainder of garlic, onion, and parsley with salt and pepper in oil. Add tomatoes and tomato paste. Cover. Simmer for 15 minutes over low heat. Mix beef mixture with rice.

Remove cabbage leaves carefully. Cut off hard ends if any. Spoon rice mixture into leaves. Fold right and left sides of leaves over filling (so filling won't escape from ends of rolls), then roll up from bottom. Fasten with poultry skewers or long wooden toothpicks. Place rolls in tomato sauce. Cover. Simmer for 15 minutes. *Makes 8 to 12 rolls.*

Fancy Green Cabbage Salad

⅓ head green cabbage
1 sweet red apple
1 tart green apple
1 carrot
2 stalks celery
2 tablespoons chopped walnuts
¼ cup raisins
¼ cup mayonnaise
¼ cup apple-cider vinegar

Remove cabbage core and shred or finely slice cabbage. Core and peel apples and cut up into small chunks. Shred carrot. Chop celery. Mix with apples, cabbage, nuts, and raisins. Mix mayonnaise and vinegar to make a dressing. Serve dressing separately. *Serves 4.*

Red Cabbage with White Wine

1 medium head red cabbage
1 medium red onion, chopped
2 tablespoons margarine or light vegetable oil
Juice of 2 lemons
¼ teaspoon freshly ground black pepper
½ cup white cooking wine

Quarter head of cabbage after removing dark outer leaves. Soak in cold water. In skillet sauté chopped onion lightly in margarine. Place cabbage quarters in pot or steamer with the lemon juice and ½ inch water. Cover and steam until heavy center leaves begin to soften slightly, about 15 minutes. Sprinkle with pepper and sautéed onion. Cover and cook for 3 to 5 minutes. Add wine. Remove from heat. Cover and let stand for 10 minutes. *Serves 4.*

Red Cabbage Excelsior Salad

¼ head red cabbage
2 carrots, shredded
4 radishes, finely chopped
1 medium red onion, finely chopped
6–8 sprigs watercress or parsley, finely chopped

2 cloves garlic, crushed
Juice of 1 lemon
¼ teaspoon freshly ground black pepper
¼ cup apple-cider vinegar

Shred or finely chop cabbage. Shred carrots. Finely chop radishes, onion, and watercress. Toss with cabbage. Crush garlic and mix with lemon juice, pepper, and vinegar to make a dressing. Serve dressing separately. *Serves 4.*

BROCCOLI RABE

Broccoli rabe (also called rape) is a member of the cabbage family. It is also having a difficult time breaking out of the kitchens of the ethnic group that loves it most, Sicilians. Vegetarian purists know it, and some black neighborhood stores sell it as a substitute when turnip or mustard greens are in short supply, but most of America has no idea how good it is!

Broccoli rabe has a slightly bitter taste but delightfully tender leaves, and when cooked properly, leaves a stimulating, fresh, brisk taste on the palate. The plant consists of a long pale-green main stem with flat bright-green leaves and a small tassel bud of little yellow flowers. Remove and discard the tough stems. Wash the leaves and tassels and tender stems thoroughly and place in a saucepan in which you have heated olive oil, garlic and some chopped onion, salt, and pepper. Place the washed rabe in, cover, simmer for five to ten minutes, and serve as an excellent side dish. It goes well with meats, spaghetti or macaroni, rice, and potatoes. Some of the chopped green leaves add a spicy touch to salad as well. Broccoli rabe is also delicious added to a chicken soup or a clear broth.

Broccoli rabe is usually sold in bunches, from late summer through the fall and winter. It is excellent as part of a winter meal. Avoid yellow or wilted broccoli rabe. One or two flowers of the bud tassel may be open, but the flower end should be green and upright, not curved over (curved over usually means overgrown).

Broccoli rabe is one of those vegetables that feels healthful while you're eating it. It is very high in vitamins A and C, with large amounts of potassium and other vital minerals. Its fiber content makes it an excellent digestive stimulant, and a clear rabe broth will chase winter colds and chills.

SOUTHERN GREENS

In areas of the United States where plantation agriculture was the society, sharecroppers, migrants, and even slaves required a cheap, fast-growing, easily cultivated, highly nutritious food to support their huge numbers. The varieties of the cabbage family known as greens became that food.

Greens include kale, collards, turnip greens, mustard greens, and kohlrabi. Kale is a crinkly moss-green leaf. Collards are like cabbages that never headed up. Turnip greens have a slightly prickly round leaf. Mustard greens have a slightly prickly ragged leaf. Kohlrabi is a hard green rounded knob from which grow stalks and leaves.

All greens have a slightly bitter taste to varying degrees, are inexpensive, and are available year-round almost everywhere. Greens are most popular in the American South, where a "mess of greens" is a mixed pot of several varieties cooked with butter, lard, or oil; garlic; onions; and a variety of pork cuts including fatback (salt pork). Greens are the central ingredient of soul food and are served with beans, rice, corn, fried fish, ham, ribs, chicken, and plenty of hot Southern gravy.

Mixed greens or a single variety can be more simply prepared by simmering or steaming them with a little garlic and oil, then serving them with a vinegar or lemon dressing or just salt and pepper. Mustard greens are especially good raw in salads, adding a spicy, earthy taste to a combination of velvety lettuce leaves.

Always select bright-green leaves free from wilt or yellowing. Avoid thick stalks and stems, flowered tips, and extremely gritty, sandy bunches. Snap off the tough bottom ends of the stalks, and wash everything several times. Break the stalks and stems into two or three evenly sized sections before cooking. Next time you're in a fresh vegetable market and you see Southern greens, buy a few. It will open up another door to the endless world of delicious fresh food.

All greens are highly nutritious. They have a very high protein and vitamin A and C content, along with remarkable amounts of calcium, potassium, and other vital minerals.

NUTRITIONAL DATA (Collard greens)

Weight (g)	Calories	Protein (g)	Carbohydrates (g)	Crude Fiber (g)	Water (g)	Fat (g)	Cholesterol (mg)
100	40	3.6	7.2	0.9	86.9	0.7	—
100	29	2.7	4.9	0.8	90.8	0.6	—

Vitamin A (IU)	Vitamin C (mg)	Thiamine B-1 (mg)	Riboflavin B-2 (mg)	Niacin (mg)	Vitamin B-6 (mg)	Vitamin B-12 (mcg)	Folic Acid (mcg)	Pantothenic Acid (mg)
6500	92	0.2	0.31	1.7	—	—	—	—
5400	46	0.14	0.2	1.2	—	—	—	—

Sodium (mg)	Potassium (mg)	Calcium (mg)	Phosphorus (mg)	Magnesium (mg)	Iron (mg)	Zinc (mg)	Copper (mg)	Manganese (mg)
43	401	203	63	57	1	—	—	—
25	234	152	39	—	0.6	—	—	—

Raw—3½ ounces
  Cooked—½ cup

NUTRITIONAL DATA *(Turnip greens)*

Weight (g)	Calories	Protein (g)	Carbohydrates (g)	Crude Fiber (g)	Water (g)	Fat (g)	Cholesterol (mg)
100	28	3	5	0.8	90.3	0.3	—
100	20	2.2	3.6	0.7	93.2	0.2	—

Vitamin A (IU)	Vitamin C (mg)	Thiamine B-1 (mg)	Riboflavin B-2 (mg)	Niacin (mg)	Vitamin B-6 (mg)	Vitamin B-12 (mcg)	Folic Acid (mcg)	Pantothenic Acid (mg)
7600	139	0.21	0.39	0.8	—	—	—	—
6300	69	0.15	0.24	0.6	—	—	—	—

Sodium (mg)	Potassium (mg)	Calcium (mg)	Phosphorus (mg)	Magnesium (mg)	Iron (mg)	Zinc (mg)	Copper (mg)	Manganese (mg)
—	—	246	58	58	1.8	—	—	—
—	—	184	37	—	1.1	—	—	—

Raw—3½ ounces
Cooked—⅔ cup

NUTRITIONAL DATA *(Mustard greens)*

Weight (g)	Calories	Protein (g)	Carbohydrates (g)	Crude Fiber (g)	Water (g)	Fat (g)	Cholesterol (mg)
100	31	3	5.6	1.1	89.5	0.5	—
100	23	2.2	4	0.9	92.6	0.4	—

Vitamin A (IU)	Vitamin C (mg)	Thiamine B-1 (mg)	Riboflavin B-2 (mg)	Niacin (mg)	Vitamin B-6 (mg)	Vitamin B-12 (mcg)	Folic Acid (mcg)	Pantothenic Acid (mg)
7000	97	0.11	0.22	0.8	—	—	—	—
5800	48	0.08	0.14	0.06	—	—	—	—

Sodium (mg)	Potassium (mg)	Calcium (mg)	Phosphorus (mg)	Magnesium (mg)	Iron (mg)	Zinc (mg)	Copper (mg)	Manganese (mg)
32	377	183	50	27	3	—	—	—
18	220	138	32	—	1.8	—	—	—

Raw—3½ ounces
Cooked—½ cup

KOHLRABI

In the wide world of Western vegetables kohlrabi is found at the mysterious fringe of foods barely known and seldom desired. Certainly at one time the soft cabbagy green kohlrabi was a standard, constant item on the tables of Central European cultural groups that seem to be less and less individualized as they become absorbed by the European and American megaculture. The kohlrabi is a late-harvest plant that produces a bulbous growth just above ground level that sprouts long-stemmed flat leaves. The flavor is often confused with cabbage and/or turnip, and although its taste is something akin to cabbage's, it is not really the same and definitely has none of the bite or sharpness of the turnip.

You will not find kohlrabi in your ordinary vegetable market. Look for neighborhoods where Hungarians, Poles, Latvians, Estonians, Czechs, Armenians, Austrians, and Germans keep their tastes and cultures, and you will find kohlrabi.

The smaller green leaves, *sans* stems, make excellent salad and go well with fresh parsley, marjoram, and French dressing. Peeled and simmered whole in a small amount of water, kohlrabi bulbs may be served plain, creamed, or cut up as part of a stir-fry or mixed vegetable dish. Raw kohlrabi bulbs can be cut up and added to salads for a refreshing crunch that increases the tactile pleasure of salad greens.

By and large the most recognition kohlrabi gets in the United States is in Southern cooking. Along with mustard and turnip greens, collards, and kale, kohlrabi is a popular and nutritious member of the greens family. It is grown plentifully and easily and offers solid nutrition. Its fame, however, comes from the way the soulful cook treats it and not from its own inherent nature. I am sure there is a Hungarian kohlrabi-lover who will disagree with me, and so be it. I have no doubt there are wonderful Old-World recipes that can make kohlrabi memorable and delicious. Unfortunately, none has emerged to keep kohlrabi from the list of endangered vegetables. Too bad. It deserves better.

Nutritionally kohlrabi is an excellent blood-cleanser and is helpful in the development and repair of gums, teeth, and bone. It contains significant amounts of calcium, phosphorus, and iron and more-than-average vegetable protein.

NUTRITIONAL DATA

Weight (g)	Calories	Protein (g)	Carbohydrates (g)	Crude Fiber (g)	Water (g)	Fat (g)	Cholesterol (mg)
100	24	1.7	5.3	1	92.2	0.1	—

Vitamin A (IU)	Vitamin C (mg)	Thiamine B-1 (mg)	Riboflavin B-2 (mg)	Niacin (mg)	Vitamin B-6 (mg)	Vitamin B-12 (mcg)	Folic Acid (mcg)	Pantothenic Acid (mg)
20	43	0.06	0.03	0.2	—	—	—	—

Sodium (mg)	Potassium (mg)	Calcium (mg)	Phosphorus (mg)	Magnesium (mg)	Iron (mg)	Zinc (mg)	Copper (mg)	Manganese (mg)
6	260	33	41	—	0.3	—	—	—

Cooked—⅔ cup

SWISS CHARD

Swiss chard is one of the most delightful leafy vegetables. It has a sweet taste and can be served as two different vegetables at the same meal. The plant is a long white rib with a leafy green top like spinach. Cut away the white ribs from the leaf, and quickly blanch them in salted water. Simmer them in a seasoned sauté of garlic, shallots, onions, and tamari or pepper. The chard rib becomes a crunchy, delicious side dish. The leaf can be cooked like spinach or can be served raw in a salad. Swiss chard is also great in a chicken soup or other broth.

Swiss chard is available from spring through winter. It is inexpensive, easy to prepare, and very versatile. It is also very easy to grow and, along with Brussels sprouts, the last thing in the garden, very often reappearing in the spring after the thaw. In the store look for very white stems and dark-green crinkly leaves. Chard should never be wilted or brown.

Like many leafy green vegetables, Swiss chard is very high in vitamin A, potassium, and other minerals. It is an excellent source of vegetable protein, easily digested, and low in fiber.

NUTRITIONAL DATA

Weight (g)	Calories	Protein (g)	Carbohydrates (g)	Crude Fiber (g)	Water (g)	Fat (g)	Cholesterol (mg)
100	25	2.4	4.6	0.8	91.1	0.3	—
100	18	1.8	3.3	0.7	93.7	0.2	—

Vitamin A (IU)	Vitamin C (mg)	Thiamine B-1 (mg)	Riboflavin B-2 (mg)	Niacin (mg)	Vitamin B-6 (mg)	Vitamin B-12 (mcg)	Folic Acid (mcg)	Pantothenic Acid (mg)
6500	32	0.06	0.17	0.5	—	—	—	—
5400	16	0.04	0.11	0.4	—	—	—	—

Sodium (mg)	Potassium (mg)	Calcium (mg)	Phosphorus (mg)	Magnesium (mg)	Iron (mg)	Zinc (mg)	Copper (mg)	Manganese (mg)
147	550	88	39	65	3.2	—	—	—
86	321	73	24	—	1.8	—	—	—

Raw—3½ ounces
  Cooked—⅗ cup

RHUBARB

Rhubarb without sweeteners is inedible. Rhubarb raw is inedible. Rhubarb leaves are poisonous. So why rhubarb? Properly cooked, the long red stems produce a soft-textured, easily digested vegetable that has a true fruit taste. Its preparation is simple, requiring only stewing or steaming with sugar. Mixed with strawberries, rhubarb is an extraordinary dessert served on shortcake with cream or baked in a pie. I love strawberry-rhubarb pie.

Most rhubarb is greenhouse-grown and is pale and tender. Farm-grown rhubarb is deep red, with thick stalks. It has more flavor but tends to be a little stringier and tougher. Although it can be used for sauces, preserves, and pickles, most rhubarb is simply stewed or baked.

Rhubarb is available from early spring through early summer. Select crisp, brightly colored medium-sized stalks without spots or dark-brown areas. Although most stores sell rhubarb without its leaves, some homegrown varieties are sold as picked, leaf and all. The green leaf should be removed before sale and certainly before leaving the store, as it is toxic; the cilia and leaf milk are both irritating and harmful.

Rhubarb offers more kinds of vital minerals and vitamins than any other vegetable. It is also an excellent aid to digestion and a gentle, effective laxative.

NUTRITIONAL DATA

Weight (g)	Calories	Protein (g)	Carbohydrates (g)	Crude Fiber (g)	Water (g)	Fat (g)	Cholesterol (mg)
137	29	0.8	7	—	128.1	0.2	—

Vitamin A (IU)	Vitamin C (mg)	Thiamine B-1 (mg)	Riboflavin B-2 (mg)	Niacin (mg)	Vitamin B-6 (mg)	Vitamin B-12 (mcg)	Folic Acid (mcg)	Pantothenic Acid (mg)
147	7	0.04	0.04	0.3	0.03	—	11	0.09

Sodium (mg)	Potassium (mg)	Calcium (mg)	Phosphorus (mg)	Magnesium (mg)	Iron (mg)	Zinc (mg)	Copper (mg)	Manganese (mg)
2	148	266	17	25	0.39	0.14	0.032	0.133

Raw—1 cup

KALE

In addition to its role as a brother to collards, kale has a life of its own. Up North kale is a crisp, delicious fall-winter vegetable with really outstanding nutritional qualities.

Kale is a member of the headless cabbage family. It has long stalks with wide, mossy-green, very crinkly leaves. The more crinkles, the better the taste. Kale has a cabbagelike flavor, with a sharpness that some people find a little too strong. It can be eaten chopped in a fresh salad or stir-fried with other vegetables. A few leaves add a nice spark to soup, and it can be cooked with pork or ham.

Kale is hardly found in supermarkets. It takes up too much room for too little money. Most fresh kale is sold at farm stands and in ethnic produce markets. When purchasing kale, make sure it is crisp and bright. Old kale is impossible. In the home garden kale often lasts all winter. It is easily cultivated and makes a beautiful fall presence when almost all of the home garden plants have finished up.

There is another variety of kale that is the most dramatic-looking of all the vegetables. It has a very crisp texture and a radiant color pattern. The small center leaves are yellow, then comes white, purple, lavender, and finally moss-green at the outer edge. It's called flowering kale and is sweeter and more delicious than green kale. Some of our customers set the kale in a low bowl of water and keep it as a centerpiece for several days before it becomes food.

Kale is nutritious and easily digestible. It is very high in vegetable protein, vitamins A and C, calcium, potassium, phosphorus, and iron.

NUTRITIONAL DATA

Weight (g)	Calories	Protein (g)	Carbohydrates (g)	Crude Fiber (g)	Water (g)	Fat (g)	Cholesterol (mg)
100	**38**	**4.2**	**6**	**1.3**	**87.5**	**0.8**	—
100	**28**	**3.2**	**4**	**1.1**	**91.2**	**0.7**	—

Vitamin A (IU)	Vitamin C (mg)	Thiamine B-1 (mg)	Riboflavin B-2 (mg)	Niacin (mg)	Vitamin B-6 (mg)	Vitamin B-12 (mcg)	Folic Acid (mcg)	Pantothenic Acid (mg)
8900	**125**	—	—	—	—	—	—	—
7400	**62**	—	—	—	—	—	—	—

Sodium (mg)	Potassium (mg)	Calcium (mg)	Phosphorus (mg)	Magnesium (mg)	Iron (mg)	Zinc (mg)	Copper (mg)	Manganese (mg)
75	**318**	**179**	**73**	**37**	**2.2**	—	—	—
43	**221**	**134**	**46**	—	**1.2**	—	—	—

Raw—3½ ounces
  Cooked—¾ cup

K A L E

Salads, Greens, and Crunchies

1. CUCUMBERS
KIRBY, GREEN, SEEDLESS

2. PARSLEY
CURLY, ITALIAN

3. WATERCRESS   4. CHIVES

5. ICEBERG LETTUCE

6. BOSTON LETTUCE   7. BIBB LETTUCE

8. LEAF LETTUCE
RED, GREEN

9. ROMAINE LETTUCE

10. CELERY   11. CELERY ROOT

12. RADISHES   13. BELGIAN ENDIVE

14. ESCAROLE   15. CHICORY

16. ARUGULA   17. DANDELIONS

18. JERUSALEM ARTICHOKES

19. SCALLIONS   20. SPROUTS

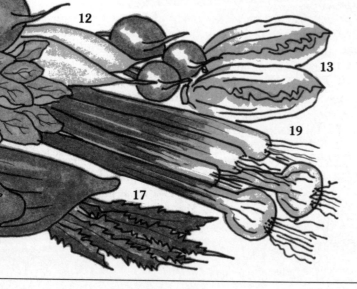

Salads should be treated with more respect. Besides their fresh, colorful appeal and varied textures, salads are the key to proper nutrition. Salad vegetables contain varying but significant degrees of vitamins and minerals and are indispensable to a balanced diet. Salad greens help maintain an alkaline balance, necessary for healthy skin, nails, and hair, clear eyes, clean breath, and healthy gums. They also help metabolize fat and carbohydrates properly, and as part of a daily diet promote vigorous intestinal activity, maintain kidney function, and help the bloodstream maintain high oxygen content.

Limp iceberg lettuce and processed bottled dressing do not a salad make. Fruit stands are abundant with four, five, and six kinds of lettuce all year. A touch of watercress, a shredded carrot, cucumber chunks, cherry tomatoes, and white radishes can turn the simplest salad bowl into a stimulating, happy part of any dinner.

Salad varieties are endless and can provide a balance to the heavier meats, sauces, breads, and pastas found on most American tables. Although some cooks, especially French ones, serve salad as a last course, I believe they should be eaten before the main course or at least along with the main part of the meal. This will reduce the hunger we all experience before mealtime, cut down on our food intake, and prepare the system for proper digestion of heavier foods that follow.

There are no real salad recipes. Anything goes. Salads can range from a completely bitter dandelion, arugula, endive, and lemon juice salad to a bowl of sweet Bibb and Boston lettuce, alfalfa sprouts, red peppers, and baby cucumbers with an apple-cider and olive-oil dressing. The trick to salads is a blend of contrasting tastes, textures, and colors. From artichoke hearts to zucchini wheels, all raw vegetables qualify for salad-makings. So do oranges, apples, pears, avocados, raisins, nuts, pineapples, lemons, limes, grapes, olives, cheeses, cold meats, croutons, and caviar.

Salad dressings are equally open to your imagination. Of course, salads and dressings have a basic form, but the combination of ingredients is limited only by imagination and willingness to experiment.

Salads are the toast of spring, when all the leaves, heads, crunchies, and buds are at their sweetest. Local sources of salad greens or homegrown varieties are far superior to those picked, packed, and shipped in from far away. Every state has a salad season, so there is a local farm source available to everyone at least once a year. The season usually occurs from March through late June or early July. Local lettuce, of any variety, has a crispness and a vitality that echoes the message of spring—new life.

During the other seasons, most stores and supermarkets do an excellent job of providing high-quality salad greens and crunchies, though some of the more specialized things like arugula, Belgian endive, chives, and Jerusalem artichokes are definitely seasonal and have an off-time. But most of the time you can choose from many things on display. Ask your fruit and vegetable man which is his freshest lettuce. Ask if he has anything special for salads and what, if anything, is grown locally.

Sometimes your local fruit stand will have what we call hearts of lettuce. These are heads of lettuce that have been trimmed down once or twice. Despite a couple of days on the stand, these hearts are almost always tender and sweet and make an excellent salad served whole, dressed with some chopped radish, scallion, and French dressing; at reduced prices they are an excellent value. It also helps out the storekeeper.

CUCUMBERS

Although cucumbers are mostly water, they are a solid, staple item at any fruit stand. I always display lots of cucumbers and am careful to select long, thin, dark-green varieties with lots of little bumps. I like to sell what I like to eat. Cucumbers are always eaten raw, and although there are many varieties, their shape rather than their flavor is the main difference among them.

There is the common long green cucumber available all year. Most of the time, however, long green cucumbers are waxed to prevent moisture loss in shipping and storage. Peel them before you eat them. Locally produced spring and summer cucumbers are not waxed, and although the skin is more difficult to digest than the white meat inside, these can be eaten unpeeled. Some long green cucumbers are half white. They are only half sweet. Avoid them.

Very long cucumbers grown in greenhouses and wrapped in plastic are seedless, burpless, and delicious. They are called gourmet cucumbers. These are also available anytime but cost slightly more than common green cucumbers (one gourmet cucumber is equal to three or four regular cucumbers, so don't let the higher price scare you). Their quality is consistent and makes them your best winter purchase.

Small, very bumpy cukes called Kirbies are very sweet and crunchy, with edible skins. Kirbies appear year-round and can be eaten whole or pickled. The Kirby cuke is what ultimately becomes the dill pickle, and tiny Kirbies become pickled gherkins. There is also a recipe for pickling large yellow cucumbers for the benefit of gardeners committed to utilizing everything their gardens produce.

Select cucumbers that are bright, firm, and mostly green, no matter the variety. Avoid cucumbers that are very white, yellow, or wrinkled or have soft spots or hollow centers. Old or overgrown cucumbers taste bitter, stale, and woody.

Except for their potassium cukes have almost no nutritional value, but their refreshing, cool taste and undeniable crunchiness make them indispensable in the salad bowl. Cucumber juice, however, is one of nature's best diuretics and kidney cleansers and is great for people who suffer from high blood pressure. Your body will love a veggie-combo drink of cucumber, carrot, and spinach juices.

NUTRITIONAL DATA

Weight (g)	Calories	Protein (g)	Carbohydrates (g)	Crude Fiber (g)	Water (g)	Fat (g)	Cholesterol (mg)
50	8	0.5	1.7	0.3	47.5	0.1	—

Vitamin A (IU)	Vitamin C (mg)	Thiamine B-1 (mg)	Riboflavin B-2 (mg)	Niacin (mg)	Vitamin B-6 (mg)	Vitamin B-12 (mcg)	Folic Acid (mcg)	Pantothenic Acid (mg)
125	6	0.01	0.02	0.1	—	—	—	—

Sodium (mg)	Potassium (mg)	Calcium (mg)	Phosphorus (mg)	Magnesium (mg)	Iron (mg)	Zinc (mg)	Copper (mg)	Manganese (mg)
3	80	13	14	6	0.6	—	—	—

Raw, with skin—½ medium

Fresh Cucumbers in Dill Marinade

½ teaspoon salt
¼ teaspoon freshly ground black pepper
2 tablespoons chopped fresh dill
3 cloves garlic, chopped
¾ cup white-wine vinegar
Juice of 1 lemon
4 green unwaxed cucumbers
4 radishes

In deep bowl combine salt, pepper, dill, and garlic with vinegar and lemon juice. Slice unpeeled cucumbers and radishes. Mix with marinade. Refrigerate for 2 to 3 hours. *Serves 4.*

Cucumber Slices

Select long dark-green summer cukes. Peel. Slice into long spears. Remove seeds by running knife blade just under seed pocket. Serve with dips or dressing.

Kirby Cuke Cooler

6–8 small Kirby cucumbers
Juice of 1 lemon
¼ teaspoon freshly ground black pepper
6–8 sprigs savory, chopped
2–3 scallions, chopped

Cut cucumbers lengthwise into halves or quarters. Place in bowl. Add lemon juice. Sprinkle with pepper, savory, and scallions. *Serves 4.*

High-Gear Cucumber Slices

Use whole unpeeled unwaxed cucumbers. Run the tines of a fork the entire length of the cucumber, penetrating the peel to the white meat. Slice crosswise and serve, or add to salads or platters for a decorative touch.

STUFFED CUCUMBERS

Cut firm, fresh, unwaxed cucumbers in half. Peel. Remove seeds with a long thin knife or apple-corer. Stuff halves with one of the following:

Whole trimmed radishes
Soft cheese
Tuna fish, crabmeat, or shrimp
Caviar
Relish

Chill. Slice.

SHREDS AND CHUNKS

2 medium cucumbers, peeled
½ head iceberg lettuce
6 sprigs parsley, chopped
6 sprigs summer savory, chopped
¼ teaspoon freshly ground black pepper
Juice of 1 lemon
1 red pepper, diced

Cut cucumbers lengthwise into spears. Remove seeds. Cut into chunks. Shred or finely slice lettuce. Mix with black pepper, parsley, and savory. Sprinkle with lemon juice and red pepper. *Serve as condiment.*

BABY PICKLES (KIRBY CUTIES)

24 Kirby cucumbers
12–16 sprigs dill
4 tablespoons mustard seed
4 cloves garlic, halved
8 tiny white onions
2 cups apple-cider vinegar
2 tablespoons pickling salt

Wash cucumbers. Divide among 4 hot sterilized quart jars. Add 3 or 4 sprigs dill, 1 tablespoon mustard seed, half a garlic clove, and 2 white onions to each jar. Combine vinegar and pickling salt in large glass, stainless steel, or enameled saucepan. Bring to boil. Fill sterilized jars, leaving ½ inch space at top. Seal (see "Canning process" entry in Glossary, page 000). Do not remove screw bands from lids during storage.

PARSLEY

Parsley is the indispensable herb. It is included in this section of salads instead of in the herb chapter because it is usually sold with salad greens, celery, and radishes. Although there are four or five kinds of parsley, it has two common forms, plain and curly. What is the difference between the two and when should you use which? The answer is simple.

Plain or flat-leaf parsley (sometimes called Italian parsley, which it is not unless it has white stems) has a stronger flavor and a longer stem. It is best used in cooking and retains its bright-green color.

Curly parsley turns dark in cooking and is best used raw in salads, garnishes, vegetable juices, cheese dips, and dressings.

Parsley may be as old as ancient Rome. Charlemagne planted it in his gardens. It grew wild along the shores of the Mediterranean from Spain to Lebanon. The Sardinians cultivated it extensively, and the Italians made it part of everything. That's where parsley stands today, a part of everything. Parsley is so commonplace we take it for granted, but a kitchen without it is no kitchen at all.

Select bright-green upright parsley. Curly parsley especially should be almost prickly in its crispness. Once you purchase it, sort out broken or wilted pieces, wash and clean it, towel-dry, and keep it refrigerated loosely in a plastic bag. The best way to keep it fresh is to use it. Parsley is an excellent additive to vegetable drinks, for both its flavor and nutrition.

Parsley is high in vitamins A and C, with significant amounts of minerals, including potassium and iron.

NUTRITIONAL DATA

Weight (g)	Calories	Protein (g)	Carbohydrates (g)	Crude Fiber (g)	Water (g)	Fat (g)	Cholesterol (mg)
100	44	3.6	8.5	1.5	85.1	0.6	—

Vitamin A (IU)	Vitamin C (mg)	Thiamine B-1 (mg)	Riboflavin B-2 (mg)	Niacin (mg)	Vitamin B-6 (mg)	Vitamin B-12 (mcg)	Folic Acid (mcg)	Pantothenic Acid (mg)
8500	172	0.12	0.26	1.2	—	—	—	—

Sodium (mg)	Potassium (mg)	Calcium (mg)	Phosphorus (mg)	Magnesium (mg)	Iron (mg)	Zinc (mg)	Copper (mg)	Manganese (mg)
45	727	203	63	41	6.2	—	—	—

Raw—3½ ounces

Garnish Galore

An excellent sprightly and tasty garnish is made from a mixture of coarsely chopped flat-leaf and curly parsley, chives or scallion greens, garlic, and red onion. More than just a colorful garnish, it adds the taste of a delightfully spicy fresh relish.

Parsley Crisps

> **1 bunch curly parsley**
> **½ cup pancake mix**
> **1 small yellow onion, finely chopped**
> **1 clove garlic, finely chopped**
> **1 egg**
> **2 tablespoons melted butter or margarine**
> **½ cup milk**
> **¼ cup stale beer**
> **¼ teaspoon freshly ground black pepper**
> **Light vegetable oil**

Wash parsley sprigs, shake dry, and chill. Mix remaining ingredients except oil and chill. Heat ¼ inch oil in pan or deep fryer until very hot but not smoking. Dip chilled parsley sprigs into batter. Allow excess to drip off. Place parsley sprigs in oil leafy ends down. Fry until browned. Remove. Drain on paper towels. Serve as an hors d'oeuvre or mixed with other tempura vegetables.

WATERCRESS

Watercress is definitely the green spice. Its taste is wide awake and its presence is definite. Watercress is found along the banks of clear running streams. It grows in every state and south of Pennsylvania can be harvested all year. Most commercial watercress is actually a large patch of wild cress that has been cared for and harvested properly. Every spring in Woodstock some of the locals from up the mountainside bring freshly gathered wild watercress to our store. After we clean and bunch it, it is a special offering for our discerning customers. The taste is very "watercress," but it has a clean and refreshing aftertaste.

Use watercress in salads, add a few sprigs to a sandwich, use it as a garnish for cold meat plates, or drop a few sprigs into a clear broth. Watercress can also be blanched and simmered in a sauté, then served with a cream sauce as a vegetable side dish.

Look for dark-green flat leaves without wilt, buds, or yellowing. The entire plant is edible, although very thick stems tend to be tough, with less flavor than the smaller hot-sweet stems and leaves.

Open the bunches after buying, wash the watercress lightly, shake it dry, wrap it loosely in a damp paper towel or put it in a loose plastic bag, and refrigerate. Though watercress is available most of the year, it has its best flavor from late April until midsummer. Late-summer watercress tends to wilt and fade quickly, so buy and use it fresh.

Watercress is consumed in small quantities, which makes it a nutritional lightweight. Proportionately, however, it contributes its share of vital minerals and vitamin A. A high chlorophyll content and clean, easily digested fiber make watercress a healthful and tasty addition to your diet. It is definitely a "live" food.

NUTRITIONAL DATA

Weight (g)	Calories	Protein (g)	Carbohydrates (g)	Crude Fiber (g)	Water (g)	Fat (g)	Cholesterol (mg)
100	19	2.2	3	0.7	93.3	0.3	—

Vitamin A (IU)	Vitamin C (mg)	Thiamine B-1 (mg)	Riboflavin B-2 (mg)	Niacin (mg)	Vitamin B-6 (mg)	Vitamin B-12 (mcg)	Folic Acid (mcg)	Pantothenic Acid (mg)
4900	79	0.08	0.16	0.9	—	—	—	—

Sodium (mg)	Potassium (mg)	Calcium (mg)	Phosphorus (mg)	Magnesium (mg)	Iron (mg)	Zinc (mg)	Copper (mg)	Manganese (mg)
52	282	151	54	20	1.7	—	—	—

Raw—3½ ounces

WATERCRESS SPICY SOUP

2 medium potatoes
1 teaspoon salt
1 small yellow onion, chopped
4 cups chicken broth
1 cup half-and-half
1 bunch watercress

Peel potatoes and cut up. Combine with salt, onion, and chicken broth. Simmer about 15 minutes. Purée hot mixture with half-and-half in blender. Add watercress a few sprigs at a time, blending at medium speed for a few seconds after each addition. Chill. *Serves 4.*

WATERCRESS-DANDELION SPRING SALAD

1 bunch watercress
¼ pound dandelions
1 Belgian endive
¼ teaspoon freshly ground black pepper
Juice of 3 oranges
Juice of 1 lemon
4 teaspoons salad oil

Wash, clean, and tear up watercress and dandelions. Slice endive and toss all three greens. Combine pepper, orange juice, lemon juice, and salad oil. Mix with greens. Chill. *Serves 3 to 4.*

CHIVES

Are chives forever confined to the limits of a sprinkle atop the baked potato and sour cream, or *le garni* floating brightly in your vichyssoise? I use chives a lot. They are the green stems of tiny, mild bulbs, less than an onion but more than a scallion. Chives add flavor to vegetable stir-fries, roast chicken, poached fish, and steamed vegetables. Their flavor is a delightful addition in salad and broth. They can be part of a sauté or chopped and mixed raw with sour cream, cream cheese, cottage cheese, cream soups, omelettes, and salad dressings.

Only the most select produce stores carry fresh chives, but a small window box or flowerpot will give you them year-round. Cut them three-quarters of the way down from the top at each picking, and they will grow new green shoots continuously. At Sunfrost we cut chives fresh from our gardens for our customers.

Chives are not a nutritional powerhouse, but they do contribute to the overall vitality of your salads and dressings with goodly amounts of minerals and vitamin A.

NUTRITIONAL DATA

Weight (g)	Calories	Protein (g)	Carbohydrates (g)	Crude Fiber (g)	Water (g)	Fat (g)	Cholesterol (mg)
10	3	0.2	0.6	0.1	9.1	trace	—

Vitamin A (IU)	Vitamin C (mg)	Thiamine B-1 (mg)	Riboflavin B-2 (mg)	Niacin (mg)	Vitamin B-6 (mg)	Vitamin B-12 (mcg)	Folic Acid (mcg)	Pantothenic Acid (mg)
581	6	0.01	0.01	0.1	—	—	—	—

Sodium (mg)	Potassium (mg)	Calcium (mg)	Phosphorus (mg)	Magnesium (mg)	Iron (mg)	Zinc (mg)	Copper (mg)	Manganese (mg)
—	25	7	4	3	0.2	—	—	—

Raw—1 tablespoon chopped

LETTUCE

How many kinds of lettuce does your local store carry? If it's only one, I know it's iceberg. America loves iceberg lettuce. No matter the price, the quality, the season, or what other varieties are available, Americans buy iceberg lettuce more than any other kind. Restaurants, supermarkets, luncheonettes, cafeterias, and most kitchens made iceberg their first and often only choice. Too bad. There are five or six other varieties available at all times, all of which are fresher, more nutritious, and for my money much tastier and more interesting.

California, Arizona, and Florida produce all of America's lettuce from September to June, but during the spring and summer local sources can provide fresh, crisp, beautiful lettuce of all varieties. But summer or winter, any fruit and vegetable market worth the name should stock five to seven varieties of lettuce. There are two kinds of lettuce: the heads and the leaves. Head lettuces include Boston butter lettuce, Bibb lettuce, and the popular, ever-present iceberg. The leaf lettuces include the red and green curly-leaf lettuces (and their straight-leaf cousins) and romaine. In some places romaine is called cos, but in this book it's romaine.

Any lettuce should be crisp and bright, without wilt or brown or yellow areas. It is often written and said that when lettuce is prepared, it should be torn by hand rather than cut with a knife unless a special recipe or form is being created, such as lettuce wedges or shredded lettuce. I have no idea why it's said and never observe this odd law in my kitchen. Maybe it has something to do with certain knife-blade metals imparting an unpleasant taste. I say keep a clean knife and cut or tear as you prefer. Lettuce in all its forms is the main ingredient of most salads. For the freshest, crispest texture the dressing should be added when the salad reaches the table or by each individual to his or her serving.

Once home, lettuce can be washed, broken up, and mixed with several other varieties. If it is kept in sealed plastic bags in the refrigerator, it will stay fresh for several days. This way the basic salad is at the ready, making preparations more convenient and leaving more time for imagination and experiments.

Lettuce, especially the leaf varieties, is high in vitamin A, riboflavin, thiamine, and vitamin C. In addition to its crispness, vibrant green color, and live taste, romaine offers the largest amounts of vitamins and minerals.

ICEBERG LETTUCE

*I*ceberg lettuce is the head head. Almost all iceberg is grown in California and shipped elsewhere. The heads are always round, with large leaves. Most merchants prefer iceberg because it requires little trimming and because most of their customers know it and want it. Most of the California variety is vacuum-packed, which removes some of the leaf moisture and creates a green, somewhat leathery leaf that lasts longer and doesn't wilt like lettuce should. Some iceberg is being shipped poly-wrapped and heat-sealed. Supermarkets love this kind of lettuce. No trimming, no waste. No self-service damage. There is also no taste, no nutrition, and no turning back. Most consumers, in choosing iceberg, look for solid, white, heavy heads, but a looser, greener, softer head is always sweeter and more tender. I never recommend iceberg lettuce as a first choice except when we have fresh-picked local leafy green iceberg. Many of our customers say, "This iceberg is too soft. I want a harder head." "Try it," I insist. I've gotten no complaints, and many of our customers now anticipate the soft iceberg season.

Soft- or hard-headed, iceberg leaves have the lettuce crunch Americans love. Iceberg adds texture and a semisharp taste to salads and sandwiches. Iceberg makes the best shredded lettuce and quarter-head wedges.

NUTRITIONAL DATA

Weight (g)	Calories	Protein (g)	Carbohydrates (g)	Crude Fiber (g)	Water (g)	Fat (g)	Cholesterol (mg)
100	13	0.9	2.9	0.5	95.5	0.1	—

Vitamin A (IU)	Vitamin C (mg)	Thiamine B-1 (mg)	Riboflavin B-2 (mg)	Niacin (mg)	Vitamin B-6 (mg)	Vitamin B-12 (mcg)	Folic Acid (mcg)	Pantothenic Acid (mg)
330	6	0.06	0.06	0.3	—	—	—	—

Sodium (mg)	Potassium (mg)	Calcium (mg)	Phosphorus (mg)	Magnesium (mg)	Iron (mg)	Zinc (mg)	Copper (mg)	Manganese (mg)
9	175	20	22	11	0.5	—	—	—

Raw—3½ ounces

BOSTON LETTUCE

Boston lettuce is a little larger than a softball. It should have bright-green soft, sweet leaves that whiten along the stalk. The inner leaves are yellower and become softer and softer and very sweet as you reach the heart. Boston is available all year round, but the very best comes from local sources in late May, June, and sometimes July (Boston lettuce grows poorly in extreme heat). Look for crisp, soft green heads with frilly-edged leaves. Sometimes Boston is grown in very black sandy soil and must be washed carefully and thoroughly. Whether sold by the head or the pound it is usually the cheapest variety. Sometimes trimmed-down Boston hearts are sold for very little money. They are excellent food and an excellent value. Boston lettuce is a fine salad-maker and provides a lovely, tasty bed for tuna, salmon, fruit, and cottage cheese. I love it with cold stewed or canned Bartlett pears.

NUTRITIONAL DATA

Weight (g)	Calories	Protein (g)	Carbohydrates (g)	Crude Fiber (g)	Water (g)	Fat (g)	Cholesterol (mg)
100	14	1.2	2.5	0.5	95.1	0.2	—

Vitamin A (IU)	Vitamin C (mg)	Thiamine B-1 (mg)	Riboflavin B-2 (mg)	Niacin (mg)	Vitamin B-6 (mg)	Vitamin B-12 (mcg)	Folic Acid (mcg)	Pantothenic Acid (mg)
970	8	0.06	0.06	0.3	—	—	—	—

Sodium (mg)	Potassium (mg)	Calcium (mg)	Phosphorus (mg)	Magnesium (mg)	Iron (mg)	Zinc (mg)	Copper (mg)	Manganese (mg)
9	264	35	26	11	2	—	—	—

Raw—3½ ounces

BIBB LETTUCE

*B*ibb lettuce is like a smaller Boston lettuce, but it is crisper, greener, and very sweet. Packed in small rectangular baskets, sixteen heads in each, Bibb is almost always sandy and needs a good washing. While the leaves may be left whole as a salad by themselves or torn and shredded to make a green leaf mix, Bibb hearts are a soft green-white and should always be eaten whole. Bibb lettuce is not a commercial item and hardly ever appears in supermarkets or small stores that wish everything came in cans. Look for Bibb from June through July in beautiful fruit markets that specialize in high-quality fresh local produce in season.

LEAF LETTUCE

*G*reen leaf lettuce never heads up. It comes with both curly- and straight-edged leaves. It is semicrisp and semisweet and makes soft, gentle salads. Many produce stores carry leaf lettuce year-round, but supermarkets seem to avoid leaf lettuce because it is fragile and so self-service can destroy more than gets sold.

Red leaf lettuce is softer than the green and has a definite sweet flavor. Though delicate and fragile, it is an excellent salad-maker. A combination of sharp leaves like endive and chicory, radishes, watercress, and scallions with the soft, sweet leaf lettuces makes exciting salads with interesting textures and contrast. I find leaf lettuces a little too soft for robust sandwiches, but they are indispensable to the family cook who prepares salads daily and wants to keep them lively and appealing.

ROMAINE LETTUCE

Romaine is the best lettuce you can eat. It is available almost everywhere all year long. There is a short-supply season in early and mid-fall; most supplies are not at their best during that period. Romaine is the only lettuce that really loves to grow in the summer heat. Good romaine is very green, with the outside leaves curling away from the center leaves. It is a very crisp lettuce with tender leaf edges and a semisweet taste and grows relatively clean. Romaine can reach a very large size and still taste young and tender, but you must avoid specimens that show a yellow leaf burst or protruding stalk, as these are overgrown and quite bitter. California and Florida ship lots of romaine on either side of summer. The California romaine seems to be a little white and the Florida variety not nearly as sweet as it should be, but in season from local sources, romaine is at its best. Caesar salad demands romaine, and sandwiches have a lot more life with romaine in the middle. The yellow-white heart is very delicious. Keep it whole and eat it whole, with a French or cheese dressing. Romaine has the highest nutritional rating of all the lettuces. It is a lettuce *extraordinaire*.

NUTRITIONAL DATA

Weight (g)	Calories	Protein (g)	Carbohydrates (g)	Crude Fiber (g)	Water (g)	Fat (g)	Cholesterol (mg)
100	18	1.3	3.5	0.7	94	0.3	—

Vitamin A (IU)	Vitamin C (mg)	Thiamine B-1 (mg)	Riboflavin B-2 (mg)	Niacin (mg)	Vitamin B-6 (mg)	Vitamin B-12 (mcg)	Folic Acid (mcg)	Pantothenic Acid (mg)
1900	18	0.05	0.08	0.4	—	—	—	—

Sodium (mg)	Potassium (mg)	Calcium (mg)	Phosphorus (mg)	Magnesium (mg)	Iron (mg)	Zinc (mg)	Copper (mg)	Manganese (mg)
9	264	68	25	11	1.4	—	—	—

Raw—3½ ounces

THE WHOLE ICEBERG

Select a firm head of iceberg lettuce. Holding the head core down, hit the head against the counter top, driving the core inward. Turn head over and remove core. Fill with water. Drain and chill. Serve head whole, turned bottom up, with core space filled with favorite salad dressing. Allow to stand several minutes before cutting into wedges.

FRENCH ICEBERG WEDGES

1 head iceberg lettuce
1 small red onion, chopped
2 cloves garlic, chopped
Juice of 1 lemon
¼ cup olive oil
¼ cup mayonnaise
1 teaspoon Worcestershire sauce
¼ teaspoon freshly ground black pepper

Core and wash lettuce according to instructions above in The Whole Iceberg. Quarter lettuce. Combine all other ingredients in bowl. Blend together until smooth. Pour over lettuce wedges. *Serves 4.*

MIXED GREEN SALAD

3–6 leaves Boston lettuce
3–4 leaves iceberg lettuce
2–3 leaves romaine lettuce
3–4 leaves leaf lettuce
3–4 bunches spinach
1 small Belgian endive
1 cucumber, peeled
1 green pepper
1 medium red onion

Trim, wash, and shred all green leaves. Slice endive and cucumber. Chop green pepper. Toss all ingredients except onion together. Slice red onion thinly and scatter rings on salad. Serve with favorite salad dressing on the side.
For a spicy flavor add 8–10 sprigs each watercress and arugula. *Serves 4.*

VELVET CRUNCH

½ head Boston lettuce
¼ head romaine lettuce
3–4 leaves Chinese cabbage

Wash all leaves. Shred lettuce and cut up Chinese cabbage. Toss. Serve with dressing on the side. *Serves 3 to 4.*

LETTUCE HEARTS

Select the hearts of Boston, romaine, or iceberg lettuce when they are about the size of a tennis ball or smaller. Remove the cores. Wash and drain. Chill. Serve with core openings filled with different dressings.

ROMAN APPLE SALAD

⅔ cup mayonnaise
2 tablespoons honey
1 teaspoon lemon juice
2 red apples, peeled and cubed
2 green apples, peeled and cubed
1 stalk celery, chopped
¼ cup chopped walnuts
¼ cup chopped dates
6 leaves romaine lettuce

Blend mayonnaise, honey, and lemon juice. Mix in all other ingredients except lettuce. Serve on lettuce leaf beds. *Serves 3 to 4.*

CELERY
AND CELERY ROOT

Between the Italians and the French, celery has been developed to a point Mother Nature never imagined. Originally a wild plant that grew along the muddy Mediterranean wetlands where the sea and land waters mingled, celery was considered by the Romans both a vegetable to be cooked with tomatoes and meats, and an herb for the relief of hangovers. The French developed a less woody, sweeter variety that we know today as table celery. Most stores carry two kinds of celery: the very green Pascal celery, which is grown mostly in California, Canada, and Michigan, and a white, tender celery, which is grown under cover to keep it white and soft. I much prefer the green for cooking and the white for eating raw.

Whatever your choice, look for bright-green leaves and white or green stalks free from brown spots. Celery should never be limp, and the core should be hard and moist at the bottom, never soft and dry. Celery is available most of the year from the celery states. Locally grown and homegrown celery rarely beats Western-grown commercial celery. Michigan also produces fine celery, but it's never quite as sweet as the California kind.

Most people treat celery as something for soup or the second basic ingredient in tuna-fish salad. I say use celery more ways. I love to stuff the stalks with cheese dips or cream cheese and olives or peanut butter. In addition to its usefulness as a salad and soup ingredient, celery adds a great touch to omelettes and is the key to the wonderful taste of a well-made Spanish omelette. Added to goose, duck, or heavy red meats while they're roasting, celery absorbs excess fats and oils. Celery sticks make excellent lunch-bag snacks and afternoon treats, although I know very few school children who will trade their afternoon Oreos, Snickers, or chocolate milk for a celery stick. Oh, well, we try.

Celery root, or celeriac, is shaped like a small knobby turnip. It can be cooked, eaten raw, or added to salads. Celery root cut in julienne slices and marinated in a mustard-curry vinaigrette is an exciting and stimulating appetizer. I enjoy celery root before my meal with a glass of white wine or a light martini. The combination awakens your palate for whatever comes next. The root must be firm and heavy, with a touch of dew. Forget celeriac if it is shriveled and woody; it is inedible.

Celery is a very alkaline food and is excellent as a digestive aid and stomach-sweetener. It provides a significant amount of calcium and, taken as a juice, provides magnesium and salt, which have a beneficial effect on the nervous system. The celery root, in the amount of a small four-root knob, contains two to five times the nutrients of the stalks.

NUTRITIONAL DATA

Weight (g)	Calories	Protein (g)	Carbohydrates (g)	Crude Fiber (g)	Water (g)	Fat (g)	Cholesterol (mg)
50	8	0.4	2	0.3	47	0.1	—
100	14	0.8	3.1	0.6	95.3	0.1	—

Vitamin A (IU)	Vitamin C (mg)	Thiamine B-1 (mg)	Riboflavin B-2 (mg)	Niacin (mg)	Vitamin B-6 (mg)	Vitamin B-12 (mcg)	Folic Acid (mcg)	Pantothenic Acid (mg)
120	5	0.02	0.02	0.2	—	—	—	—
230	6	0.02	0.03	0.3	—	—	—	—

Sodium (mg)	Potassium (mg)	Calcium (mg)	Phosphorus (mg)	Magnesium (mg)	Iron (mg)	Zinc (mg)	Copper (mg)	Manganese (mg)
63	170	20	14	11	0.2	—	—	—
88	239	31	22	—	0.2	—	—	—

Raw—1 stalk
  Cooked—⅘ cup diced

C E L E R Y   A N D   C E L E R Y   R O O T

GENTLE CRUNCH SALAD

>   ¼ cup vegetable oil
>   ¼ cup vinegar
>   2 cloves garlic, chopped
>   ½ small red onion, chopped
>   2 stalks celery, chopped
>   8 string beans, cut up
>   4 Kirby cucumbers, sliced
>   4 radishes, sliced
>   1 medium Jerusalem artichoke, cubed
>   ¼ teaspoon freshly ground black pepper

Combine oil, vinegar, garlic, and onion in blender to make dressing. Mix with remaining ingredients except for pepper. Sprinkle with black pepper.
   Serves 2 to 3.

CELERIAC LE HOT

>   1 celery root
>   Juice of 2 lemons
>   Dash Tabasco sauce
>   ¼ cup grainy mustard
>   ¼ cup mayonnaise

Peel celery root and slice into thin strips. Marinate in lemon juice in the refrigerator. Mix Tabasco sauce, mustard, and mayonnaise with lemon juice marinade. Pour mixture over celery root. Serve as appetizer.

BRAISED CELERY

>   1 bunch celery
>   1 small yellow onion, chopped
>   6–8 sprigs flat-leaf parsley, chopped
>   Light sesame oil
>   3–4 scallions, chopped
>   2–3 teaspoons tamari sauce
>   Dash garlic powder

Clean and trim celery leaves. Cut stalks into 4- to 6-inch pieces. Sauté onion and parsley in oil. Wash celery and place wet pieces into sauté mixture. Add ¼ cup water and the chopped scallions. Sprinkle with tamari sauce, extra dash sesame oil, and dash garlic powder. Cover. Simmer for 5 minutes. Remove cover. Increase heat and cook celery, turning, until slightly browned. *Serves 4 to 6.*

CELERIAC IN CREAM SAUCE

2 celery roots
4 tablespoons butter
1 cup milk
¼ cup instant-blending flour

Peel celery roots and cut into strips or wedges. Steam until just soft. Melt butter and slowly add milk. Lower heat. Add flour 1 teaspoon at a time, stirring constantly. Cook, stirring, until sauce thickens. Drain celery root. Pour on cream sauce. *Serves 2 to 4.*

RADISHES

Everybody has seen radishes carved in elaborate rosettes that always seem to be left on the plate after the party. Radishes deserve better. At one time they were the "food of kings." Today we dip them and munch them and toss them in salad. One hardly ever thinks of eating radishes.

Actually, there are two different kinds of radishes. One kind includes the common small round red radish, the red and white radish, and the white "icicle" radish. These radishes are used raw in salads, hors d'oeuvres, and decorations.

The other kind of radish includes the Japanese radish, which is white and two to three feet long, the rose-colored Chinese long radish, and a group of thick-skinned black radishes with white flesh. These radishes can be grated, chopped, and sliced for salads and appetizers. They can also be cooked, producing a tangy, creamy vegetable with many of the taste properties of a yam or a small sweet turnip.

All radishes must be firm and bright when purchased. Fresh radishes should be purchased with bright-green leaves and plenty of them. I never sold radishes in a plastic bag in my life. They look artificial and taste like wooden beads. Demand fresh. You'll get it.

Radishes are excellent companions to avocados, creating a delightful contrast between the soft, gentle avocado and the crunchy, hot radish. Radish wheels will activate your salads and stimulate your appetite and taste buds. They also make an interesting salad or condiment as themselves *au vinaigrette*.

Fresh radishes are usually available in major markets most of the year, particularly the larger cooking varieties. Fresh garden radishes are best from May until July and once again in late summer and early fall. I like to munch on rosy spring radishes with guacamole and shredded lettuce.

Nutritionally, radishes contain excellent and balanced nutrients. Unfortunately, raw radish nutrition requires you to eat ten or more of them. "Please pass the radishes, Mom!"? Cooked large radishes, on the other hand, can be eaten in amounts significant enough to be worthwhile. The Far East cultures that specialize in abundant nutritionally efficient food have made the radish a major cooking item in the daily diet.

NUTRITIONAL DATA

Weight (g)	Calories	Protein (g)	Carbohydrates (g)	Crude Fiber (g)	Water (g)	Fat (g)	Cholesterol (mg)
100	17	1	3.6	0.7	94.5	0.1	—
100	19	0.9	4.2	0.7	94.1	0.1	—

Vitamin A (IU)	Vitamin C (mg)	Thiamine B-1 (mg)	Riboflavin B-2 (mg)	Niacin (mg)	Vitamin B-6 (mg)	Vitamin B-12 (mcg)	Folic Acid (mcg)	Pantothenic Acid (mg)
10	26	0.03	0.03	0.3	—	—	—	—
10	32	0.03	0.02	0.4	—	—	—	—

Sodium (mg)	Potassium (mg)	Calcium (mg)	Phosphorus (mg)	Magnesium (mg)	Iron (mg)	Zinc (mg)	Copper (mg)	Manganese (mg)
18	322	30	31	15	1	—	—	—
—	180	35	26	—	0.6	—	—	—

Red, raw—10 small
   Oriental, raw—3½ ounces

Fresh Radish Relish

> **12 red radishes**
> **3–4 icicle radishes**
> **1 medium black radish, peeled**
> **¼ pound Chinese radish**
> **2–3 stalks celery**
> **½ seedless cucumber or 1 garden cucumber,**
>   **peeled and seeded**
> **3–4 scallions (including tops)**
> **1 small red onion**
> **2 cloves garlic or 2 shallots**
> **6–8 sprigs curly parsley**
> **¼ teaspoon salt**
> **¼ teaspoon freshly ground black pepper**
> **Juice of 1 lemon**

Finely chop all ingredients except for salt, pepper, and lemon juice. Mix well. Add salt, pepper, and lemon juice. Serve with sliced tomatoes, avocado halves, cheese, or cold seafood salads; or sprinkle on green lettuce salads, baked potatoes, cottage cheese, or sour cream. *Serves 4.*

Radish Hors d'oeuvres

Some of the larger radish varieties, such as the black and Chinese radishes and some of the bigger icicle radishes, can be cut into wedges, sticks, and chunks as well as slices. Their colorful skins and crisp textures make them excellent companions to celery, carrots, cukes, and raw vegetable appetizers. These larger radishes have definite flavor and spiciness, but less of the burn some folks dislike in the fiery little red radishes. Cut-up large radishes are delicious served with guacamole, mayonnaise, mustard, salsa, crackers, bread sticks, and fresh bread and butter. Any buffet table improves with the inclusion of unique varieties of properly presented radishes.

Radishes Vinaigrette

This recipe may include as few as one variety of radish or as many as can be had.

> **12 red radishes**
> **3–4 icicle radishes**
> **1 medium black radish, peeled**
> **¼ pound Chinese radish**

2 cloves garlic
10 sprigs curly parsley
Juice of 1 lemon
2 tablespoons apple-cider vinegar
2 tablespoons olive oil
¼ teaspoon salt
¼ teaspoon freshly ground black pepper
1 small yellow onion

Coarsely chop all radishes. Finely chop garlic and parsley. Mix with radishes. Add lemon juice, vinegar, olive oil, salt, and pepper. Slice the onion into thin rings. Toss with marinated radishes. Chill for 3 to 4 hours. Serve with green salad, cottage cheese, sour cream, or warm bread and butter. *Serves 4.*

BELGIAN ENDIVE

Belgian endive, or witloof ("whiteleaf"), is a broad-leaved chicory plant grown indoors in darkness in heat-producing beds of peat, soil, and manure. This produces a rapid-growing plant without chlorophyll and high in sweetness and flavor. When harvested, Belgian endive looks like a small, tightly wrapped, white-husked ear of corn. These endive plants are packed in small waxed-paper–lined wooden boxes of twelve pounds each. They are usually sold under refrigeration for four or five dollars a pound, or about seventy-five cents each. Belgian endive is one of the world's finest eating plants, beautiful to see and feel and an original addition to any table.

Belgian endive is usually sliced and eaten raw in salads, or eaten leaf by leaf with dips and sauces, but it can also be braised or wrapped with ham, then simmered and served with a cheese sauce or gravy.

The best-quality Belgian endive arrives in late winter, spring, and early summer. Make sure it is fresh, with a minimum of green, no brown spots at all, and no wilted edges.

Belgian endive is a high alkaline vegetable with significant amounts of vitamin A and potassium.

ESCAROLE AND CHICORY

*E*scarole and chicory are basically the same plant, but with different shapes and flavors. They are internationally known as endive, but American custom and labeling calls each by its own name. Your fruit man buys them this way, and so should you.

Escarole is a flat white-stalked plant with broad, fleshy green leaves. It is slightly sharp, sometimes even semisweet. Escarole is a favorite of Italian cooks as a companion to the chicken in chicken soup. Escarole can also be added to salads as torn leaves, or chopped fine and added to a lettuce mix for tacos, burritos, and Mexican salads. It can also be braised, mixed with stir-fried vegetables, or blanched and served with a light butter-cream sauce. Escarole makes an excellent bed for tuna, salmon, crab, and shrimp salad.

Chicory is a wild-looking green plant with narrow raggedy leaves and a bitter taste. Although chicory will never gain popularity as a vegetable enjoyed for itself, it is an excellent addition to a salad. It can also be used with its fellow bitter-leaved greens, dandelions and arugula, to form a wickedly delicious salad, dressed with olive oil, salt, fresh black pepper, and a splash of fresh orange juice.

You are not expected to run wild over chicory and escarole, now or ever. But new tastes and food discoveries will keep your table fresh and more delicious. I promise you that your salads and soups will be more delicious and that you'll feel better the more often you select bright, fresh, healthful vegetables like escarole and chicory.

Nutritionally escarole and chicory are well balanced, with large amounts of vitamin A and vital minerals.

NUTRITIONAL DATA *(Escarole)*

Weight (g)	Calories	Protein (g)	Carbohydrates (g)	Crude Fiber (g)	Water (g)	Fat (g)	Cholesterol (mg)
100	20	1.7	4.1	0.9	93.1	0.1	—

Vitamin A (IU)	Vitamin C (mg)	Thiamine B-1 (mg)	Riboflavin B-2 (mg)	Niacin (mg)	Vitamin B-6 (mg)	Vitamin B-12 (mcg)	Folic Acid (mcg)	Pantothenic Acid (mg)
3300	10	0.07	0.14	0.5	—	—	—	—

Sodium (mg)	Potassium (mg)	Calcium (mg)	Phosphorus (mg)	Magnesium (mg)	Iron (mg)	Zinc (mg)	Copper (mg)	Manganese (mg)
14	294	81	54	10	1.7	—	—	—

Escarole, raw—20 long leaves

NUTRITIONAL DATA *(Chicory)*

Weight (g)	Calories	Protein (g)	Carbohydrates (g)	Crude Fiber (g)	Water (g)	Fat (g)	Cholesterol (mg)
100	20	1.8	3.8	0.8	—	0.3	—

Vitamin A (IU)	Vitamin C (mg)	Thiamine B-1 (mg)	Riboflavin B-2 (mg)	Niacin (mg)	Vitamin B-6 (mg)	Vitamin B-12 (mcg)	Folic Acid (mcg)	Pantothenic Acid (mg)
4000	22	0.06	0.1	0.5	—	—	—	—

Sodium (mg)	Potassium (mg)	Calcium (mg)	Phosphorus (mg)	Magnesium (mg)	Iron (mg)	Zinc (mg)	Copper (mg)	Manganese (mg)
—	420	86	40	13	0.9	—	—	—

Chicory, raw—30–40 small leaves

ESCAROLE SOLO

   1 small head escarole
   2–3 cloves garlic
   1 small yellow onion
   Light vegetable oil
   ¼ teaspoon dried oregano
   ¼ teaspoon salt
   ¼ teaspoon freshly ground black pepper
   Freshly grated Parmesan cheese

Separate, wash, and break escarole leaves into pieces, removing any tough ends. Let stand in cold water. Chop garlic and onion and gently brown in oil with oregano and salt. Lower heat and add barely drained escarole. Cover. Simmer for 8 to 10 minutes. Sprinkle with black pepper and grated Parmesan cheese. Serve as a dinner vegetable or before the entrée. *Serves 3 to 4.*

ESCAROLE NOODLES

   Light vegetable oil
   1 medium head escarole
   1 clove garlic, chopped
   ½ small yellow onion, chopped
   ½ teaspoon salt
   ¼ teaspoon freshly ground black pepper
   1 pound wide flat noodles
   1 cup light cream or milk
   ¼ pound butter or margarine
   ¼ pound freshly grated Parmesan or Romano cheese

Bring to a boil a large pot of water with 1 tablespoon vegetable oil to cook noodles. While water heats, separate and wash escarole leaves, removing any tough white ends. Cut up into 1- to 2-inch pieces. Let stand in cold water. Heat thin layer of vegetable oil in bottom of deep pot. Add garlic, onion, salt, and pepper. Just before garlic and onion brown, lower heat and add barely drained wet escarole. Cover pot.

   Cook noodles in boiling water. Heat cream and melt butter in it. Cover and keep hot. Drain cooked noodles thoroughly. Add to escarole. Mix. Cover. Slowly add cheese to cream and butter, stirring constantly. When mixture thickens, pour over noodles and escarole. Serve in small bowls. *Serves 4.*

ESCAROLE BROTH

    1 small head escarole
    2 cloves garlic
    Light vegetable oil
    ½ teaspoon salt
    ¼ teaspoon freshly ground black pepper
    2 quarts chicken stock

Separate, wash, and break escarole leaves into pieces, removing any tough ends. Let stand in cold water. Slice garlic and gently brown in oil with salt and pepper. Remove garlic and add drained wet escarole. Cover and simmer over low heat for 3 to 4 minutes. Heat chicken stock. Add escarole to broth. Simmer.
    Serves 4 to 6.

WICKED CHICORY SALAD

    1 small head chicory
    6–8 bunchlets dandelion greens
    6–8 bunchlets or 16–20 leaves arugula
    ½ teaspoon salt
    ¼ teaspoon freshly ground black pepper
    ¼ cup olive oil
    Juice of 1 orange

Separate, clean, and wash chicory, dandelions, and arugula. Drain. Break apart and mix. Toss with salt, pepper, olive oil, and orange juice. Serve chilled.
    Serves 4 to 6.

ARUGULA

Arugula is an Americanized version of the Italian salad herb *rucola* but is the same plant the French call *roquette,* or rocket. By any name arugula is a distinctive, pungent little leaf that will definitely make any salad bowl jump. Almost too strong to go it alone, arugula loves to mix it up with soft lettuce or its wild cousins dandelions and chicory. It goes well with onions and scallions and is at its best with basil, garlic, tomatoes, olive oil, and oregano.

Most fruit stands or supermarkets never carry arugula. It can be grown, however, in almost any kind of soil. After sowing the seed, the seedlings should be thinned. The entire plant should be picked soon thereafter at a height of eight to ten inches and definitely before flowering. Arugula can be dried and used as a seasoning or frozen and thawed for a cooking flavor or herb. Only fresh arugula makes fresh salad, so plant small beds. Keep sowing seed from spring through fall, and you will maintain a constant five- or six-month supply. South of the Mason-Dixon, arugula can be grown and enjoyed year-round. Look for seed packets labeled "roquette."

Learn to use arugula, and your whole green world of salads, tomatoes, and light summer eating will improve immensely. Select bright-green leaves with no wilt or yellowing. Wash, dry, and keep wrapped in the refrigerator.

Specific nutritional data for arugula seem unavailable, but as a member of the spicy green-leaved group it probably contains vital minerals and vitamin A. More importantly, arugula will increase the taste appeal of your salads, which are vitally important to overall good digestion and nutrition.

LEMON ARUGULA

1 bunch arugula
Juice of 1 lemon
Dash onion powder
Dash garlic powder
¼ teaspoon freshly ground black pepper

Wash, de-stem, and tear up arugula. Mix with remaining ingredients. Toss. *Serves 2.*

DANDELIONS

You will almost never find dandelion greens in your local supermarket. The very best dandelions are gathered in early spring from wild meadows. These wild springtime plants, picked before flowering, are a delectable, exciting bittersweet salad green. Some Italian and French fruit and vegetable markets offer dandelions in the springtime and again in the first days of fall, following the frost. They should always be bright green, crisp, and washed thoroughly.

Springtime was a dandelion festival in my house. My mother served them steamed with garlic and olive oil or as a salad, and most children would rather die than eat them. I never really liked the cooked version, but the raw salads tasted good to me then and still do today.

Dandelion flowers provide Western bees their nectar for a thick, tasty dandelion honey. Dandelions make excellent tea and wine. The roots are dried and roasted for a clever coffee substitute, and I have read that the roots also contain a latex-type milk that the Russians have used to produce a synthetic rubber. Imagine all of this from a pesty, ill-smelling yellow flower that interrupts the serene green of our front lawns!

Dandelions are extraordinarily nutritious. They have a relatively large amount of protein and massive amounts of vitamins A and C, potassium, calcium, and iron.

NUTRITIONAL DATA

Weight (g)	Calories	Protein (g)	Carbohydrates (g)	Crude Fiber (g)	Water (g)	Fat (g)	Cholesterol (mg)
100	45	2.7	9.2	1.6	85.6	0.7	—
100	33	2	6.4	1.3	89.8	0.6	—

Vitamin A (IU)	Vitamin C (mg)	Thiamine B-1 (mg)	Riboflavin B-2 (mg)	Niacin (mg)	Vitamin B-6 (mg)	Vitamin B-12 (mcg)	Folic Acid (mcg)	Pantothenic Acid (mg)
14000	35	0.19	0.26	—	—	—	—	—
11700	18	0.13	0.16	—	—	—	—	—

Sodium (mg)	Potassium (mg)	Calcium (mg)	Phosphorus (mg)	Magnesium (mg)	Iron (mg)	Zinc (mg)	Copper (mg)	Manganese (mg)
76	397	187	66	36	3.1	—	—	—
44	232	140	42	—	1.8	—	—	—

Raw—3½ ounces
Cooked—½ cup

ITALIAN-AMERICAN SALAD

6–8 bunchlets or 16–20 leaves arugula
4–6 bunchlets dandelion greens
  (Belgian endive may be substituted)
½ bunch watercress
2 cloves garlic, finely chopped
¼ teaspoon freshly ground black pepper
Juice of 1 lemon
3–4 tablespoons olive oil

Separate, clean, and wash arugula, dandelions, and watercress. Break into pieces and mix with finely chopped garlic, pepper, lemon juice, and olive oil. Toss.
  Serves 4.

SICILIAN SALAD

6–8 bunchlets or 16–20 leaves arugula
6–8 bunchlets dandelion greens
1 small red onion
2 cloves garlic
¼ cup freshly grated Parmesan cheese
¼ cup red-wine vinegar
¼ cup olive oil
¼ teaspoon freshly ground black pepper

Separate, clean, and wash arugula and dandelions; leaves should remain whole. Finely chop onion and garlic. Mix with greens. Add remaining ingredients. Toss.
  Serves 4.

NORTHERN ITALIAN SALAD

6–8 bunchlets or 16–20 leaves arugula
6–8 bunchlets dandelion greens
6–8 escarole or chicory leaves
2 cloves garlic
¼ cup white-wine vinegar
¼ cup olive oil
¼ teaspoon freshly ground black pepper

Separate, clean, and wash arugula, dandelions, and escarole. Break into pieces and mix together. Chop garlic. Mix all ingredients together. Toss.
  Serves 4 to 6.

JERUSALEM ARTICHOKES

The Jerusalem artichoke, or sunchoke, is a plant-producing tuber of the sunflower family. Unknown and underused, the Jerusalem artichoke has nothing to do with Jerusalem or artichokes. It is nevertheless a versatile vegetable with many of the nutritional properties of the sunflower.

Jerusalem artichokes are available almost year-round and should be firm and bright. Although they come in many shapes, there are basically two kinds: the whitish, roundish Western sunchoke and the Midwestern or Northeastern elongated red Jerusalem artichoke. Inside they all have the same color, taste, and texture. Some of the knobbier, convoluted varieties require a good scrubbing with a vegetable brush to be really grit-free.

Jerusalem artichokes are at their best in your salad, cut in chunks or slices. They add crunch, spiciness without heat, and a clean, crisp splash of white. Jerusalem artichokes can also be sliced and eaten raw with an olive oil and garlic dressing, steamed and served with a cream sauce, or blanched, dipped in batter, and fried. They are an excellent addition to soup and mix well with any stir-fry or tempura mix. Combined with raw string beans, parsley, carrots, and juiced, Jerusalem artichokes contribute to a high-energy vegetable juice that aids adrenal production and calcium metabolism.

By themselves Jerusalem artichokes are an easily digested source of vegetable protein, with hardly any calories and high amounts of iron and niacin.

NUTRITIONAL DATA

Weight (g)	Calories	Protein (g)	Carbohydrates (g)	Crude Fiber (g)	Water (g)	Fat (g)	Cholesterol (mg)
100	35	2.3	16.7	0.8	79.8	0.1	—

Vitamin A (IU)	Vitamin C (mg)	Thiamine B-1 (mg)	Riboflavin B-2 (mg)	Niacin (mg)	Vitamin B-6 (mg)	Vitamin B-12 (mcg)	Folic Acid (mcg)	Pantothenic Acid (mg)
20	4	0.2	0.06	1.3	—	—	—	—

Sodium (mg)	Potassium (mg)	Calcium (mg)	Phosphorus (mg)	Magnesium (mg)	Iron (mg)	Zinc (mg)	Copper (mg)	Manganese (mg)
—	—	14	78	11	3.4	—	—	—

Raw—4 small

SCALLIONS

A scallion is not an onion, not a true bulb, not a shallot. It is sometimes called a green onion or a spring onion. Whatever the scallion is, it is a delightful, zesty addition to both cooked and raw recipes.

Scallions can be served raw with cheeses and dips, along with carrot, cucumber, and celery sticks. They can be chopped and mixed in a salad or sauté. Scallions are an excellent substitute for onions when a milder taste is desired. The chopped scallion top is excellent added to an eggplant stuffing or dropped in a soup just before serving.

Look for straight bright-green scallions with bright white bulbs, and stems without wilt or rot. Curved scallions indicate age and toughness. Early summer produces local scallions as long as thirty inches, with a large round white bulb. These are scallions at their finest, with a fresh, almost sweet, onion flavor. Fresh scallions are highly desirable and should be enjoyed whenever available. Most scallions are the straight bulbless variety, grown and shipped from California year-round. The root hairs and tops of the leaves are clipped for shipping. Unless very fresh and straight, these commercial scallions do not have the same brisk, tangy flavor of the fresh-picked local varieties.

Scallions will never win any nutrition awards, although they are high in potassium and niacin. They do, however, liven up the salad and soup bowl and make fresh vegetables tastier and more appetizing.

NUTRITIONAL DATA

Weight (g)	Calories	Protein (g)	Carbohydrates (g)	Crude Fiber (g)	Water (g)	Fat (g)	Cholesterol (mg)
100	45	1.1	10.5	1.0	87.6	0.2	—

Vitamin A (IU)	Vitamin C (mg)	Thiamine B-1 (mg)	Riboflavin B-2 (mg)	Niacin (mg)	Vitamin B-6 (mg)	Vitamin B-12 (mcg)	Folic Acid (mcg)	Pantothenic Acid (mg)
trace	25	0.05	0.04	0.4	—	—	—	—

Sodium (mg)	Potassium (mg)	Calcium (mg)	Phosphorus (mg)	Magnesium (mg)	Iron (mg)	Zinc (mg)	Copper (mg)	Manganese (mg)
5	231	40	39	—	0.6	—	—	—

Raw—5 medium

WHOLE BARBECUED SCALLIONS

Large round-bulb scallions, trimmed
Olive oil
1 teaspoon garlic powder
1 teaspoon onion powder
1 teaspoon salt

*R*emove root hairs from scallion bulbs. Blanch bulb ends of scallions in boiling water.

Place 2 inches of olive oil in cup. Mix together in a bowl the garlic powder, onion powder, salt, and pepper. Dip scallion bulbs in olive oil, roll in seasonings, and place on grill. Turn often until browned.

CRISPY FRIED SCALLIONS

Round- or long-bulb scallions, trimmed
1 cup bread crumbs
2 cloves garlic, finely chopped
1 small onion, finely chopped
6 sprigs parsley, finely chopped
¼ teaspoon dried oregano
¼ teaspoon salt
¼ teaspoon freshly ground black pepper
2 eggs, lightly beaten
Light vegetable oil

*R*emove root hairs from scallion bulbs. Mix bread crumbs, garlic, onion, parsley, oregano, salt, and pepper. Blanch scallions in boiling water to just above white bulb. Discard unblanched greens. Heat ¼ inch vegetable oil in frying pan. Dip bulbs in egg, roll in bread crumbs, and place in hot oil. Turn when one side is crisp. Drain on paper towels.

SCALLION OMELETTE

8–10 scallions (including green tops), trimmed
6 eggs, lightly beaten
½ cup milk
2 tablespoons mayonnaise
¼ teaspoon salt

¼ teaspoon freshly ground black pepper
4 shallots
3–4 sprigs flat-leaf parsley, chopped
2–3 sprigs savory, chopped
4 tablespoons butter or margarine

Remove root hairs from scallion bulbs. Beat eggs, milk, mayonnaise, salt, and pepper. Chop scallion bulbs along with shallots, parsley, and savory. Sauté in butter. When barely softened, add egg mixture. Cover and cook over medium heat.

Coarsely chop scallion greens and sprinkle over half-cooked omelette. Fold over. Cover and continue cooking for 2 to 3 minutes. Turn folded omelette over and cook until done. *Serves 3 to 4.*

WHOLE BRAISED SCALLIONS

12 scallions, trimmed
3–4 ounces sesame or walnut oil
2 cloves garlic, chopped
½ teaspoon salt
¼ teaspoon freshly ground black pepper
2 tablespoons tamari sauce

Remove root hairs from scallions and blanch bulb ends. Heat oil, garlic, salt, and pepper. Place whole scallions in pan. Brown one side and turn. Sprinkle with tamari sauce. Brown other side.

SPROUTS

Whether you buy them or sprout them, sprouts are incredibly nutritious and delicious. One of the most popular and tasty specimens is the alfalfa sprout (which is a nutritional powerhouse). It is eaten raw in salads and sandwiches, or as a topping for a cheese or tuna melt or even a hamburger.

Radish sprouts are slightly less nutritious but a lot more peppery and flavorful. They are, of course, the sprouts of the radish seed and will enhance salads, sandwiches, and raw vegetable dishes. Radish- and alfalfa-sprout seeds can be sprouted together for a tasty combination.

The mung bean (or bean sprout) does everything other sprouts do but also cooks up well as part of a mixed stir-fry or Chinese vegetable dish. Bean sprouts can also be dropped in soup or a clear chicken or beef broth. I enjoy bean sprouts raw or eaten with avocado and tomato.

When buying sprouts, inspect the container carefully. Sprouts should always be fluffy and white. Tiny specks of green, with bits of the seed husk present, are excellent signs of freshness. Soggy brown sprouts, or containers with milky fluid, should be avoided. Overage sprouts have no taste or nutritional value; they are worthless.

Alfalfa sprouts have a very high protein and zinc content. Mung-bean sprouts also have significant protein and serious mineral content. Radish sprouts are yet to be recognized for their nutritional values, but it stands to reason that their mineral content would be as significant or more so than that of the adult root. For taste, texture, and nutrition sprouts are a vital addition to your salads and raw vegetable diet.

NUTRITIONAL DATA

Weight (g)	Calories	Protein (g)	Carbohydrates (g)	Crude Fiber (g)	Water (g)	Fat (g)	Cholesterol (mg)
100	35	3.8	6.6	0.7	88.8	0.2	—
100	41	5.1	5.5	1.7	88.3	0.6	—

Vitamin A (IU)	Vitamin C (mg)	Thiamine B-1 (mg)	Riboflavin B-2 (mg)	Niacin (mg)	Vitamin B-6 (mg)	Vitamin B-12 (mcg)	Folic Acid (mcg)	Pantothenic Acid (mg)
20	19	0.13	0.13	0.8	—	—	—	—
—	16	0.14	0.21	—	—	—	—	—

Sodium (mg)	Potassium (mg)	Calcium (mg)	Phosphorus (mg)	Magnesium (mg)	Iron (mg)	Zinc (mg)	Copper (mg)	Manganese (mg)
5	223	19	64	—	1.3	—	—	—
—	—	28	—	—	1.4	1	—	—

Mung bean, raw—3½ ounces
Alfalfa, raw—3½ ounces

SALAD DRESSINGS

*T*oo often a beautiful, delicious, healthy salad is drowned by a combination of creams, fluids, juices, spices, seasonings, sugar, cheese, and salt called a salad dressing. Lighten up, folks. A subtle use of dressing will enhance the salad, not overwhelm it. I make all my own dressings, each one tailored to the salad of the meal. If you use the same dressing all the time, all your salads will taste the same. How boring. Salads that contain sharp, spicy greens should be dressed with a mild, sweetish taste. Gentle-tasting salads like iceberg and Boston lettuce ones taste even better with a more piquant dressing made with vinegar, lemon, pepper, and such.

All ingredients qualify in salad dressing, from the usual oil and vinegar to tamari, Worcestershire sauce, pineapple and orange juices, yogurt, and liquefied cheeses. The trick is balance and subtlety. And *never dress the whole bowl.* Allow each person to dress his or her own salad to taste. Leftover salad keeps fresher naked anyway.

OIL AND VINEGAR DRESSING

> 1 small yellow onion, finely chopped
> 2 cloves garlic, finely chopped
> ½ cup olive oil
> ¼ cup apple-cider vinegar
> 1 teaspoon chopped fresh basil or ¼ teaspoon dried basil
> ¼ teaspoon dried oregano
> ¼ teaspoon salt
> Dash freshly ground black pepper

Combine all ingredients and shake thoroughly. Serve separately, in cruet. *Makes 6 to 8 ounces.*

CREAMY GARLIC DRESSING

> 1 small red onion, chopped
> 6 cloves garlic, chopped
> 4 sprigs curly parsley, chopped
> ¾ cup olive oil

¼ cup white-wine vinegar
¼ cup freshly grated Parmesan cheese
1 tablespoon lemon juice
¼ teaspoon freshly ground black pepper
Dash salt

Combine all ingredients in blender. Blend slowly until smooth. Serve separately, in cruet. *Makes 10 to 12 ounces.*

COUNTRY ITALIAN DRESSING

24 fresh basil leaves
1 small red onion, finely chopped
6 cloves garlic, chopped
3–4 sprigs oregano, chopped
¾ cup olive oil
¼ cup red-wine vinegar
¼ teaspoon dried oregano
¼ teaspoon freshly ground black pepper
Dash salt

Wash and finely chop basil leaves. Combine with remaining ingredients and shake thoroughly. Serve separately, in cruet. *Makes 8 to 10 ounces.*

LEMON LIGHT DRESSING

Juice and grated peel of 3 lemons
½ cup olive oil
¼ teaspoon chopped fresh or dried tarragon
¼ teaspoon chopped fresh or dried savory
¼ teaspoon freshly ground black pepper

Combine all ingredients and shake thoroughly. Serve separately, in cruet. *Makes 4 to 6 ounces.*

YOGURT–LA MAYONNAISE–DILL DRESSING

3 radishes, peeled and crushed
½ cup yogurt
¼ cup mayonnaise

2 tablespoons chopped fresh dill
1 tablespoon lemon juice
¼ teaspoon salt
¼ teaspoon freshly ground black pepper

Combine all ingredients and stir thoroughly. Serve separately, in cruet. *Makes 6 ounces.*

A FRENCH WAY

1 small yellow onion, finely chopped
2 cloves garlic, crushed
½ cup catsup
¼ cup mayonnaise
1 tablespoon Worcestershire sauce
Juice of 1 lemon
¼ teaspoon salt
⅛ teaspoon freshly ground black pepper

Combine all ingredients and stir until smooth. Serve separately, in cruet. *Makes 6 to 8 ounces.*

A JAPANESE WAY

4 red radishes or 2 ounces Chinese radish,
   peeled and crushed
½ teaspoon crushed fresh ginger
½ cup tamari sauce
Juice of 1 lemon
¼ cup sesame oil
1 tablespoon grated horseradish

Combine all ingredients in blender. Blend slowly until smooth. Serve with shrimp, sushi, salads, and cold vegetables. *Makes 6 to 8 ounces.*

MEXICAN CEVICHE

2 medium very red tomatoes, finely chopped,
   with their juice
1 green pepper, finely diced

1 jalapeño or cherry pepper, finely chopped
1 large white or red onion, finely chopped
4–6 cloves garlic, finely chopped
10–12 sprigs coriander (cilantro), chopped (leaves and stems)
Juice of 12 limes
½ teaspoon salt
¼ teaspoon freshly ground black pepper

Combine all ingredients and mix well. Makes an excellent marinade for seafood, especially lobster, shrimp, and squid. *Makes 8 to 10 ounces.*

Tomatoes

1. FLORIDA
2. MEXICAN
3. CALIFORNIA
4. GREENHOUSE
5. BEEFSTEAK
6. RUTGERS
7. BIG BOY
8. ROMA PLUM
9. TEAR DROP PLUM
10. CHERRY
11. TINY TIM CHERRY
12. SUN-RAY
13. ITALIAN PEPPER

*F*rom their early days, when they were known as poison apples, through their French reputation as "love apples," and their Italian heritage as "golden apples," the tomato has always received singular attention in the world of fruits and vegetables. Tomatoes were originally grown in Peru as small round yellow fruit. Through Spanish explorers, the Peruvian seeds were sent to the then Spanish kingdom of Naples—before and since a center of Italian culture and cuisine. The rest is history. The tomato is definitely the backbone of Italian cooking, and no other culture has spread the use and pleasure of a vegetable more than Italians have done for tomatoes. Once the gardeners and cooks reached tomato perfection somewhere around 1950, the businessmen and scientists took over. It is no joke that the agribusiness goal is to create square self-stacking tomatoes that don't bruise or rot. Taste and nutrition seem to be of little importance to the new economic tomatoheads.

When you are shopping, please don't squeeze the tomatoes. If you must touch them at all—and no one does at Sunfrost—handle them tenderly and with respect for their bruisability. (The reason that supermarkets have artificial tomatoes is because self-service destroys real tomatoes.) Shop at a good fruit stand where the clerks will select hard, ripe, perfect specimens to your taste. If they don't, complain. They will next time. A good fruit and vegetable market should have pink, red, and very red tomatoes on hand at all times. I always suggest allowing tomatoes to ripen at home. That way you will always have solid, ripe, unbruised tomatoes on hand. *Tomatoes should never be refrigerated except to chill them immediately before eating.* Refrigeration kills the flavor, the texture, and any hope of full-ripened maturity.

Tomatoes are not the most nutritious things in the world, but they do contain large amounts of vitamin A and potassium. Besides, what else do you put on a sandwich?

NUTRITIONAL DATA

Weight (g)	Calories	Protein (g)	Carbohydrates (g)	Crude Fiber (g)	Water (g)	Fat (g)	Cholesterol (mg)
150	33	1.6	7	0.8	140.3	0.3	—
100	26	1.3	5.5	0.6	92.4	0.2	—

Vitamin A (IU)	Vitamin C (mg)	Thiamine B-1 (mg)	Riboflavin B-2 (mg)	Niacin (mg)	Vitamin B-6 (mg)	Vitamin B-12 (mcg)	Folic Acid (mcg)	Pantothenic Acid (mg)
1350	34	0.09	0.06	1	—	—	—	—
1000	24	0.07	0.05	0.8	—	—	—	—

Sodium (mg)	Potassium (mg)	Calcium (mg)	Phosphorus (mg)	Magnesium (mg)	Iron (mg)	Zinc (mg)	Copper (mg)	Manganese (mg)
4	366	20	40	21	0.8	—	—	—
4	287	15	32	—	0.6	—	—	—

Raw—1 medium
  Boiled—½ cup

WINTER TOMATOES

Although there are many varieties of tomatoes, we will divide them into two classes: winter tomatoes and summer tomatoes.

Winter tomatoes appear in the stores throughout the United States from December until June. Some are imported, some grown locally in our Southern states, and all are shipped great distances north and east to satisfy America's demand for summertime food all year long. It is to our credit as a society that we can satisfy our food desires even if the greater goals of peace and prosperity still seem a far piece away. Nevertheless, by hook or by truck fresh red tomatoes can be had for the asking in the far reaches of snowbound North Dakota in the middle of January. Winter tomatoes come from Florida, Mexico, California, and greenhouses here and in Europe.

FLORIDA TOMATOES

Florida tomatoes are the basic supermarket tomato. They are hard and thick, never turn red, and always disappoint our taste. Still, we buy them by the thousands of tons. Florida grows a great tomato. I've eaten vine-ripened tomatoes in Florida in January, and they taste wonderful. The trouble is that Florida tomatoes for shipping are all picked green at one time by mechanical pickers. They are stored in huge tubs, refrigerated, and shipped in large quantities to repacking houses around the country, where they are sized, gassed with ethylene to turn their skins red, and packed loose in twenty-five-pound boxes for retailers and supermarket buying agents. Florida tomatoes grown for shipping outside the state almost never fully ripen, never sweeten up, never turn red inside, and never taste wonderful. To make matters worse all supermarkets and many retailers refrigerate their stock. That is the final tomato insult.

If you really need a tomato and Florida tomatoes are all you can buy, buy carefully. Look for really red ones that are redder at the blossom end. This means they are ripening naturally. Take them home, and let them turn red all over. Let them soften. Then eat 'em. If they still seem tasteless, don't buy them anymore. Wait for summer.

MEXICAN TOMATOES

Mexican tomatoes are flattish and deep pink, often with definite ribs and a dark stem end. The most important thing to know about them is that Mexican tomatoes are picked only when they're ripe. They are packed in two single layers in twenty-pound boxes, not railroad tubs, and shipped daily. Cheap labor makes this hand-picking red-crop process possible. A Mexican tomato will ripen deep red right to the middle and will taste like a fresh-picked tomato because it is. Of course, if the retailer refrigerates his stock or if the tomatoes sit on a truck all day in January, their sweet, juicy tomato taste will be history.

Properly handled, Mexican tomatoes have the reddest meat, the sweetest juice, and the "realest" taste of all the tomatoes you can buy all winter. *Gracias, Mexico. Su tomates estas muy deliciosas.*

CALIFORNIA TOMATOES

California tomatoes have a very bright red color and are flattish and heavy. They have large, solid meat areas with tender walls. They are juicy but are still firm and not watery. California picks red tomatoes and leaves the green ones to ripen for another day's harvest. California tomatoes are shipped in twenty-pound wooden or corrugated boxes, and despite long-distance shipping and small-weight packaging, California tomatoes are always reasonably priced and always high quality.

If any people in the world have the business of fresh fruit and vegetables down cold, it is the California growers. The California grower never comes to market—only the fruits of his labor do. The California system is based on growers' cooperatives, which sell enormous produce tonnage to packers and shippers, who grade, label, package, and ship all over the world. The produce is sent to agents called receivers, who place carload or trailer-load lots up for auction at distribution points. In the United States these auctions take place in all major metropolitan areas. Buying agents or brokers bid and buy for their chain-store, wholesale, and large retail clients and add their commission to the purchase price. The produce is then trucked to wholesale houses, warehouses, and stores. It is a miracle of modern merchandising that beautiful fragile things like tomatoes can be planted, cultivated, picked, packed, shipped, sold, and sold again and come to us as perfect produce. I have met only a few California growers, but every time I open a package of California produce, especially tomatoes, I mentally tip my hat in a personal, respectful way to the world's best farmer: the California grower.

During the late spring, when the first local salad greens begin to appear and tomatoes are in demand, the California tomatoes arrive right on time. They are big, red, firm, juicy, and flavorful. You must, however, buy them two or three days ahead of time and allow them to ripen, stem ends up, to a deep red in your home. California tomatoes are completely versatile. They are firm enough to cut up for salads, tasty enough for slicing for sandwiches or by themselves on a plate. Their vine-ripened flavor makes them perfect for sauces and gravies. Look for California tomatoes in May and June, and have a good tomato time.

GREENHOUSE TOMATOES

Greenhouse tomatoes are also excellent. The domestic varieties appear in good fruit stores in late April and keep on coming until June or early July. Holland exports and we import tons of very large, very red greenhouse tomatoes in single-layer wooden trays. Each tomato is individually cradled in a cardboard cell and retains its bright-green stem. The American greenhouse tomato is also picked and shipped with its stem still intact. Both the American variety and the Dutch import are picture-perfect tomatoes. They must, however, be purchased a few days in advance and ripened fully at home. When you slice them, they will be red to the middle, with that tomato aroma and unique, tangy tomato taste. Greenhouse tomatoes have a higher juice content than field-grown varieties and a thinner skin. This makes them more fragile, and greater care must be taken in their handling. They are more expensive and must be handled with an appreciation for the tomato miracle that they are.

Greenhouse tomatoes are more averse to cold temperatures than all other tomato varieties. Refrigeration will make them lose their flavor and become watery.

They are excellent sliced for an Italian-style garlic-basil salad, and their large size and real flavor makes them perfect for an all-American sandwich.

SUMMER TOMATOES

Summer tomatoes are no problem. They are easy to grow, and they grow anywhere. I have tomatoes in my garden that grew from the culled tomatoes I plowed under last fall. Nevertheless supermarkets still continue to sell hard, manipulated, artificially ripened tomatoes. But the roadside stands, the farm markets, and all fruit and vegetable stores are full of delicious vine-ripened fresh-picked tomatoes from local sources. The local tomato varieties are endless, but there are some basic ones you can easily recognize, intelligently select, and enjoy.

BEEFSTEAK TOMATOES

The beefsteak tomato looks like a slab of steak when you slice it. If it doesn't, it isn't. Beefsteaks are clumsily shaped, often growing in a curved convolution. They are deep orange-red rather than maroon-red and are sometimes streaked with green at the top end. When you slice a beefsteak, it is almost all meat, with small seed pockets, thick walls, and thick juice. If the beefsteak has very yellow shoulders (the bumpy area around the stem socket), it will never fully ripen; the yellow meat is inedible. Look for very large, clumsy, heavy tomatoes with a large blossom-end scar (the larger blossoms seem to produce the very best flavored beefsteaks). Beefsteaks are at their best sliced, with basil, garlic, oregano, chopped onion, pepper, and a splash of vinegar and good olive oil. Slice up a hot loaf of Italian bread, crack a jug of red cellar wine, and you can tell everyone to call you later.

Once again, look for beauty in the clumps, lumps, and bumps—and *please* don't squeeze the beefsteaks.

RUTGERS TOMATOES

The Rutgers tomato is a university graduate. The New Jersey State University at Rutgers has an internationally famous horticulture department that developed this variety as a more beautiful, slightly smaller, redder version of the famous Jersey beefsteak. They succeeded in producing a medium to large round, very red tomato with a lot of inner meat and a thin edible skin. The Rutgers is cultivated throughout the United States and appears in almost every seed catalog.

The Rutgers is also very durable and has excellent tomato flavor. It can be sliced, quartered, chopped, or made into sauce. It is an excellent stuffing tomato. The Rutgers ripens easily and evenly, and as red as it gets outside, the inside is the same. And with its fresh field-ripe flavor, the Rutgers is excellent for canning, as either whole tomatoes or cooked sauce. Look for Rutgers from early July through mid-August. The bigger sizes appear first, so if you're into peeling and canning, get the early ones.

BIG BOY TOMATOES

Big Boy tomatoes are relatively new and very popular. The Big Boy is large, flat, oval and a deep red when it ripens. Like the beefsteak it is almost all meat, with thick walls, small seed pockets, and thick juice. It is a handsome tomato, without the clumps and free form of a beefsteak. It probably has the classic profile for a large tomato. It also tastes great. Slice it for its best flavor, but it is a solid tomato and can be cut and served any way you like.

Big Boys arrive in mid-August. They are a big, heavy tomato that requires a lot of growing time, and the plants can only produce for about three to four weeks. Big Boys are gone by early September except for the last slim pickings, which are smaller in size but have excellent flavor. Big Boys are delicious in all the ways we like tomatoes. Their deep-red color and thick texture make them an excellent choice for canning.

In addition to the beefsteak, Rutgers, and Big Boys, there are several popular versions of the Star variety: Star Fire, Jet Star, Star Boy, Jet Fire, and Fire Boy. They are more or less versions and hybrids of the big three and come in orange, orange-red, and red colors, round and flat shapes, and mostly firm textures. Look for bright colors, firm textures, and that tomato fragrance, and no matter the name, you're going to get a great tomato.

PLUM TOMATOES

Summertime also brings plum tomatoes. Most plum tomatoes you will find are versions of the Roma, a large tear-drop-shaped, firm, deep-red Italian tomato. They are excellent raw or in salads. They blanch and peel easily and cook up into a thick, aromatic tomato sauce. Romas are available almost all year from Mexico, Florida, and California, but the zesty, saucy kind that ripens on the vine and fills the store with its special aroma arrives in August and keeps coming well into September. Add a few Romas to a stir-fry, steam a few and mix them with scrambled eggs, or just sprinkle them with a little salt and pepper and eat like fruit. There is no way a fresh-picked Roma plum tomato is not great.

CHERRY TOMATOES

Cherry tomatoes seem to be everywhere. Except for the local vine-ripened varieties the best ones come from Mexico, where, once again, they are picked red and shipped daily. I find Florida cherry tomatoes tasteless. Most California cherry tomatoes are glassy and watery except for a variety shipped under the Wilson label. These are great. Once you start eating, it's one after another, a wonderful combination of pepper and sugar. I get the same taste kick from cherry tomatoes from local sources when they start to appear in August. There is another variety of very small cherry tomato called Tiny Tim. These are big as large marbles or smaller. They are bright, bright red and have a sweet, deep tomato flavor that is even more compelling than that of their bigger cousins.

Cherry tomatoes should never be too big, and they should always be red, not orange. Look for those that still retain their stems and have not been refrigerated. Avoid those with yellow spots or a depressed dark stem end.

Although cherry tomatoes are in the stores all year long, the local varieties are a summertime thing that happens in August and September. Until then it's Mexico or California Wilsons or bust.

SUN-RAY TOMATOES

One of my favorite summer foods is the Sun-Ray tomato. Little-known and hardly available, it is a wonderfully clean acid-free tomato that still retains its unique tomato flavor and texture. The Sun-Ray is a deep, bright yellow color, like a school bus. It is usually medium to large, round, and smoothly shaped. Sun-Rays have thin skin, thick meat, solid seed pockets, and thick juice. They go well in salads, especially an all-tomato salad, sandwiches, and sliced platters. The Sun-Ray tomato is especially helpful in a high-alkaline diet as part of a cleansing and detoxification program.

There are no commercial varieties of Sun-Ray tomatoes that I know of, except that I think Holland is beginning to develop an acid-free variety for export. Local crops appear at farm stands and great fruit stands—never in supermarkets—in late August and early September.

ITALIAN PEPPER TOMATOES

Jimmy DeNova is an Italian farmer in upstate New York who grows specialty items for the Italian trade. He cultivates basil, oregano, Italian parsley, arugula, cardoons, long and very tiny eggplants, incredibly curved pale-green cucumber squash, dandelions, Sicilian bush basil, broccoli rabe, and a pimiento-shaped deep-orange tomato with a thick green stem called a Pepper tomato. I think it is the best tomato you can eat.

The Italian Pepper tomato has thick tomato meat with very few seeds. The sweet tomato juice is locked within the meat, not sloshing around the watery seed pockets as in so many varieties. The Pepper tomato tastes like it's already been salted and peppered. The shape is longish with a pointed end and has a slight curve. It looks a lot like those little red or gold horns many Italian people wear on a chain around their necks.

You can slice the Pepper tomato or cut it up into pieces for salad. I also like to peel one, hold it by its stem like a Popsicle, and eat it while dipping it in a seasoning. I love making a production out of rare and beautiful food.

Good as it is raw, the Pepper tomato is out of this world in tomato sauce. It blanches and peels easily. Its low seed content and firm flesh creates a thick, deep, dark sauce that sucks up all the seasoning you can handle. It is the perfect tomato for an old-fashioned macaroni meat gravy, the kind that slowly, slowly cooks for hours and hours with a lamb shank, a beef shank, a pork bone, or meatballs simmering along with the garlic, parsley, oregano, basil, pepper, and onions. (Along with Marconi, Fermi, Da Vinci, Michelangelo, Columbus, Gucci, and Joe DiMaggio, the Italians have given us spaghetti sauce.) No matter how you create your version of the classic sauce, it all starts and ends with the tomato, and the Italian Pepper tomato does it best.

So where do you find this wonderful tomato if you don't have an upstate New York farmer? Look for Italians, the old-fashioned kind where the old men wear hats and play *bocce* and the women wear black and can cook the pants off any group of women anywhere in the world. Look in their stores. Look in late August in roadside markets run by Italian families, where everyone works, right down to the little kids. Look in city markets, in stores that hang cheeses and salami in their windows. They're there. It just takes more effort to find a rare anything.

ITALIAN TOMATO SAUCE

12 ripe medium tomatoes
1 head garlic
2 large yellow onions
¼ teaspoon salt
½ teaspoon freshly ground black pepper
1 teaspoon dried oregano
4–6 basil stems with leaves, chopped, or 1 tablespoon
   dried basil
½ cup olive oil

Wash, de-stem, blanch, and peel tomatoes. Place in deep pot, cover, and simmer. Chop garlic and onions. Place salt, pepper, oregano, and chopped basil in ungreased sauté pan. Let seasonings char ever so slightly as pan heats. Add oil, garlic, and onions. Cover and sauté until just about to brown. Combine with tomatoes. Stir. Continue to simmer 3 to 4 hours, stirring thoroughly from the bottom every 15 minutes. If sauce remains watery, skim off liquid or simmer without cover until excess liquid evaporates, then re-cover. *Makes 1 quart.*

ITALIAN GRAVY

For thick tomato gravy add 1 6½-ounce can tomato paste to sauce and/or add bits of pan-browned pork, veal or beef that contains small bones, neck bones or ribs, cooked meatballs, *braccioles,* or sausage.

MEXICAN SALSA

12 ripe red tomatoes
3–4 cloves garlic, chopped
1 large sweet onion, chopped

**12–15 sprigs coriander (cilantro), chopped
  (stems and leaves)**
6–8 jalapeño peppers, chopped
1 small green pepper, finely chopped
2 tablespoons honey
1 teaspoon cornstarch or arrowroot

Wash, de-stem, blanch, and peel tomatoes. Press through strainer, discarding seeds and dry pulp. Chop garlic, onion, coriander, and peppers and combine with strained tomatoes. Add honey and cornstarch. Blend until smooth. Keep chilled. *Makes 1 quart.*

CANNED WHOLE TOMATOES

Prepare canning jars (see "Canning process" entry in Glossary, page 000). as per instructions. Wash, de-stem, quickly blanch, and peel each tomato, allowing 6 to 8 whole tomatoes per quart jar. Pack in wide-mouth sterilized jars with a large fresh basil leaf in each, leaving ½ inch space at top. Close and process 40 minutes in boiling water. Store in cool place.

FRIED GREEN TOMATOES

4 unbruised green tomatoes, sliced
Olive oil
2 cloves garlic, sliced
½ teaspoon dried oregano
¼ teaspoon salt
¼ teaspoon freshly ground black pepper

Wash, de-stem, and slice tomatoes. Place them in heated lightly oiled pan. Add remaining ingredients and fry. Turn. Fry. Drain on paper towels. Serve as side dish or as an excellent tomato sandwich filling.

HOMEMADE CATSUP

12 very ripe tomatoes
1 tablespoon sugar
1 small yellow onion, finely chopped
2 cloves garlic, crushed
8 sprigs parsley, finely chopped

2–3 sprigs tarragon, chopped
Juice of 1 lemon
1 tablespoon grated fresh horseradish
1 tablespoon vinegar
¼ teaspoon dried oregano
¼ teaspoon salt
¼ teaspoon freshly ground black pepper

Wash, de-stem, blanch, and peel tomatoes. Strain to remove seeds. Cook tomatoes with sugar over medium heat in shallow pan. When they begin to thicken, cool. Combine with remaining ingredients in blender or food processor until smooth. *Makes 24 ounces.*

TOMATO RELISH

4 large firm tomatoes, coarsely chopped
1 medium green pepper, finely chopped
2 jalapeño or cherry peppers, de-seeded and finely chopped
4 radishes, chopped
1 small red onion, finely chopped
4 cloves garlic, finely chopped
8–10 basil leaves, chopped
Juice of 2 limes
½ teaspoon dried oregano

Combine all ingredients and mix well. Chill. *Makes 2 cups.*

TOMATO AND GREEN-PEPPER SALAD

4 firm red medium tomatoes
2 crisp medium green peppers
¼ teaspoon salt
¼ teaspoon freshly ground black pepper
¼ cup olive oil
1 tablespoon dried oregano

Wash, de-stem, and cut tomatoes into wedges. Seed peppers and cut into wedges. Combine with salt, pepper, olive oil, and oregano. Serve very cold. *Makes 2 cups.*

TOMATOES ITALIANO

2 very large ripe beefsteak tomatoes
¼ cup olive oil
24 basil leaves, chopped
4 cloves garlic, finely chopped
1 medium red onion, finely chopped
6–8 sprigs flat-leaf parsley, chopped
¼ teaspoon salt
½ teaspoon freshly ground black pepper
2 large cloves garlic, sliced
2 teaspoons dried oregano

Wash, de-stem, and thickly slice tomatoes. Fan out the slices on a plate. Drip the olive oil over the slices. Chop basil and sprinkle it on tomatoes along with chopped garlic, onion, parsley, salt, and pepper. Lay the garlic slices on top of the tomato slices. Sprinkle with oregano. Serve at room temperature.

TOMATO-LEMON SALAD

3–4 firm red medium tomatoes
Juice and grated peel of 1 lemon
½ teaspoon freshly ground black pepper
¼ teaspoon salt

Wash, de-stem, and cut tomatoes into wedges. Place in bowl. Sprinkle with grated lemon peel. Add salt and pepper and lightly sprinkle with lemon juice. Serve cold.

STUFFED TOMATOES

Select flat, firm red tomatoes. Wash and remove stems and center cores. Stuff with tuna, crab, shrimp, chicken, or turkey salad or a combination of chopped cucumber, celery, and radishes mixed with lemon juice, mayonnaise or yogurt, salt, and pepper.

HUEVOS RANCHEROS

1 yellow onion, chopped
2–3 cloves garlic, chopped

1 green pepper, chopped
3–4 sprigs parsley or coriander (cilantro), chopped
¼ teaspoon salt
Dash freshly ground black pepper
Light vegetable oil or margarine
3–4 tomatoes, blanched, peeled, and crushed
2 eggs

Sauté onion, garlic, green pepper, parsley, salt, and black pepper. Add crushed tomatoes. Mix. Cover and simmer. Break eggs into a cup. When tomatoes and other ingredients soften, push to one side of pan, increase heat slightly, and add eggs. Cover. When eggs are cooked, spoon tomato sauce onto plate. Serve eggs on top, with toast or tortillas. *Serves 2.*

CHICKEN SALSA

1 large Spanish onion
3–4 cloves garlic
1 medium zucchini
6–8 sprigs parsley
1 green pepper
Light vegetable oil
¼ teaspoon salt
¼ teaspoon freshly ground black pepper
1 tablespoon dried oregano
4 tomatoes, peeled and cut up
6 pieces chicken

Chop onion, garlic, zucchini, and parsley. Slice green pepper. Heat vegetable oil with garlic, salt, black pepper, and oregano. Add tomatoes, green pepper, onion, zucchini, and parsley. Cover and simmer over medium heat for 15 to 20 minutes. Add chicken. Cover and simmer over low heat for 40 to 50 minutes. Serve from pan. *Serves 3 to 4.*

STEWED TOMATOES

Per large ripe tomato:

1 scallion, chopped
½ stalk celery, chopped

1 small yellow onion, chopped
1 clove garlic, chopped
2 tablespoons cooked rice
Light vegetable oil
½ teaspoon freshly grated cheese

Sauté all ingredients except tomato and grated cheese. Wash, de-stem, and stuff tomatoes. Place in baking pan. Cover with lid or foil. Bake in preheated 325° F. oven for 20 minutes. Sprinkle with grated cheese.

GREEK TOMATOES

2–3 firm red medium tomatoes
6 artichoke hearts, halved
1 seedless cucumber, diced
12–15 Greek olives, chopped
3–4 sprigs savory, chopped
Juice of 1 lemon
½ cup red-wine vinegar
¼ cup olive oil
½ teaspoon freshly ground black pepper
¼ teaspoon salt

Wash, core, and cut tomatoes into wedges. Combine all ingredients. Chill for 3 to 4 hours. Serve with hot bread, butter, and cheese board.   *Serves 4.*

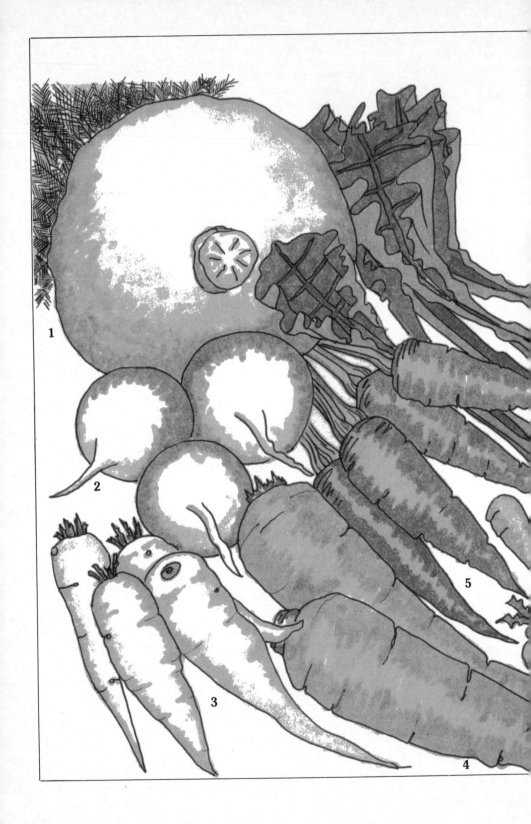

ROOTS

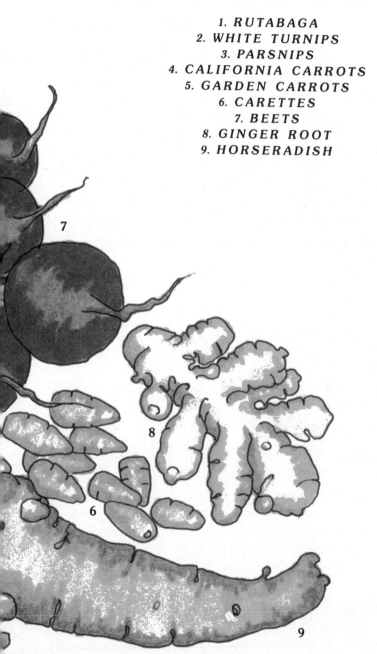

1. RUTABAGA
2. WHITE TURNIPS
3. PARSNIPS
4. CALIFORNIA CARROTS
5. GARDEN CARROTS
6. CARETTES
7. BEETS
8. GINGER ROOT
9. HORSERADISH

*R*oots are the forgotten children of vegetables. After carrots and a few canned beets most of us never get to the root of the vegetable matter. The reason lies somewhere in the beginning of our present culture patterns. The East Asian world thought roots were food grown underground in the domain of the devil and wouldn't eat them. Throughout history the upper class and aristocracy considered roots peasant fodder. Today, because of low production, poor distribution, and lack of retailing interest, the root foods do not pop into mind when customers shop for produce or plan meals and menus.

Except for carrots and, to a lesser degree, beets the root vegetables are never prominently displayed or merchandised. More's the pity. The more we know about the earth's abundance, the more we learn to recognize *all the fresh foods* in our stores and markets, the more versatile we will be in our ability to feed ourselves with pleasure and imagination. There are no *awful* natural foods, not even the roots. If we continually turn up our noses at the unfamiliar, we will continue to limit fresh food resources for ourselves. It is how we perceive these foods and how we prepare them that make them delicious and exciting or tasteless and dull.

I expect no one will ever start a rutabaga rage. But how nice it is to know what is available and when and what one can do with every item in the endless variety of earth's bounty.

RUTABAGAS

*A*lthough turnips do grow around the world in an impressive array of shapes and colors, we shall deal with the only two turnip varieties found in our supermarkets and fruit and vegetable stands. Little can be said about yellow turnips, called rutabagas, except that most people dislike them, have never used them, and don't expect to. The rutabaga is a dense yellow and purple globe that makes its way to market in October and lasts throughout the winter. Rutabagas sell slowly, so they arrive heavily waxed, which protects them from shriveling and waste. There are, however, delicious uses for the rutabaga. One excellent one is to cut it up and place the pieces around a pot roast. Or it makes an excellent addition to vegetable soup, imparting a nice gentle tang to the overall flavor. Rutabagas can be mashed with butter and cinnamon, blanched for tempura, or mixed with peas and carrots, for instance, as a spicy side vegetable. Rutabagas are also an excellent cooking companion to roast duck, absorbing grease and adding their own definite tang to the game flavor.

Always select heavy, hard, bright-looking rutabagas. Avoid woody, dry specimens or those of pale color. The top half of the rutabaga is purple, and fresh specimens should be brightly covered with that color.

Rutabagas have a large amount of calcium and vitamin A. They are easily digested, and their high carbohydrate levels make them a high-energy, nonfattening food.

NUTRITIONAL DATA

Weight (g)	Calories	Protein (g)	Carbohydrates (g)	Crude Fiber (g)	Water (g)	Fat (g)	Cholesterol (mg)
100	35	0.9	8.2	1.1	90.2	0.1	—

Vitamin A (IU)	Vitamin C (mg)	Thiamine B-1 (mg)	Riboflavin B-2 (mg)	Niacin (mg)	Vitamin B-6 (mg)	Vitamin B-12 (mcg)	Folic Acid (mcg)	Pantothenic Acid (mg)
550	26	0.06	0.06	0.8	—	—	—	—

Sodium (mg)	Potassium (mg)	Calcium (mg)	Phosphorus (mg)	Magnesium (mg)	Iron (mg)	Zinc (mg)	Copper (mg)	Manganese (mg)
4	167	59	31	—	0.3	—	—	—

Cooked—½ cup diced

Sweet Turnip Mash

1 medium rutabaga
6 tablespoons maple syrup
6 tablespoons butter
1 tablespoon brown sugar

Peel, cut up, and cook rutabaga in 2 inches boiling water until soft. Coarsely mash softened rutabaga with maple syrup and 2 tablespoons butter. Place in shallow baking pan. Cut up 4 tablespoons butter over top. Sprinkle with brown sugar. Bake in a preheated 350° F. oven for 15 to 20 minutes, then broil quickly for 2 minutes. *Serves 4.*

Turnip-Potato-Vegetable Soup

1 medium rutabaga
2 large white potatoes
2 medium carrots
2 quarts chicken stock
10 small white boiling onions
¼ teaspoon salt
Dash freshly ground black pepper
½ pound string beans
8–10 sprigs flat-leaf parsley
2 stalks celery

Peel and cube rutabaga and potatoes. Peel and dice carrots. Place in pot with chicken stock and onions. Add salt and pepper. Simmer for 15 to 20 minutes. Trim and cut up string beans. Chop parsley and celery. Add when potatoes begin to soften. Cook for 10 to 15 minutes. *Serves 4 to 6.*

White Turnips

White turnips are first cousins to the yellow but are smaller and are usually grown for their greens, which are becoming increasingly popular. They are sometimes round but usually flattened, with a purple top. By themselves white turnips have several uses, both cooked and raw. They can be cut up and added to salads, sliced and served with avocado or dip, added either in small pieces or shredded to soup for a slight sharp flavor, or cooked as a hot vegetable by themselves or with carrots.

Turnips go very well with leg of lamb and game meats. Cooked separately, they make an excellent companion dish to turnip greens. White turnips also make excellent additions to salad, imparting a fresh, crunchy, spicy texture.

Select firm, fresh turnips without rust or shriveling. Store them in a plastic bag in your refrigerator, and they'll keep for weeks.

Although white turnips are eaten chiefly for their flavor and natural spice, they have a large amount of potassium and are excellent digestive aids.

NUTRITIONAL DATA

Weight (g)	Calories	Protein (g)	Carbohydrates (g)	Crude Fiber (g)	Water (g)	Fat (g)	Cholesterol (mg)
100	23	0.8	4.9	0.9	93.6	0.2	—

Vitamin A (IU)	Vitamin C (mg)	Thiamine B-1 (mg)	Riboflavin B-2 (mg)	Niacin (mg)	Vitamin B-6 (mg)	Vitamin B-12 (mcg)	Folic Acid (mcg)	Pantothenic Acid (mg)
trace	22	0.04	0.05	0.3	—	—	—	—

Sodium (mg)	Potassium (mg)	Calcium (mg)	Phosphorus (mg)	Magnesium (mg)	Iron (mg)	Zinc (mg)	Copper (mg)	Manganese (mg)
34	188	35	24	—	0.4	—	—	—

Cooked—⅔ cup diced

PARSNIPS

Since the Middle Ages the potato has replaced the parsnip as a filling high-starch, high-protein vegetable. There was a time, however, when the sweet, nutty, aromatic flavor of parsnips was a popular table delight for everyone from emperors to peasants.

Today parsnips are chiefly appreciated by ethnic groups whose cultural habits have roots in those lands lost somewhere between Germany and Russia. Soup aficionados consider a hard white fresh parsnip as indispensable as any ingredient. I enjoy fresh parsnips as a vegetable, but few others do. Perhaps our American tastes demand our vegetables have their spiciness added on in the cooking, rather than as an inherent "pepper," as in the parsnip.

The best parsnips are picked after the frost, sometimes after a hard freeze. The best varieties have a high crown with a deep impression from which the greens grow. Parsnip greens, like carrot greens, are inedible, and most commercial parsnip-growers have developed a crew-cut variety with little green. There are also round parsnips with not quite the same tang or flavor.

Steamed parsnips are delicious served with salt, pepper, and butter. Cut lengthwise and quickly blanched, parsnips are excellent with roast meats and celery. Small pieces of raw parsnip add texture and a tingly taste to mixed green salad.

Parsnips have a high potassium content, some useable protein, and nonfattening carbohydrates.

NUTRITIONAL DATA

Weight (g)	Calories	Protein (g)	Carbohydrates (g)	Crude Fiber (g)	Water (g)	Fat (g)	Cholesterol (mg)
100	66	1.5	14.9	2	82.2	0.5	—

Vitamin A (IU)	Vitamin C (mg)	Thiamine B-1 (mg)	Riboflavin B-2 (mg)	Niacin (mg)	Vitamin B-6 (mg)	Vitamin B-12 (mcg)	Folic Acid (mcg)	Pantothenic Acid (mg)
30	10	0.07	0.08	0.1	—	—	—	—

Sodium (mg)	Potassium (mg)	Calcium (mg)	Phosphorus (mg)	Magnesium (mg)	Iron (mg)	Zinc (mg)	Copper (mg)	Manganese (mg)
8	379	45	62	—	0.6	—	—	—

Cooked—½ cup diced

CARROTS

Carrots are one of the most important, nutritious, and versatile vegetables available. The traditional bunches of carrots with the bushy green tops have almost disappeared from our stores except the farm stand in season. Actually, once carrots are picked, the tops rob the edible root of its moisture and nutritional quality. Most carrots are sold in look-alike one- to five-pound cello bags, which keep the carrots firm and fresh.

There are many differences in carrots. They grow short and stubby, long and thin, fat, twisted, stumpy, hairy, and hairless. The best carrots by far, however, are grown in California. Even very large California carrots tend to be sweeter and more tender than any other kind. Carrots require alkaline soil for optimum sweetness and soft earth for tender, easy growth. California growers supply both. Even the large central core of California carrots is more worthwhile and tastier than the whole of many of the other carrots available. So read your labels and ask your clerk. Medium-size carrots, especially locally grown varieties, are also excellent and tender. I especially dislike Florida carrots. They always taste bitter and go limp quickly.

Look for bright, shiny carrots with smaller cores and wholesome tips. Their crowns might be slightly green, red, or even purple, but the roots themselves must be deep orange and free from bruise, rot, or black skin.

There are also very small carrots packed in cello bags, called Carettes. These are either the immature roots, dug up in the early thinning process during cultivation, or actual dwarf carrots. There is no way to tell the difference until you taste them. Dwarfs are definitely better tasting and sweeter.

Carrots can be stewed, steamed, stir-fried, souped, juiced, baked, grated, shredded, or cut up raw into sticks, wheels, or chunks for salads and dips. Carrot juice is a delicious source of energy and a cleansing agent of little residue and easy assimilation.

Carrots are an excellent source of many vital nutrients. They contain four times the Minimum Daily Requirement of vitamin A. They are also high in vitamin E, vital minerals, and nonfattening carbohydrates.

NUTRITIONAL DATA

Weight (g)	Calories	Protein (g)	Carbohydrates (g)	Crude Fiber (g)	Water (g)	Fat (g)	Cholesterol (mg)
100	42	1.1	9.7	1	88.2	0.2	—
100	31	0.9	7.1	1	91.2	0.2	—

Vitamin A (IU)	Vitamin C (mg)	Thiamine B-1 (mg)	Riboflavin B-2 (mg)	Niacin (mg)	Vitamin B-6 (mg)	Vitamin B-12 (mcg)	Folic Acid (mcg)	Pantothenic Acid (mg)
11000	8	0.06	0.05	0.6	—	—	—	—
10500	6	0.05	0.05	0.5	—	—	—	—

Sodium (mg)	Potassium (mg)	Calcium (mg)	Phosphorus (mg)	Magnesium (mg)	Iron (mg)	Zinc (mg)	Copper (mg)	Manganese (mg)
47	341	37	36	23	0.7	—	—	—
33	222	33	31	—	0.6	—	—	—

Raw—1 large
  Cooked—2/3 cup

CARROTS 'N' CREAM

4 large carrots
¾ cup light cream or half-and-half
6 tablespoons butter
6 tablespoons instant-mixing flour
Dash freshly ground black pepper

Peel and slice carrots. Place in pan with 1½ inches water. Cover and cook until just soft. Drain. Cover.

Heat cream and butter together until butter melts. Stir gently. Lower heat. Add flour 1 teaspoon at a time, stirring constantly. Cook, stirring, until sauce thickens. If too thick, add water 1 tablespoon at a time. Place carrots in bowl. Pour on cream sauce. Season. *Serves 4.*

BUTTERED CARROTS

4 large carrots
4 tablespoons butter
Sprinkle of chopped fresh savory
Dash salt
Dash freshly ground black pepper

Peel carrots. Crinkle-cut, julienne, or slice and place in 1½ inches water. Cover. Cook until just soft. Drain, keeping 6–8 tablespoons water, and mix in butter. Sprinkle with chopped savory, salt, and pepper. Cover. Simmer for 2 minutes. *Serves 4.*

HONEY-BUTTER CARETTES

12–15 freshly dug carrot thinnings or 1 cello bag Carettes
¼ cup honey
4 tablespoons melted butter
3–4 sprigs curly parsley, chopped

Wash Carettes. Steam in 1 inch water. Drain. Spoon honey and melted butter over Carettes. Allow to stand over very low heat until honey melts. Serve in white or blue bowl with chopped parsley as a garnish. *Serves 4.*

Maple Carrots

    4 large carrots
    ¼ cup maple syrup
    2 tablespoons butter

*P*eel, slice, and cook carrots in 1½ inches water until just soft. Drain. Spoon maple syrup and butter over carrots. Cover. Simmer over low heat for 1 to 2 minutes. *Serves 4.*

Grated and Shredded Carrots

*Y*our ordinary kitchen grater is an easy way to grate or flake fresh carrots. Shredded carrots can be created with a counter-top hand shredder or food processor.

Freshly grated or shredded carrots are sweet eating and add a colorful and interesting aspect to salad plates, cold meat plates, cream soups, and cold vegetable platters.

Shredded Carrot Salad

    2 medium carrots
    2 medium beets
    2 medium white turnips
    ¼ cup olive oil
    2 tablespoons lemon juice
    1 clove garlic, crushed
    ¼ teaspoon salt
    ¼ teaspoon freshly ground black pepper

*P*eel and shred carrots, beets, and turnips. In blender at low speed combine olive oil, lemon juice, garlic, salt, and pepper. Toss vegetables lightly, adding dressing to taste. *Serves to 6.*

Coconut Carrots

    4 large carrots
    4 tablespoons butter
    8–10 tablespoons grated fresh coconut
       or shredded baking coconut

*P*eel and slice carrots lengthwise into quarters. Boil until barely soft. Butter shallow baking pan and melt remaining butter. Place carrots in pan. Spoon butter and coconut over carrots. Place in preheated 350° F. oven for 20 minutes.

Serves 4.

BEETS

*B*eets come three ways. They arrive first in midspring as little ruby-green leaves with tiny beet roots. These are the beet-crop thinnings and are a delicious, wonderfully digested cooked vegetable. Lightly tossed with lemon juice and pepper, beet thinnings are also a bright, tasty salad.

The classic bunch of beets with its dark-purple roots and crisp wide emerald leaves is actually two vegetables in one, the roots and the leaves. Beet roots should always be cooked whole, with an inch or so of the stem left attached. Peeling and slicing is done afterward. Shredded or julienned raw beets are a great salad ingredient that imparts a sweetness and dramatic red that is colorful and inspirational to the appetite.

Fresh beet tops can be prepared several ways, all of which are highly nutritious and pleasing to the vegetable-leaf lover. Remove the red stem from the green leaf before cooking. Use little water and any kind of spicy seasoning, from simple black pepper to curry powder and mixed garden herbs. The leaves can be shredded and added to fresh garden salads. A salad of beet tops, lemon juice, olive oil, and pepper is a great salad taste. Fresh beets are available from here and there year-round, but the best beet comes from local sources at the beginning of spring into the deep summer. Look for beet roots that are hard, with fresh tails and no cuts. The leaves should be crisp and fanlike, holding up erect on their stems. Loose beets without tops become soft quickly. Wash them, store them in plastic bags, and refrigerate them until you use them.

Beets contain ten per cent sucrose carbohydrates, many times more than any other vegetable, an excellent source of nonfattening energy. They are high in vital mineral content, especially potassium and iron. Beets are also an excellent juicing vegetable, especially good when mixed with carrot and celery or string beans. The combination is high-energy and very refreshing. Beets provide a diuretic action and, eaten as a vegetable, act as a body broom. Don't let the scarlet urine scare you. It's only beets.

NUTRITIONAL DATA *(Beet root)*

Weight (g)	Calories	Protein (g)	Carbohydrates (g)	Crude Fiber (g)	Water (g)	Fat (g)	Cholesterol (mg)
100	43	1.6	9.9	0.8	87.3	0.1	—
83	27	0.9	6	0.7	75.5	0.1	—

Vitamin A (IU)	Vitamin C (mg)	Thiamine B-1 (mg)	Riboflavin B-2 (mg)	Niacin (mg)	Vitamin B-6 (mg)	Vitamin B-12 (mcg)	Folic Acid (mcg)	Pantothenic Acid (mg)
20	10	0.03	0.05	0.4	—	—	—	—
17	5	0.03	0.03	0.3	—	—	—	—

Sodium (mg)	Potassium (mg)	Calcium (mg)	Phosphorus (mg)	Magnesium (mg)	Iron (mg)	Zinc (mg)	Copper (mg)	Manganese (mg)
60	335	16	33	27	0.7	—	—	—
36	172	12	19	—	0.4	—	—	—

Raw—2 medium
   Cooked—½ cup diced

NUTRITIONAL DATA *(Beet greens)*

Weight (g)	Calories	Protein (g)	Carbohydrates (g)	Crude Fiber (g)	Water (g)	Fat (g)	Cholesterol (mg)
100	24	2.2	4.6	1.3	90.9	0.3	—
100	18	1.7	3.3	1.1	93.6	0.2	—

Vitamin A (IU)	Vitamin C (mg)	Thiamine B-1 (mg)	Riboflavin B-2 (mg)	Niacin (mg)	Vitamin B-6 (mg)	Vitamin B-12 (mcg)	Folic Acid (mcg)	Pantothenic Acid (mg)
6100	30	0.1	0.22	0.4	—	—	—	—
5100	15	0.07	0.15	0.3	—	—	—	—

Sodium (mg)	Potassium (mg)	Calcium (mg)	Phosphorus (mg)	Magnesium (mg)	Iron (mg)	Zinc (mg)	Copper (mg)	Manganese (mg)
130	570	119	40	106	3.3	—	—	—
76	332	99	25	—	1.9	—	—	—

Raw—3½ ounces
   Cooked—½ cup

Marinated Sliced Beets

6 medium beets
1 small Spanish onion
6–8 sprigs curly parsley, chopped
¼ cup lemon juice
½ cup white-wine vinegar
2 bay leaves
8–10 peppercorns
¼ teaspoon salt
¼ teaspoon freshly ground black pepper

Remove beet tops, leaving tails and 1½ inches of stem attached (reserve tops for another use). Place beets in water and boil until just soft. Drain. Cool. Peel and slice. Thinly slice onion. Mix beets and onion with remaining ingredients. Refrigerate for 2 to 3 hours. Serve with salad, cottage cheese, or sour cream. *Serves 4.*

Beets in Butter

4–6 medium beets
4 tablespoons butter

Remove beet tops, leaving tails and 1½ inches of stem attached. Steam or boil until soft. Drain and cool. Add butter. Cover. Let stand over very low heat for 1 to 2 minutes. *Serves 4.*

Cold Beet Borscht

4–6 medium beets
1¼–1½ cups sour cream

Remove beet tops, leaving tails and 1½ inches of stem attached. Steam or boil until soft. Remove beets; simmer beet water until it begins to thicken. Peel and slice, dice, or quarter beets. Return to beet water and chill. Serve with generous dollops of sour cream. *Serves 4.*

BEET TOPS

1 pound beet thinnings or 1 bunch beet tops
2–3 cloves garlic
1 small yellow onion
¼ cup sesame oil
¼ cup tamari sauce

De-stem and thoroughly wash beet leaves. Let stand in cold water. Chop garlic and onion and sauté in sesame oil. Add wet beet leaves. Sprinkle with tamari sauce. Cover. Let simmer over low heat for 8 to 10 minutes. Stir. Serve with rice or baked potatoes and main events. *Serves 2.*

BEET-TOP SALAD

1 pound beet thinnings or 1 bunch beet tops
½ small yellow onion, chopped
2 cloves garlic, crushed
1 teaspoon chopped fresh dill, marjoram, or savory
¼ cup lemon juice
2 tablespoons white-wine vinegar

De-stem and wash beet leaves thoroughly. Drain. Tear up leaves. Combine remaining ingredients and serve separately as dressing.

The salad can be varied by the addition of spinach, Boston lettuce, Belgian endive, or watercress. *Serves 2 to 3.*

GINGER

Ginger has come out of the Oriental kitchen and into the contemporary American scene as a stimulating and clean natural spice. Actually, ginger has been part of the American scene since Colonial times, being the essential ingredient in gingerbread cookies and houses and in New England candies and ice cream. The Continental Army listed ginger as standard issue for the troops.

Earlier than all of this, ginger had made its mark from China across

India to Greece and Italy. The Italians made it part of their daily bread. Some parts of southern Italy were so accustomed to ginger that it bore no specific name. It was simply identified by a word describing its flavor, something close to "strong."

Most of the fresh ginger sold in stores in America is known as Jamaican ginger but was probably imported from Puerto Rico, where it grows year-round. When you select your roots, or "hands," look for hard, knobby specimens. They are usually beige, fat, and clumpish, and no two are ever alike. Avoid wrinkles, mold, and a spongy texture, and keep ginger wrapped in the refrigerator.

Ginger is the ingredient essential to all curry combinations. It is an excellent addition grated into soups, salads, sautés, stuffings, and marinated and ground-meat preparations such as meat loaf and hamburgers. I like to grate some over a roasting chicken that I have already been basting with butter, oregano, pepper, and honey. A few thin slices of ginger added to a vegetable stir-fry will show up in the eating as spicy bits to be remembered and enjoyed. Ginger is also a surprisingly delightful ingredient in any vegetable-juice combination. The sweetness of carrot juice in contrast with ginger heat is a pleasant taste experience.

Ginger is not one of the world's leading nutritious items. It is basically a tasty root that enhances the appetite appeal of the food it escorts. Ginger, however, contains goodly amounts of potassium, iron, and phosphorus. Unfortunately, nobody eats enough ginger at one time for its nutritional contribution to have measurable impact. Use it for the love of it.

NUTRITIONAL DATA

Weight (g)	Calories	Protein (g)	Carbohydrates (g)	Crude Fiber (g)	Water (g)	Fat (g)	Cholesterol (mg)
100	49	1.4	9.5	1.1	87	1	—

Vitamin A (IU)	Vitamin C (mg)	Thiamine B-1 (mg)	Riboflavin B-2 (mg)	Niacin (mg)	Vitamin B-6 (mg)	Vitamin B-12 (mcg)	Folic Acid (mcg)	Pantothenic Acid (mg)
10	4	0.02	0.04	0.7	—	—	—	—

Sodium (mg)	Potassium (mg)	Calcium (mg)	Phosphorus (mg)	Magnesium (mg)	Iron (mg)	Zinc (mg)	Copper (mg)	Manganese (mg)
6	264	23	36	—	2.1	—	—	—

Raw—3½ ounces

GINGER SAUTÉ

¼ cup olive oil
4 tablespoons butter
2 ounces fresh ginger, grated
1 small yellow onion, finely chopped
2 cloves garlic, crushed
6–8 sprigs parsley, chopped
¼ teaspoon freshly ground black pepper

Heat olive oil and butter. Sauté remaining ingredients until brown. This makes an excellent sauté base for lamb pieces, browned beef, cube steak, boneless chicken, shrimp, and mushrooms.

GINGER CURRY

2 ounces fresh ginger, grated
¼ cup olive oil
¼ cup walnut oil
¼ teaspoon curry powder
Dash chili powder
¼ teaspoon freshly ground black pepper

Combine all ingredients and mix well. This makes a sauté base for meats, vegetables, or seafood. It can also be brushed on shish kebabs or barbecued meats.

GINGER MARINADE

4 ounces fresh ginger, grated
1 small yellow onion, finely chopped
2 cloves garlic, chopped
¼ cup lime juice
¼ cup red-wine vinegar
¼ teaspoon salt
¼ teaspoon freshly ground black pepper

Combine all ingredients. Add cubed meats for shish kebab, seafood for seafood salad, mushrooms, hard-boiled eggs, or raw broccoli and cauliflower; or serve as a spicy dressing for avocado and other salads.

HORSERADISH

Years ago I knew horseradish in two forms. Like most of us I knew it grated in jars and small crocks, served with clams, shrimp, and oysters. My other horseradish experience came in small damp burlap bags filled with fresh-picked whole phallic-shaped roots that were sold in great quantities several times a year during the Jewish holy-day seasons. Even today on the Lower East Side of Manhattan grizzled old Jewish men in flat caps and white coats hawk pickles, herring, and freshly ground horseradish from their open-air stalls.

Sometimes sliced and eaten raw in small epicurean circles as an appetizer, horseradish is at its best freshly ground. It can sustain a touch of salt and distilled vinegar, which enhances its bite, but little more is ever required. Fresh-ground horseradish is a wonderful sauce for shellfish, thickly sliced cold German meats like liverwurst and goose pâté, and English boiled beef. Horseradish can also be added to a meat marinade or a barbecue sauce, or it can be grated over roasts and poultry headed for the oven.

The convenience of freshly ground jarred horseradish and the lack of a hand grinder in most kitchens have caused the noble root to disappear from most stores and markets. I assure you, however, that a freshly grated fresh-picked horseradish root is far superior to any jarred variety. Look for fresh horseradish in early spring and late fall. Grate up a batch, and turn yourself on to this unique and ancient root. Evidence of its use has been found in the tombs of the Egyptian pharaohs from as far back as the Twelfth Dynasty. Select hard whitish roots with a large burled, brownish top knob. Avoid horseradish that is very dark in color, sprouted, or not rock-hard.

As much as I tout horseradish for its taste and form, I make no case for its nutritional value. Suffice it to say its use increases the taste appeal and consumption of the more nutritionally powerful meats and fish it dines with.

HERBS

1. LEAF BASIL
2. SICILIAN BUSH BASIL
3. THYME
4. MARJORAM
5. SAGE
6. CORIANDER
7. TARRAGON
8. DILL
9. OREGANO
10. MINT
11. COMFREY
12. ROSEMARY
13. SAVORY

Most fruit men don't keep a heavy mental inventory of herbs. They know that they exist and have a vague idea of their uses and essences, but by and large retail produce people leave herbs to the herbalists and the spice companies that package herbs in small tins with large profits.

Herbs are small green plants with aromatic, nutritional, and flavorful properties that can elevate even the simplest cooking to new horizons. Fresh herbs can be picked and used immediately, cut and refrigerated for weeks, or snipped, dried, and used for months or even longer. Herbs produce flowers, most often late in their growing season, which have a more intense version of the leaves' flavor.

Most stores do not carry many fresh herbs, if any at all, and it may be necessary for you to grow your own supply. Herbs can be grown in windowsill pots or boxes, as part of annual gardens or perennial borders, or planted in complicated ornamental patterns that have been the pride of kings. Learn to use herbs in your cooking, and you will develop subtleties and flavors that will turn the usual into a deliciously unique cuisine.

Although there are many varieties of herbs, from the simplest and most common to the most esoteric and magical varieties, the following pages include those most easily available and most useful in the home kitchen.

Herbs are easy to grow but difficult to germinate (basil excepted), so look for them in small pots in the perennial plant section of a good farm stand or greenhouse. Herbs can be picked all summer, then cut back at harvest time and mulched. Basil needs annual planting and actually is treated more like a leaf vegetable or salad.

Fresh-cut herbs can be wrapped in a damp paper towel and kept in your refrigerator for weeks. Herbs that are harvested in large amounts at season's end should be kept unwashed, gathered into small bunches with an elastic band, and hung stem ends up until dry and crisp. Simply brushing the dry bunch into your open hand will release highly aromatic crumblings full of the herbs' flavorful essence. Several dried herbs can be mixed to your own taste, kept in jars, and used as your favorite seasoning for salads, poultry, fish, meats, and soups.

There are many nutritional and healthful properties to fresh herbs, but my purpose in this book is to offer information about fruits and vegetables from their origins to the kitchen table. As such, I will only include those basic herbs most used in cooking and seasoning.

The wealth of information necessary to understand the healing and health powers of herbs is beyond my scope as a fruit man. I recognize herbal magic and from time to time as you read on about specific herbs, you will find information I have gathered from experience and study. There are excellent books, however, which focus directly on the full range of herbal nutrition, and I recommend their use.

This book will help you identify, purchase, and maintain the fresh herbs you choose, whether for flavor or nutrition.

BASIL

*B*asil is the king of herbs. It has been the favorite herb of the Mediterranean for centuries, and as they did with so many other fresh fruits and vegetables, the Italians made it happen.

There are two general classes of basil: bush basil and leaf basil. Bush basil has smallish pointed leaves that form a definite small round bush. Though mostly green, bush basil can be violet or violet-tinged. Bush basil is said to be sweeter than leaf basil, but I believe that is not so.

Leaf basil grows on long stems with opposite oval leaves. Like many herbs, the plant grows always from within the crotch formed by the opposite leaves. When picking leaf basil, pinch the stalk at a crotch, forcing the plant to grow thicker rather than taller. This also inhibits the production of the flowers, which sprout from the end of each stem when it reaches about eighteen inches. Inhibiting flower production lengthens the productive life of the plant, retains early plant sweetness, and produces a thicker, darker leaf. Leaf basil can be crinkly like lettuce, smooth-leafed, violet, or variegated. It is all good, with basically the same flavor.

Two weeks before frost is expected, allow the basil plant to flower. Pull the whole plant, wash the root end, and hang it upside down to dry. The dried plant can be used by brushing the leaves and flowers into your hand, or all the leaves and flowers can be removed and stored in a jar for winter use until fresh basil appears next season.

When buying basil, be aware that it is seasonal and that basil out of season is always expensive. Basil is best from local sources from June until the first frost. Dry, frozen, or oil-packed basil may have to do in January, but there is nothing like fresh-picked garden basil. We pick basil every day it grows at Sunfrost. Sometimes on a weekend one of our crew picks basil three or four times a day to meet the demand for the real thing.

Basil can be used in the preparation of poultry, lamb, pork, and seafood. It has a delightful taste in salads and pasta soups. But basil is best with tomatoes and tomato-salad dressing. When you think of anything tomato, especially Mediterranean-style tomato, always think of basil. Basil, tomatoes, olive oil, and garlic are always superb together, whether cooked or made into an uncooked sauce or eaten fresh as salad.

Basil is also the main ingredient of one of the finest sauce recipes ever created, called pesto. The fresh basil leaves are ground with Romano and Sardo cheeses, pignola nuts and walnuts, salt, olive oil, and fresh garlic. Pesto can be served on pasta, noodles, or squash and never needs cooking.

THYME

Thyme contains an essential oil that is highly aromatic and beneficial. Ancient Greeks considered thyme an aphrodisiac and generously sprinkled their homes with thyme leaves in preparation for intimate romantic gatherings. More commonly used as a seasoning, thyme has a definite taste of its own, and its use should be careful. A light sprinkling or small amount of fresh chopped leaf will delightfully enhance meat stews, meat loaf, hamburger, baked fish, or poultry stuffings. A small *bouquet garni* of thyme, bay leaf, and parsley gently brushed into a soup or chowder will add a delightful mischievous taste.

Thyme has tiny narrow leaves that grow on woody little twigs and can be carefully pinched off to encourage tight, bushy growth. There is a wilder variety called creeping thyme that is very popular. It has a light lemon scent and is excellent with fish, especially when it's stuffed and poached. Wild thyme has a rounder leaf and a softer twig. Dried thyme, sage, and rosemary make an excellent garden herb mix that will find its way into many of your recipes.

Marjoram

Marjoram has a small, flat, roundish leaf sprouting from tough, woody, twiglike branches. It produces a round pod from which sprout highly aromatic knots of purple or white flowers. Fresh marjoram loves eggs and omelettes and poached fish. In its dried state it is excellent sprinkled on meat roasts or broiled chops.

Marjoram is very close to oregano in its flavor. The essential difference is the spiciness of each. Italian, Greek, and North African recipes call for oregano almost always, while the more northern cuisines of France and England favor marjoram. I recommend oregano for all recipes that are created with gusto, passion, and broad strokes of flavor. Marjoram is definitely more subtle, a pastel that tenderly enhances light-flavored foods, light-colored meats, puddinglike casseroles, and cream soups. Marjoram is also highly aromatic and, hung to dry, will perfume the area with a clean, feminine fragrance. Marjoram bunches will minimize cooking odors that remain trapped in tightly sealed modern winter kitchens.

Sage

"The sage in bloom is like perfume" is not just a line from a Texas country song. Bright-purple flowers spring from the base of the top leaves of the sage plant, filling the garden or kitchen with a pervasive, arousing fragrance. But watch out for sage. Sage is never to be taken lightly or used heavily. The sage leaf is highly pungent. It is best used with high-fat meats and poultry—pork, goose, duck, and sausages. Almost all sausages are sage-flavored. So is scrapple. Goat cheese sometimes has a fine coat of chopped sage, and there are several English cheese favorites that feature layers of cheese with thin layers of finely chopped or ground sage leaves. Almost all prepared poultry seasonings contain sage leaf, as it is an excellent herb in a well-seasoned poultry stuffing, especially one for turkey. Sage also forms a wonderful flavor combination with cranberries.

Sage should be used often but sparingly, as it never holds back its flavor. When dried, sage can be mixed with thyme and rosemary for a useful, friendly herb seasoning. Tied in a small *bouquet garni* with thyme and marjoram, it can be brushed on roast meats or fowl in basting, dipped into soup as it simmers, or gently stirred in a gravy or sauce. This threesome gives powerful but pleasant herbal enhancement to your kitchen creations.

CORIANDER

The Latin Americans call it cilantro. It is the essential seasoning in all of Spanish Latin America, and that's where you find coriander used most. There is no salsa, *ceviche,* soup, dressing, or gravy without it.

Coriander is a definite taste. The same recipe with coriander and without it are two totally different experiences. The unprepared taster will be surprised by coriander, maybe unpleasantly. Once appreciated, however, coriander, or "hot parsley," will be an absolute requirement for that unique piquance that no other herb quite gives. Coriander is rare in most retail stores. It germinates with slight difficulty but grows bountifully and easily.

Coriander is also a very adaptable herb. It will retain its strong identity with game, pork, or lightly spiced meats like sausage, but it will back off a little when used with chicken, fish, or young lamb. All curry combinations use coriander leaves or seeds as part of the heat, yet it will remain quiet but identifiable in mashed potatoes or an omelette.

Coriander has two kinds of leaves on each stem: a small lacy leaf resembling the leaf of the carrot, to which it is related, and a wider, rounder, flat fan-shaped leaf. It also bears two kinds of seeds. This probably accounts for the unique and ubiquitous flavor of this internationally popular little sprig. Coriander has thin, wispy stems and most often is sold leaves, stems, and small roots. Coriander is always better smaller.

Ask your retailer to carry coriander. It is becoming increasingly popular, and if more fruit and vegetable stores would carry it, they would do themselves and their customers a solid flavor. In our store we make it part of our Latin American–inspired items, along with our homemade salsa, guacamole salad, tropical fruits, and Mexican fruit drinks.

TARRAGON

The French claim it as their own, and real tarragon is actually called French tarragon. The Italians never use it (they say it tastes like licorice), and the Americans almost never use it (they say it is tasteless). Blame it on the Russians. The problem is that American seed-sellers back off from selling French tarragon seeds because only a very small percentage of them ever germinate. True tarragon must be propagated by root or stem cuttings. Russian tarragon germinates quite easily from seed, but Russian tarragon has no flavor. Americans, used to getting what we want, have seen no purpose in planting, buying, or using an herb with no taste.

French tarragon is available from garden centers, roadside stands, and fruit markets that concentrate on the spring garden business of annuals, perennials, and herbs.

Assuming you've found your source, you'll find the true tarragon plant has a flavor that seems to be most at home in light sauces, egg dishes, cream sauces, and wine vinegar. Subtle foods like seafood, veal, and small fowl are delightfully enhanced by the artful use of tarragon. Tarragon is also great in soups and either fresh or dried is a delicious salad sprinkle.

Tarragon grows into large plants up to three feet high. The longish, slender leaves can be picked and used fresh, or small branches can be clipped and easily hung and dried for winter use.

DILL

No herb needs timely picking more than dill. The small, fine, feathery dill plant is a delightful taste addition to spring salads of cukes and early lettuce. Fresh early dill, chopped and sprinkled on baked potatoes with sour cream, is a wonderful treat, as is fresh dill, cottage cheese, and tomatoes. The midstage dill plant has a stronger yet gentle flavor that is a wonderful taste in soup and seafood chowder. The mature plant produces a large flat flower containing the dill seeds. These flower heads, cut with long stems, are dried and the seeds gathered for pickling. Their sharp, incisive flavor creates the famous flavor of the dill pickle.

In the fall the tall dill plants contain all three stages of dill. The entire plant can be pulled by its roots, hung, and dried in a warm, dry area. It will retain its fresh bright-green color, and you can select the fine, midsize, or seed part of the plant for your specific needs. Most good fruit and vegetable markets carry big bundles of harvested dill through September and October. It is never expensive and always a nice addition to the general atmosphere of a properly kept kitchen.

OREGANO

Oregano is the Italian herb and should not be confused with its milder cousin marjoram. Used in other cultures for native recipes, it is always used in recipes called Italian style. The flowers of the oregano plant are the most pungent part and in their dried state create that distinct taste so popular in pizza and Italian sauces. There are some varieties of oregano, however, whose smaller dusky-green leaves are highly aromatic and very flavorful. Fresh oregano adds a spicy freshness to salads, and a few sprigs in lentil soup, *pasta e fagiole* (not pasta fazool), or minestrone will raise these popular recipes several notches in the flavor department. Fresh oregano is absolutely best on a freshly sliced beefsteak tomato, with generous slices of garlic, a sprinkle of olive oil, and freshly ground pepper.

Dried oregano can be generously mixed with melted butter and basted on a roasting chicken to create a crisp, flavorful crust that will surprise and please almost every palate. It also creates a new deliciousness when sprinkled generously on roast meats or seafood. My grandfather used to split open a loaf of fresh, hot Italian bread, butter it, and sprinkle it with home-pressed olive oil, freshly ground pepper, chopped garlic, and dried oregano. It was a midafternoon feasting that I will never forget.

MINT

There are over two thousand varieties of mint. However, there are only two kinds sold at your favorite fruit and vegetable stand: spearmint and peppermint. Spearmint is the oldest version of all mints, and while the leaves closely resemble those of peppermint, the flavor does not. Spearmint throws a flower tassel, while peppermint develops a round purple flower much like clover.

Peppermint has more oils and a very strong taste of menthol not found in spearmint. Spearmint grows on green stems and peppermint on purple. If you have the choice, choose peppermint. If not, you will probably not be able to tell the difference if you are simply using mint for mixed drinks and fruit ades.

Mint grows easily. Whether cut, crushed, rolled, or left whole, it mixes well with meats, melons, fruit drinks, salad vegetables, and teas. Fresh mint leaves crushed with sugar and a few drops of water make an excellent flavoring for refreshing summer drinks. Mint leaves wrapped around whole garlic cloves that are stuffed into tiny pockets throughout a leg of lamb will create a taste permeating the meat that most people find a superb addition to the lamb flavor.

COMFREY

Comfrey is the most nutritionally healthful plant and the most easily grown. A simple root cutting will produce enormous plants up to four feet high and just as wide. The rampant tubers run wild, producing offshoot plants at random. The long hairy leaves are rich in iron and B vitamins, and the chemical properties of the plant promote rapid healing of injured bone and muscle tissue. Comfrey is an excellent detoxicant, especially effective on liver and lung tissues, creating richer, cleaner blood cells that carry greater amounts of oxygen and nourishment. Comfrey tea, comfrey syrup, and the comfrey leaf blended in fruit juice are refreshing and stimulating. It is far superior in its fresh form and therefore only seasonally available in most areas.

ROSEMARY

*F*resh rosemary is difficult to germinate and requires hot days and cool nights to grow to its maximum height of four to five feet. It is difficult to find and often expensive, probably because it has not become an American taste. The long, narrow leaf contains essential oils, including tannic acid and resin. Its taste, however, is far from bitter and imparts a soothing quality to strong-flavored meats. Italians use rosemary with a passion, stuffing it generously into preparations of lamb legs and shanks, pork knuckles, rabbit, and shellfish. It goes well with meat dishes cooked in wine and is totally compatible with garlic. No wonder the Italians love it.

Rosemary gives an excellent balance to sweet jellies and is compatible with veal and seafood, although I find it a little strong used with light meats or fish. It is an excellent additive to dressings and sauces and an active ingredient for marinades and in the preparation of barbecued meats. Along with sage and thyme rosemary makes a useful dried herb mix to be sprinkled into soups and gravies, but, I add, use it with caution and sensitivity. Fresh rosemary chopped into any rice recipe is excellent and will make the taste of those little grains jump up and dance.

SAVORY

Savory is the salad herb. It is named for the seasons: winter savory and summer savory, with much the same taste. I find summer savory better tasting and softer textured. In both their dried and fresh forms, the savories create excellent salad dressings mixed with yogurt, mayonnaise, avocado, garlic, and cheese, or simply with lemon juice. Savory can also be used in sausage or ground meat, but both the summer and winter varieties seem to be at home mostly with garden greens and fresh bean recipes.

I recommend dried summer savory mixed with oregano, thyme, a bit of sage, and marjoram in a shaker bottle to be kept on the table with other condiments. The mixture is an excellent sprinkle for winter salads, when the greens do not have the same sparkle and pep as the spring and summer varieties. The savory, along with the rest of the mix, increases the overall flavor and pleasure of a January or February "fresh" green garden salad.

Most retailers do not carry savory, but retailers of spring annuals, perennials, and herbs can supply you with tiny potted plants that grow easily to one foot high or more even in poor soil or a sunny windowsill pot. Keep a little savory on hand. It is a surprisingly effective plant for so unheralded a fresh-food item.

POTATOES

1. YAMS
2. IDAHO BAKERS
3. MAINE
4. CALIFORNIA RUSSET
5. NEW #1 WHITE
6. NEW #2 WHITE
7. NEW #1 RED
8. NEW RED 'BABY BLISS'
9. PEA
10. NEW WHITE RUSSET
11. RED SWEET
12. WHITE SWEET

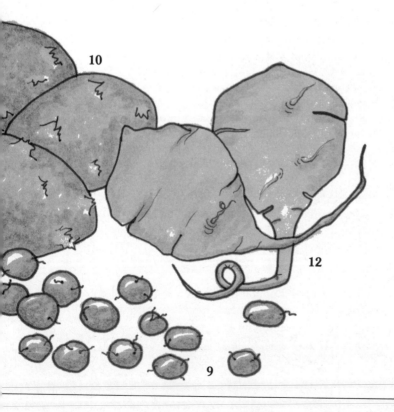

*P*otatoes are the noble beasts of the fruit business. They are just what they look like, a strong, sturdy, powerfully nutritious, efficient food. I've always liked the heft and feel of a fresh-dug, lightly cured potato of any variety. There is no better food. Maybe that's why the potato became a basic food for every country in the Western world.

Fifty years ago, when the wave of ethnic immigration was at its height in America, Ireland moved most of its population to America just because of potatoes. Poles, Ukrainians, Germans, Jews of all nations, Hungarians, Russians, and thousands of families from lost places like Latvia and Estonia came to America and created a national food pastime. America's affair with the potato comes in many forms: chips, sticks, puffs, crinkles, ridges, home fries, french fries, and fresh frozen. The potato also comes wrapped in silver foil next to a piece of nasty-looking roast beef and thirty-eight dead peas!

Real fresh potatoes have nothing to do with any of those things that start out as potatoes but end up processed, packaged, and priced to death. The more you mess with potatoes, the more you pay and the less you get. Nothing you buy in any processed form tastes anything like fresh-dug potatoes.

You can have really good fresh potatoes almost all year round. Potatoes are great keepers and store well but not forever. So look for fresh inventories and new varieties in the stores and markets you shop. Potatoes are grown in every state. Some states produce several crops of early, then later, varieties. In New York first potatoes start showing up by September. These are the early new potatoes, almost transparent-skinned, bright, hard things bursting with vitality and life. Later on in October the thick-skinned dirty "old" potatoes are finally harvested and stored in deep barns three-quarters underground for winter packing. The same scene is going on all over the Northeast. In Maine the famous Aroostook potato is being gathered from its deep brown bed to be bagged and buried for months and months before it goes to winter market. In Idaho the famous Idaho Baker is packed in fifty-pound boxes for as long as nine months. And in between the Northern seasons Florida, Texas, California, and Oregon are picking and shipping all over America to every little town and store.

Potatoes were once considered unfit food for anything other than cattle and the lowliest peasant. Throughout history the aristocracy ignored the potato as food and left it to, or forced it on, the peasant class. How foolish. The protein-rich, mineral-laden potato is a tremendously nutritious food. Perhaps the potato is the reason that we have so few of the aristocracy left, while the peasants have survived to become the teachers, farmers, engineers, scientists, and peasants of our modern society.

Potatoes grow everywhere, but they started in Peru long before the first Incas. Today the Andean mountain people of Peru grow over two hundred varieties of potatoes, from tiny black ones to giant red tubers with stripes and polka dots, and every color in between. Once potatoes were the main food of the Western world; today potato consumption is down almost forty per cent from its peak early in this century. I remember potatoes that came to our store in rough one-hundred-pound burlap bags by the carload. Today potatoes come fifty pounds to a thick paper sack or cardboard box as part of a mixed load of vegetables. I recall women who bought at least five pounds a day, every day, sometimes more. Today customers buy four potatoes for a meal or three pounds a week. One-third of all potato consumption in the United States today is of processed potatoes in one form or another. I have a feeling that that fraction is increasing at a dramatic rate. The fast-food industry is consuming frozen, precut, partially cooked potatoes for "fat" fries as fast as they can be supplied. At nine cents per pound of potatoes, prepared frozen and shipped to be refried and sold at eighty-nine cents for two ounces, the fast-food boys are piling up big profits. At that rate they could give the hamburger away and still make money.

Potatoes, sweet potatoes, and yams are in no way related. The true potato is South American in origin. The sweet potato is Asian, and the yam originated in Africa, where it grows quite large and is a mainstay of the African diet. (More peasants?) Although potatoes, sweet potatoes, and yams are often substituted for one another on our American tables, the true potato is significantly higher in nutritional qualities although all three are recommended as regular beneficial additions to our daily diets. Potatoes are almost ten per cent usable protein. They supply important amounts of potassium, calcium, and phosphorus. They have more iron than milk, which, by the way, is heavy in solid fats, which are difficult for the body to digest and utilize. Potatoes are low in salt and high in starch. Potato starch is highly digestible, is easily assimilated, and does not produce body fat. The butter, oil, sour cream, mayonnaise, and dressing we smother potatoes in does! The calories in potatoes are one-quarter that in bread and more useful.

Fresh potato harvests arrive in our stores all the time from different parts of the country except in the dead of winter. The first fresh potatoes arrive from Florida in March or early April. The harvest moves north all spring, summer, and fall and eastward from California from the beginning of May. Some new potatoes have a green tinge to the skin, and some people think this is poisonous. Potatoes are a member of the deadly nightshade family, as are tomatoes and eggplants. The chemical solanine in pure form and massive amounts is deadly serious, but in potatoes, there is not enough in a carload to disturb anything in our chemical balance. What the green does indicate, however, is that the potatoes have been out of their bags and on display too long. New potatoes need darkness and are at their best when used fresh. Buy Idaho potatoes, Maine potatoes, and western russets all winter. The rest of the year look for fresh new inventories with light, bright skins in all colors and sizes. Learn to prepare potatoes in new and interesting ways, and you will enlarge your family's fresh-food pleasure and overall nutritional intake.

Good-quality yams and sweet potatoes are available year-round, except in midsummer, at reasonable prices. They are at their best, however, in the fall during harvest time. During September, October, and November yams and sweets are thin-skinned, moist, and very succulent. Yams and sweet potatoes are cured for storage (sort of dried out in hot rooms) and for most of the year lose some of that harvest succulence. They are, like all fresh fruits and vegetables, at their best when they are at their freshest. Look for hard bright-skinned specimens without wrinkles, gray spots, soft spots, or a woody feel.

Yams and sweet potatoes are two of the most easily digested foods, and their gentle ways combined with their powerful nutrition make them highly desirable foods. Since there is no information available on the specific nutritional values of each potato variety, the following tables indicate the general levels of potato nutrition.

NUTRITIONAL DATA

Weight (g)	Calories	Protein (g)	Carbohydrates (g)	Crude Fiber (g)	Water (g)	Fat (g)	Cholesterol (mg)
100	76	2.1	17.1	0.5	79.8	0.1	—
100	95	2.6	21.1	0.6	75.1	0.1	—
100	76	2.1	17.1	0.5	79.8	0.1	—

Vitamin A (IU)	Vitamin C (mg)	Thiamine B-1 (mg)	Riboflavin B-2 (mg)	Niacin (mg)	Vitamin B-6 (mg)	Vitamin B-12 (mcg)	Folic Acid (mcg)	Pantothenic Acid (mg)
trace	20	0.1	0.04	1.5	—	—	—	—
trace	20	0.1	0.04	1.7	—	—	—	—
trace	16	0.09	0.04	1.5	—	—	—	—

Sodium (mg)	Potassium (mg)	Calcium (mg)	Phosphorus (mg)	Magnesium (mg)	Iron (mg)	Zinc (mg)	Copper (mg)	Manganese (mg)
3	407	7	53	34	0.6	—	—	—
4	503	9	65	—	0.7	—	—	—
3	407	7	53	—	0.6	—	—	—

White, raw—1 medium
White, baked—1 medium
White, boiled, with skin—1 medium

YAMS

Yams and sweet potatoes look alike and taste alike. Yams come in different skin shades from brown to dull red, and the meat inside can be red, bright orange, or yellow. Just as potatoes are the mainstay food of the Western world, yams are the basic diet of the African, West Indian, and some of the South Pacific world. They grow from two to four pounds each on the average, with some specimens close to one hundred pounds.

In the United States yams can only be grown in the deep South, where they are a heavy favorite among the black African descendants of the original Africans who carried their yam cuttings with them from Africa to the West Indian Islands in the holds of the wretched slave ships. Yams, like okra, are one of the very few items of African life to reach the new world.

Yams differ from sweets and potatoes in the way they grow. Yams grow on plants and hang from plant stems. Their weight causes the stems to bend to the soil, and the yams become partially embedded like an exposed underground tuber. They are, however, truly above-ground tubers and are sometimes called the air potato. Although the yam is botanically very different from the sweet potato, most produce managers, clerks, and even experienced fruit men cannot tell the difference. Yams are never red or orange. Sweet potatoes are. Yams are always bigger, smoother skinned, and heavier than sweets.

Yams can be baked, boiled, steamed, mashed, blanched and fried, candied, and made into a sweet pie filling with a taste and texture very near that of pumpkin pie. Yam soup is an African dish, and yams can even be distilled for a potent alcoholic drink.

Look for smooth-skinned brightly colored specimens without dark spots or wrinkles.

NUTRITIONAL DATA

Weight (g)	Calories	Protein (g)	Carbohydrates (g)		Crude Fiber (g)	Water (g)	Fat (g)	Cholesterol (mg)
200	210	4.8	48.2		1.8	—	0.4	—

Vitamin A (IU)	Vitamin C (mg)	Thiamine B-1 (mg)	Riboflavin B-2 (mg)	Niacin (mg)	Vitamin B-6 (mg)	Vitamin B-12 (mcg)	Folic Acid (mcg)	Pantothenic Acid (mg)
trace	18	0.18	0.08	1.2	—	—	—	—

Sodium (mg)	Potassium (mg)	Calcium (mg)	Phosphorus (mg)	Magnesium (mg)	Iron (mg)	Zinc (mg)	Copper (mg)	Manganese (mg)
—	—	8	100	—	1.2	—	—	—

Cooked—1 cup

IDAHO BAKERS

Idaho baking potatoes are the best potatoes in the world. The very best specimens have been taken to the best soils of France, South America, England, and our own southern United States, and nothing happened. Potatoes, yes, but nothing like the thick-skinned oblong, flattish tuber that bakes into a glistening white, fluffy morsel with that unique crisp, flaky skin.

Idaho potatoes grow in the mineral-rich volcanic-ash soil left behind in the Idaho foothills thousands and thousands of years ago. They are harvested in late summer and early fall, and most are packed in fifty-pound boxes, stored, and shipped all over the United States for the next nine to ten months. They are always graded, sorted, and packed by size and shape so that when you purchase them, you are assured of even cooking.

There are many potatoes sold as baking potatoes that are grown and shipped all over the country, but the Idaho baking potato is unique and unequalled for quality, taste, and texture. They are packed 70, 80, 90, 100, and 120 potatoes to the box. The 120-size is a most delectable potato that bakes quickly and makes a delightful personal serving. The 70-size is enormous, takes seventy-five to ninety minutes to bake, and can compete with a lobster or steak for room on a plate. I've always sold 90- or 100-size Idahos. They are impressive on the plate, delicious, and certainly filling. Definitely your money's worth.

An average-sized Idaho potato should be lightly scrubbed, then placed in a preheated 350° F. oven for forty minutes. Then increase the temperature to 450° F. for ten minutes more. This makes the skin crisp and flaky. If you like faster cooking and you're not into microwaved food, insert two common ten-penny nails in each potato. Idaho potatoes also make excellent fried potato skins and french fries. When making french fries, wash the whole potato either peeled or unpeeled and dry it. Don't wash the cut pieces. Idaho potatoes store very well in a cool, dark place and keep for months.

Its high quality and rich growing soil give the Idaho Baker the highest nutritional ranking, especially in mineral content, of all the potatoes.

MAINE POTATOES

As the Idaho potato is to the oven, the Maine potato is to the stove top. Maine potatoes, also called round whites, and most of their Northeast cousins are harvested in mid- or late fall and have a very dark thick skin. They are usually dirty and require peeling or, at the least, a thorough and complete scrubbing. In the trade Maine potatoes are called old potatoes because they are left long underground, often well past the frost, and are sold all fall, winter, and spring into June. They are also called Aroostook potatoes after the county in Maine that is the center of that state's potato industry.

These knobby big brown tubers are excellent in soup, make the best mashed potatoes because of their smooth, moist texture, and fry quickly either sliced of julienned. Cut up, they are wonderful companions to roast meats, chicken, pork, and lamb. There is no doubt that this common, endlessly useful Maine variety is a direct descendant of the potato that kept most of the Western world alive for centuries.

Maine potatoes keep well but will sprout even under ideal storage conditions. Once sprouted, Maine potatoes can be cut into sections containing one or more eyes and then planted in a six-inch trough and covered with soft soil. During the growing season hoe up the trenches into long mounds. This causes smaller green plant growth and greater potato development. Harvest in early September, and you will enjoy potatoes as never before.

When buying Maine potatoes, look for the first harvest beginning in early October. Fresh, nutritious supplies are available most of the winter into late February. Select unsprouted, largish, flat potatoes free from wrinkles, cuts, or green areas.

CALIFORNIA RUSSET POTATOES

California produces a russet potato with a thick, dimpled, leathery skin that makes it ideal for baking. It has a very short season—about three weeks in late summer and September. After the summer Idaho Baker drought and just before the new Idaho harvest the California "leather-back" is a welcome sight in the produce department. This unique and delicious variety is hardly ever found in supermarkets, but most select and knowledgeable fruit-stand vendors will feature this dark-skinned long oval potato during its brief availability. Like the Idaho, it is an excellent baker and makes delicious fried potato skins. Look for clean, bright, russet-colored potatoes that feel heavy for their size. When baking them, select potatoes of even size to make sure they all cook in the same time.

NEW WHITE POTATOES

New white potatoes are early-harvest potatoes. They have a long oval shape and very transparent pale skins, often flaky. They are medium-sized and very clean. Sometimes they have a slight green coloring, a solanine residue that is eliminated in cooking and is never harmful. They are marketed in two size grades. The #1 is a U.S.D.A. grading that indicates a minimum size, absence of diseased or damaged specimens, and overall quality.

New potatoes first appear in the markets in May from Florida, then in late June and July from California. Then the Carolina and Virginia crops are picked, and finally the first harvests from the Midwest and Northeast appear at the tail end of summer and beginning of fall.

New white potatoes, like all early-harvest vegetables, cook very quickly. They are excellent with roast meats and fowl and need not be blanched first. New potatoes can also be served without peeling. They make excellent potato salad but are a little too moist for fries and mashed. New potatoes sliced and blanched will add a delicious dimension to salads and cold plates.

New white potatoes #2 are small round potatoes with the same characteristics as #1, only much smaller. Usually the size of golf balls, they make excellent parsley potatoes and go well with roasts and chicken. They are always served in their skins, which are totally edible and scarcely noticeable. The skins of these baby potatoes lend an interesting nutty flavor to the tubers as well as preserve their nutrition complete. Every customer who has discovered these tiny morsels in our store has since made them a regular part of his or her vegetable variety, eagerly awaiting each season's new arrival.

NEW RED POTATOES

New red potatoes are a red-skinned variety of potato grown and harvested at the same time as their new white cousins. They are round, not oval, their skins are smooth, they have no solanine residue, and they tend to taste a touch sweeter than the oval white potato. The skins are slightly heavier but are edible and delicious.

The new red potato is the very best boiling potato. They can be cooked with their jackets on; they are easily removed once cooled if desired. New red potatoes can be broiled with roasts, again with skins intact; sliced and home fried; or sliced, blanched, and chilled to be served cold as part of a vegetable salad.

New red potatoes are first harvested in Florida and are available in April and early May. Florida red potatoes have a flaky, damp skin and must be eaten fresh to be at their best. The California red potato follows soon after and is a harder, shiny, bright potato that is less moist and keeps better. Northeast red potatoes have dusty jackets and very white insides. They are harder and drier and keep longer and fresher than the Florida or California varieties. Look for these during the fall harvest, not only in the Northeast but also in all the states except in the far West and deep South.

As do their white cousins, new red potatoes come in two U.S.D.A. grades. The larger potatoes for general use are #1. New red potatoes #2 are called Baby Bliss. These pink potato morsels are one of the most delightful and prettiest dishes you can put on your table. Boiled whole and dressed with butter, garlic powder or salt, and much chopped curly parsley, then served in a white bowl, they are one of the finest combinations of color, taste, texture, and nutrition you can create. Baby Bliss parsley potatoes are simple to prepare and incredibly inexpensive. They will become one of your family's favorite foods. Baby Bliss are readily available in spring and early summer from Florida and California and arrive again in the fall from almost all the other states except those in the deep South.

PEA POTATOES

You hardly ever see them anywhere. But if you're lucky enough to live near a knowledgeable farm stand, ask them for pea potatoes, or Peewees. They are tiny, tiny white or red potatoes about the size of small marbles. Naturally, they're hard to pick and sort, but some farmers do take the time to cull them out and bring them to their markets.

Pea potatoes are always cooked whole and can be mixed with peas or green beans; dropped into soups; sprinkled around roasts or fowl; boiled, buttered, and parslied; or blanched and mixed with stir-fried vegetables.

Pea potatoes are never shipped anywhere. They are only available from local or homegrown sources at roadside markets, farm stands, and great fruit and vegetable stores like Sunfrost Farms. Even then they are rare. So in early fall ask around. You may get lucky and enjoy something especially sweet and delicious.

NEW RUSSET POTATOES

The russet potato always features a textured or deeper skin that crusts up in baking and seals the moist potato tissue within. The skin of the new russet is not as deeply formed as in the mature russet, so new russets are better for boiling, roasting, and making potato salad or cold sliced platters.

A truly freshly dug russet is really copper-colored, and the skin is formed by a coating of tiny coppery grits, or surface cells. It is definitely not a whole skin or peel.

New russets are grown in California and shipped in early summer, but there are many local varieties available in late August and early September. New russets are long ovals in shape, have a flaky or ricy texture, and are extremely versatile in the oven, the roaster, boiling water, or hot oil.

Look for them in early fall at your local market or roadside stand. Their taste is extremely clean and wholesome, and your digestive system and protein process will love them.

SWEET POTATOES

Aside from the obvious color differences, there are several other distinctions between the two sweet potato relations, red and white. Red sweet potatoes grow larger than the white and seem to keep their firmness and moisture content a good deal longer. White sweet potatoes tend to be a bit less moist and a bit more fibrous. The white sweet potato, especially when first picked, has a sweet, creamy taste that aficionados look for in early fall when white sweets are first harvested. Both white and red sweet potatoes can be baked, fried, mashed, puréed, or candied, or they can be used in breads, biscuits, and the Southern United States' favorite, sweet potato pie. Mostly, however, the sweet potato is baked, split, buttered, and eaten out of its skin with delight and healthful benefit.

Look for bright-skinned unbruised sweet potatoes with firm ends and no soft spots. If you plan to bake your sweets, select them all of the same size. I always recommend the long thin specimens for baking because they cook faster and more uniformly while retaining most of their moistness. A 7- or 8-inch sweet potato about 1½ to 2 inches thick will cook in thirty-five minutes at 350° F.

Red sweet potatoes are available almost all year round, but they are at their best in fall and early winter. White sweets seem to arrive in early fall and disappear, except for brief cameo appearances in select stores in early summer, when they are harvested in the deep Southern states and shipped north and west.

NUTRITIONAL DATA

Weight (g)	Calories	Protein (g)	Carbohydrates (g)	Crude Fiber (g)	Water (g)	Fat (g)	Cholesterol (mg)
180	254	3.8	58.5	1.6	114.7	0.9	—
180	172	2.6	39.8	0.6	127.1	1.3	—

Vitamin A (IU)	Vitamin C (mg)	Thiamine B-1 (mg)	Riboflavin B-2 (mg)	Niacin (mg)	Vitamin B-6 (mg)	Vitamin B-12 (mcg)	Folic Acid (mcg)	Pantothenic Acid (mg)
14600	40	0.16	0.13	1.3	—	—	—	—
11940	26	0.14	0.09	0.9	—	—	—	—

Sodium (mg)	Potassium (mg)	Calcium (mg)	Phosphorus (mg)	Magnesium (mg)	Iron (mg)	Zinc (mg)	Copper (mg)	Manganese (mg)
22	540	72	104	—	1.6	—	—	—
15	367	48	71	—	1.1	—	—	—

Baked—1 large
  Boiled—1 large

Baked Potatoes

Select medium baking potatoes of the same size. Preheat oven to 350° F. Puncture top of each potato with fork or cut X about ½ inch deep. Place on rack near back of oven. Bake for 40 minutes. Raise oven to 425° F. for 5 to 8 minutes before serving.

Stuffed Baked Potatoes

Puncture top of each baked potato with fork. Cut lengthwise in half. Scoop out potato and mash each with one of the following:

1 tablespoon chopped onion or scallion
2 tablespoons butter
1 tablespoon cottage cheese
Dash salt and pepper

1 tablespoon chopped chives
2 tablespoons butter
1 tablespoon sour cream
Dash salt and pepper

Chopped cooked broccoli
2 tablespoons butter
Grated cheddar cheese
Dash salt and pepper

2 tablespoons onion sautéed in butter
2 tablespoons tomato sauce

Replace mashed mixture in skins. Place in preheated 400° F. oven and bake for 5 to 10 minutes. Serve in skins.

Peasant Potatoes

4 large red potatoes, unpeeled, or white potatoes, peeled

Quarter and boil the potatoes until just soft. Drain. Serve with cabbage, sauerbraten, heavy red meats, or sour cream or cottage cheese.

POTATO SKINS

> 4–6 large white potatoes
> Light vegetable oil
> ¼ teaspoon salt
> ¼ teaspoon freshly ground black pepper
> 2 cloves garlic, chopped

Wash and peel potatoes thickly, leaving considerable potato attached to peel. Pat skins dry. Heat oil about ⅛ inch deep. Add salt, pepper, and chopped garlic. Add potato skins. Brown one side, then turn. Drain on paper towels. Serve salted or spiced with herb seasoning. *Serves 4.*

COUNTRY MASHED POTATOES

> 2½ pounds Maine potatoes or new white potatoes
> 1 cup milk
> 6 tablespoons butter
> 6–8 sprigs parsley, chopped
> ½ teaspoon freshly ground black pepper

Peel, cut up, and boil potatoes until soft but firm. Drain. Cover. Heat milk and butter together until butter melts. Break up potatoes with fork or masher. Add milk-and-butter mixture slowly while mashing. Add parsley and pepper, mashing until desired texture is reached. *Serves 4 to 6.*

POTATO SOUP

> 3 tablespoons butter or margarine, or 3–4 small salt
>   pork cubes, or 3 tablespoons bacon drippings
> 3–4 medium white potatoes, peeled and cubed
> 2 stalks celery, chopped
> 1 small yellow onion, chopped
> 2 scallions, chopped
> 1 shallot, chopped
> ½ teaspoon salt
> ¼ teaspoon freshly ground black pepper
> 2 tablespoons flour
> 2 cups milk
> 1 tablespoon chopped parsley, chives, or savory

Melt butter in large saucepan (if salt pork is used, sauté and retain drippings). Add potatoes, celery, onion, scallions, shallot, salt, pepper, and 1 cup water. Cover. Cook over medium heat for 10 to 15 minutes until just soft. Combine flour and a small amount of the milk until smooth. Add to potato mixture slowly. Add remaining milk. Cook over medium heat until soup begins to bubble. Garnish with chopped parsley. *Serves 3 to 4.*

BUTTER-CREAM POTATOES

4–6 medium Maine potatoes, peeled and cubed;
   or new red or russet potatoes, unpeeled, scrubbed and cubed
1 medium carrot, diced
½–¾ cup shelled peas
6 tablespoons butter or margarine
¾ cup milk, half-and-half, or light cream
Dash salt
Dash freshly ground black pepper
¼ cup flour

*B*oil potatoes until barely soft. Drain. Cover. Boil diced carrot in ½ inch water. Add peas just before carrot softens. Drain. Cover. Melt butter and slowly add milk. Add salt and pepper. Add flour 1 teaspoon at a time, stirring constantly. Cook, stirring, until sauce thickens. If too thick, sauce may be thinned by adding tablespoons of potato water. Place potatoes in bowl. Pour butter-cream sauce over and top with peas and carrots. *Serves 4 to 6.*

LEFTOVER FRIES

Leftover potatoes (roast, boiled, or baked)
Light vegetable oil
2–3 cloves garlic, chopped
6–8 sprigs parsley, chopped
1 small Spanish onion, chopped
¼ teaspoon salt
¼ teaspoon freshly ground black pepper
¼ cup bread crumbs

*T*hickly slice potatoes. Heat a thin layer of vegetable oil in a sauté pan. Sauté garlic, parsley, onion, salt, and pepper. When hot and just turning brown, add potato slices. Sprinkle with bread crumbs. Brown potatoes on one side. Turn and brown other side. Drain on paper towels.

ROAST POTATOES

> #2 red or white new potatoes, whole and unpeeled,
>   or #1 new Maine or russet potatoes, cut up
>   and peeled or left unpeeled
> 4 tablespoons butter or margarine, melted
> 1 tablespoon dried onion flakes
> 1 tablespoon dried parsley
> 1 tablespoon garlic powder
> 2 tablespoons salt
> 1 teaspoon freshly ground black pepper

Place potatoes in boiling water just long enough barely to soften. Drain. Cool. Combine remaining ingredients in flattish bowl. Roll potatoes to coat. Place around roast for last 15 minutes of cooking time. These potatoes can accompany any roast meat or fowl.

TIN FOIL BARBECUE

Wash and puncture potatoes. Wrap in foil, making sure water droplets cling to potatoes. Place directly on grill or coals, turning every 5 to 10 minutes for 1 to 1¼ hours.

POTATO-ONION SHISH KEBABS

> #2 red or white new potatoes
> White boiling onions
> 4 ounces butter or margarine, melted
> ¼ cup Worcestershire sauce
> ¼ cup catsup
> ½ teaspoon dried oregano
> ½ teaspoon salt
> ¼ teaspoon freshly ground black pepper

Boil potatoes until fork just barely penetrates. Drain. Cool. Drop onions into boiling water just long enough to soften. Drain. Cool. Combine remaining ingredients to make sauce. Alternate potatoes and onions on skewers. Place on grill at medium height. Brush with sauce while cooking. Turn often.

POTATO PANCAKES

2 cups cold mashed potatoes or grated leftover baked potatoes
1 teaspoon chopped chives or scallion greens
1 egg
2 tablespoons flour
Dash salt
¼ teaspoon freshly ground black pepper
¼ cup bread crumbs or cornmeal
3 tablespoons butter or margarine

Combine all ingredients except butter and bread crumbs. Mix until blended. Shape into pancakes or patties. Dust with bread crumbs. Heat griddle or skillet and fry patties in hot butter until brown on one side. Turn and brown other sides. Drain on paper towels. *Makes 4 to 6 patties.*

DOWN-HOME FRIES

4–5 medium round potatoes, sliced but not peeled
2 tablespoons finely chopped parsley
1 cup flour or cornmeal
1 tablespoon dried onion flakes
½ teaspoon garlic powder
½ teaspoon salt
½ teaspoon freshly ground black pepper
Light vegetable oil

Wash and thickly slice potatoes. Place in boiling water until just barely soft. Drain. Cool. Combine remaining ingredients except oil in flattish bowl while heating thin layer of oil in skillet. Coat potato skins with seasoned flour. Fry until golden brown on one side. Turn. Remove when crisp. *Serves 4.*

SHALLOW-FRY FRENCH FRIES

4–6 russet, Idaho, or #1 new white potatoes
Light vegetable oil
½ teaspoon garlic powder
½ teaspoon onion powder
¼ teaspoon dried oregano
1 teaspoon dried onion flakes

3–4 cloves garlic, finely chopped
¼ teaspoon salt
¼ teaspoon freshly ground black pepper

Peel and wash potatoes. Dry thoroughly. Cut potatoes into long, thin strips. Spread out on paper towels to absorb any moisture.

Heat ⅛ inch oil until hot enough that it pops when a drop of water is sprinkled in. Season oil with garlic powder, onion powder, oregano, and dried onion flakes. Place as many potatoes as will fit in single layer in pan. Add proportionate amount of chopped garlic, salt, and pepper. Turn potatoes with fork as they brown. Control heat to keep the oil from smoking. Add more oil slowly if necessary between batches. Remove when golden brown. Drain on paper towels. *Serves 4.*

ALL-AMERICAN POTATO SALAD

4–6 medium #1 Maine, new red, new white, or russet potatoes
2 stalks celery, chopped
1 medium yellow onion, chopped
2 cloves garlic, chopped
6 tablespoons finely chopped curly parsley
6–8 tablespoons mayonnaise
2 tablespoons grainy mustard
½ teaspoon salt
¼ teaspoon freshly ground black pepper
1 small red pepper, diced

Peel and cube potatoes. Cover with water and boil until they begin to soften. Drain and chill. Combine remaining ingredients except for red pepper. Mix with cold potatoes. Top with diced red pepper. Serve in blue bowl. *Serves 4.*

POTATO-CHEESE CASSEROLE

3–4 medium potatoes
4 tablespoons butter
½ cup half-and-half or light cream
8 ounces yellow cheddar, Monterey jack, or havarti cheese
2 tablespoons chopped parsley
2 tablespoons chopped dill
Dash paprika

Peel, cube, and boil potatoes. Drain and place in casserole dish. Melt butter in half-and-half. Cut cheese into small cubes. Mix all ingredients with potatoes except for paprika. Cover and bake in preheated 350° F. oven for 20 to 25 minutes. Uncover. Sprinkle with paprika and return to oven at 425° F. for 3 to 5 minutes. *Serves 4.*

PARSLEY POTATOES

16–20 small #2 new white or red potatoes
20–24 sprigs curly parsley, chopped
¼ pound butter or margarine
½ teaspoon salt
¼ teaspoon freshly ground black pepper

Wash potatoes and boil whole. Drain. Add remaining ingredients. Mix. Let stand over low heat for 1 to 2 minutes. Serve red potatoes in white or glass bowl. Serve white potatoes in blue or brown bowl. *Serves 3 to 4.*

POTATO OMELETTE

2–3 medium red or white potatoes
6 eggs
½ red pepper, chopped
1 small yellow onion, chopped
6–8 sprigs parsley, chopped
¼ teaspoon salt
Dash freshly ground black pepper
4 tablespoons butter or margarine
3–4 scallions, chopped

Wash and dice potatoes. Boil until soft. Beat eggs. Sauté all ingredients except potatoes and scallions in butter. When tender, lower heat and add beaten eggs. Cover. Allow to cook three-quarters through. Add potatoes and scallions. Fold over. Cook without cover. Turn and complete cooking. Serve from platter, with lettuce, sliced tomato, and sliced Spanish onion. *Serves 4.*

CANDIED YAMS

> 4 medium yams or red sweet potatoes
> 4 tablespoons butter or margarine
> ½ cup maple syrup
> ¼ cup brown sugar
> ½ cup chopped walnuts

Boil yams until soft. Drain. Cool. Peel. Combine butter, maple syrup, brown sugar, and ¼ cup water. Bring to a boil. Thickly slice yams and lay in shallow baking pan. Pour maple mixture over yam slices. Sprinkle with nuts. Bake in preheated 350° F. oven for 20 minutes.

BAKED YAMS

Select yams of any size. A round yam or sweet potato takes longer to bake than a long yam or sweet potato. An average 8-ounce yam bakes in 30 to 40 minutes in a preheated 350° F. oven. After baking, split the skin and push in from either end, creating a perfect pocket for butter, honey, maple syrup, chopped nuts, raisins and dates in butter, or just plain nothing.

WHITE SWEET POTATOES

These potatoes may be scrubbed, boiled whole, and allowed to cool, then peeled and served with butter, honey, maple syrup, raisins cooked in water to make a thick syrup, or hot orange marmalade. They may also be baked for 30 minutes in a preheated 350° F. oven, then at 425° F. for 5 minutes more.

MASHED YAMS WITH APRICOT PURÉE

> 2 pounds yams or red sweet potatoes
> 1 pound fresh apricots, 1 small can apricots, or 15–18 dried apricots
> ½ cup orange juice
> 2 tablespoons honey or maple syrup
> Cinnamon

Boil yams until soft. Drain. Cool. Peel. Remove stones from fresh apricots and steam until very soft; dried apricots must also be steamed until soft. Purée apricots, orange juice, and honey in blender until smooth. Mash yams and place in loaf pan. Pour apricot purée over yams and bake in a preheated 350° F. oven for 20 minutes. Sprinkle with cinnamon. *Serves 4 to 6.*

FRIED YAM CIRCLES

4–6 large round yams or red sweet potatoes
4 tablespoons butter or margarine
¼ cup brown sugar

Boil yams until soft. Drain. Cool. Peel. Slice into thick circles. Heat butter in skillet. Place yam slices in skillet. Sprinkle with brown sugar as they brown. Serve hot, with eggs, grits, sausages, greens, or turkey. *Serves 6 to 8.*

BREADED YAM CAKES

2 cups mashed cooked yams
¼ cup flour
2 tablespoons brown sugar
Bread crumbs or cornmeal
4 tablespoons butter

Mix yams with flour and brown sugar. Form into patties or cakes. Dust with bread crumbs. Heat butter in skillet. Cook yam cakes over medium heat until brown. *Serves 4.*

WINTER SQUASH

1. HUBBARD
2. BUTTERNUT
3. ACORN   4. BUTTERCUP
5. GOLDEN NUGGET   6. SPAGHETTI
7. SUGAR PIE PUMPKIN
8. JACK-O'-LANTERN
PUMPKIN

*L*ong before the time of supermarkets, refrigeration, and year-round fresh produce it was imperative that the family fall harvest include food that would see the folks through long nonproductive winter months. In addition to grains, salted or smoked meats, and cheese, winter food supplies included apples, potatoes, and a wide variety of colorful garden squashes with hard, thick rinds that preserved and protected the fruit inside. (Even zucchinis, if allowed to reach maturity, will develop a thick skin that will preserve them through much of the winter.) These winter squashes—relatives of melons, gourds, and cucumbers—created themselves in different sizes, shapes, and colors, yet all retain the same basic characteristics of taste and nutrition and can be prepared in pretty much the same fashion.

Winter squashes are available in abundance throughout the fall and are the least expensive of all vegetables during their harvest period. Throughout the harvest months of September, October, and November many roadside and country farm stands feature the complete variety of winter squashes for as little as ten cents a pound. It is well worth while to stock up on several baskets of mixed varieties and let your culinary imagination prepare them for your family throughout the winter. It's also a good idea to take a ride to the country for a day or weekend.

Although winter squashes can be individually prepared in a variety of recipes from soup to bread, all kinds can be prepared in the same manner. The squash is split in half, quartered, or cut into pieces; seeded; placed cut side down in one-half inch of water, and baked at 350° F. until almost soft. The squash is then removed and sprinkled with cinnamon, brown sugar, maple syrup, honey, tamari sauce, chopped bacon, or butter and cheese and broiled or baked at 450° F. for just a few minutes. Smaller squashes such as acorn or Golden Nugget become individual servings, while larger squashes like hubbard, spaghetti and big butternut can be cut and served in sections. Winter squashes can also be stuffed with sautéed vegetables or rice and baked.

Some winter squashes like hubbard, big butternut, and the small Sugar Pie pumpkins make excellent pie fillings for traditional pumpkin pie. All squash seeds can be removed, seasoned, and roasted for a highly nutritious and delicious family snack, but pumpkin seeds are the biggest and the best-tasting and give you the most for your money.

Squashes seem to hybridize themselves with incredible ease; their forms and shapes seem endless. The most popular and widespread varieties are acorn, butternut, and hubbard squashes. There are many other varieties, however, that appear in local areas from surrounding farms and never leave town.

NUTRITIONAL DATA

Weight (g)	Calories	Protein (g)	Carbohydrates (g)	Crude Fiber (g)	Water (g)	Fat (g)	Cholesterol (mg)
100	63	1.8	15.4	1.8	81.4	0.4	—
100	38	1.1	9.2	1.4	88.8	0.3	—

Vitamin A (IU)	Vitamin C (mg)	Thiamine B-1 (mg)	Riboflavin B-2 (mg)	Niacin (mg)	Vitamin B-6 (mg)	Vitamin B-12 (mcg)	Folic Acid (mcg)	Pantothenic Acid (mg)
4200	13	0.05	0.13	0.7	—	—	—	—
3500	8	0.04	0.10	0.4	—	—	—	—

Sodium (mg)	Potassium (mg)	Calcium (mg)	Phosphorus (mg)	Magnesium (mg)	Iron (mg)	Zinc (mg)	Copper (mg)	Manganese (mg)
1	461	28	48	—	0.8	—	—	—
1	258	20	32	—	0.5	—	—	—

Baked—½ cup
 Boiled—⅖ cup mashed

HUBBARD SQUASH

The hubbard squash is the least pretty of the winter squash family. It has a hard, thick rind with bumpy skin and comes in blue, gray, green, or orange. The flesh, however, is always orange and very sweet-flavored. An unbruised fresh hubbard can be stored in a quiet, cool place for months and still taste sweet and good. An average hubbard squash can weigh twenty pounds. Most fruit-stand vendors will cut and sell hubbards in halves or quarters, and even cut they keep well in your refrigerator.

Hubbard squash can be baked and seasoned sweetly or spicily. It can be baked and added to soup, or blanched and added to a vegetable mix. But I think hubbard is best in pumpkin pie. The hubbard squash has a thicker, firmer texture than fresh pumpkin and "sets up" easier. It has the same flavor as pie pumpkin but requires less sugar. (For my dough a great pumpkin pie is in the spice, the cream, the crust, and the baker.) Hubbard squash also sells for less money than pie pumpkins.

Hubbard squash is a highly nutritious food. It is easily digested, with little residue. It delivers enormous quantities of vitamin A, niacin, and phosphorus. Its vegetable protein content is high, along with its vitamin B-1 and B-2 content. More hubbard squash during the winter months will definitely help a family's health and food variety.

NUTRITIONAL DATA

Weight (g)	Calories	Protein (g)	Carbohydrates (g)	Crude Fiber (g)	Water (g)	Fat (g)	Cholesterol (mg)
205	103	3.7	24	3.7	174.5	0.8	—
245	74	2.7	16.9	3.4	223.2	0.7	—

Vitamin A (IU)	Vitamin C (mg)	Thiamine B-1 (mg)	Riboflavin B-2 (mg)	Niacin (mg)	Vitamin B-6 (mg)	Vitamin B-12 (mcg)	Folic Acid (mcg)	Pantothenic Acid (mg)
9840	21	0.1	0.27	1.4	—	—	—	—
10050	15	0.1	0.25	1	—	—	—	—

Sodium (mg)	Potassium (mg)	Calcium (mg)	Phosphorus (mg)	Magnesium (mg)	Iron (mg)	Zinc (mg)	Copper (mg)	Manganese (mg)
2	556	49	80	—	1.6	—	—	—
2	372	42	64	—	1.2	—	—	—

Baked—1 cup mashed
  Boiled—1 cup mashed

BUTTERNUT SQUASH

*I*f you have not made butternut squash one of your family's fall and winter foods, you have passed up one of the world's great food gifts. Butternut squash is cheap. It keeps for months and months. It is easy to prepare. It is smooth, creamy, and delicious. And it's beautiful.

Butternut squash has a large, round fleshy bottom that encloses the seeds, and a cylindrical upper part that is solid flesh. The skin is smooth, with a deep butterscotch color. The best-tasting butternut has no traces of green. The flesh is deep orange, with a distinctive butterscotch flavor that most people find surprisingly delicious. I am never surprised anymore. I love butternut squash and cook it often, always simply.

Very small butternuts (like small zucchini, small potatoes, small onions, small peas) are very sweet. They make excellent single servings cut in half, cooked until soft, then baked with a topping of butter and maple syrup. Sometimes I add chopped walnuts. Oh, God, is that good! The large butternuts can be served in sections after baking, or the flesh can be removed from the shell, then mashed or cut up, and browned in the oven or broiler.

Butternut squash is available year-round from Florida and the Southern farm states. There are plentiful supplies, but prices tend to climb higher as winter drags on. It really pays to buy a bushel or two in the fall during the harvest and keep them all winter long. You'll love the idea when you see them in the store for eighty-nine cents a pound and you have them at home for a dime.

In addition to its superb texture and taste, the butternut squash delivers extraordinary nutrition. It is high in vegetable protein and extremely high in vitamin A, niacin, calcium, potassium, phosphorus, and iron. There is every reason to make butternut squash a regular family member.

NUTRITIONAL DATA

Weight (g)	Calories	Protein (g)	Carbohydrates (g)	Crude Fiber (g)	Water (g)	Fat (g)	Cholesterol (mg)
205	139	3.7	35.9	3.7	163.2	0.2	—
245	100	2.7	25.5	3.4	215.1	0.2	—

Vitamin A (IU)	Vitamin C (mg)	Thiamine B-1 (mg)	Riboflavin B-2 (mg)	Niacin (mg)	Vitamin B-6 (mg)	Vitamin B-12 (mcg)	Folic Acid (mcg)	Pantothenic Acid (mg)
13120	16	0.1	0.27	1.4	—	—	—	—
13230	12	0.1	0.25	1	—	—	—	—

Sodium (mg)	Potassium (mg)	Calcium (mg)	Phosphorus (mg)	Magnesium (mg)	Iron (mg)	Zinc (mg)	Copper (mg)	Manganese (mg)
2	1248	82	148	—	2.1	—	—	—
2	835	71	120	—	1.7	—	—	—

Baked—1 cup mashed
Boiled—1 cup mashed

ACORN SQUASH

Shaped like a giant acorn with a definite pointed end, the acorn squash is the most famous of all the winter squashes. Its popularity is responsible for its year-round availability, although most folks tend to minimize its use in the summer, when there is such an abundance of fresh green and leafy vegetables.

Both the green and golden varieties of acorn squash have a definite nut taste to their flavor, although the golden variety tends to be a little sweeter and the green more moist. Both are at their best when they have hard shiny skins, although late in the storage season they tend to have a softer dull surface due to partial dehydration. As I've said, I recommend a trip to the country during harvest time to bring home several bushels of mixed squash, especially acorn. It is so good.

Select green acorn squashes with a splash of orange or yellow, but golden acorns are best when they are completely orange. Cook acorn squashes like other varieties, simply and in the oven. Other than that, acorn squash requires little more than a touch of butter or cinnamon to enhance its natural good taste. A baked acorn-squash half with sprightly seasonings makes a delightful single-serving vegetable for any wholesome meal.

Acorn squash is loaded with minerals and contains higher-than-average amounts of vitamins A, C, and B. It is easily digested and low in carbohydrates.

NUTRITIONAL DATA

Weight (g)	Calories	Protein (g)	Carbohydrates (g)	Crude Fiber (g)	Water (g)	Fat (g)	Cholesterol (mg)
195	86	3	21.8	3.5	161.7	0.2	—
245	83	2.9	20.6	3.4	219.8	0.2	—

Vitamin A (IU)	Vitamin C (mg)	Thiamine B-1 (mg)	Riboflavin B-2 (mg)	Niacin (mg)	Vitamin B-6 (mg)	Vitamin B-12 (mcg)	Folic Acid (mcg)	Pantothenic Acid (mg)
2180	20	0.08	0.20	1.1	—	—	—	—
2700	20	0.1	0.25	1	—	—	—	—

Sodium (mg)	Potassium (mg)	Calcium (mg)	Phosphorus (mg)	Magnesium (mg)	Iron (mg)	Zinc (mg)	Copper (mg)	Manganese (mg)
2	749	61	45	—	1.7	—	—	—
2	659	69	49	—	2	—	—	—

Baked—½ medium
  Boiled—1 cup mashed

BUTTERCUP SQUASH

*T*he Buttercup squash is not found everywhere, though it should be. It has a shape like a fat small wheel. It is dark green with white stripes and a gray protrusion emerging from the blossom end. The skin is covered with raised rough spots. Buttercups are best when they are hard and semiglossy. Buttercup flesh is yellow, almost pale, but is sweet and smooth textured. Buttercups retain their natural moisture and are excellent when stuffed and baked whole.

GOLDEN NUGGET SQUASH

*C*losely related to the acorn variety, the Golden Nugget is nevertheless its own squash. It is very sweet, deep-orange in color inside and out, and mostly available in small, single-serving sizes, which makes it delightful to serve. Golden Nuggets can be opened like small pumpkins (which is exactly what they look like), scooped clean, brushed inside with butter and seasonings, and baked whole. They can also be split, baked cut side down in water, and prepared exactly like acorn squash. Golden Nugget squash does not keep as well as the other winter varieties. Its golden color is evidence of its ripened maturity, and it really tastes best eaten fresh.

When buying Golden Nugget squash, select hard, shiny specimens without soft or wrinkled ends. They are best when the stem is green and still flexible.

SPAGHETTI SQUASH

Spaghetti squash has to be one of the easiest squashes to prepare and the most fun to serve. Select large white spaghetti squashes with tender skin and little yellowing. Spaghetti squash is at its best in the summer and by late fall begins to harden up. By winter the skin is a tough shell, the stem hard and dry, and the meat less moist but stringier. After cutting the squash in half and removing the seeds, place it cut side down in a half inch of water and bake at 350° F. until it is soft. After it cools, scoop out the soft white meat with a fork into a bowl. *Voila!* Spaghetti in squash form, or squash in spaghetti form. Add your tomato sauce or butter and cheese and enjoy. Spaghetti squash is also excellent served with a fresh pesto sauce or mixed with shrimp, clams, and tomato sauce and rebaked.

SUGAR PIE PUMPKINS

The Sugar Pie pumpkin is different from the American Jack-o'-Lantern pumpkin. The Sugar Pie has a thicker, deeper-colored flesh with a tight, smallish seed pocket. Its taste is sweeter and its texture firmer. Although it can be cooked like a squash, it is best when used for pumpkin-pie filling. Sugar Pie pumpkins also make delicious soup. Select hard, shiny pumpkins that feel dense and heavy.

Sugar Pie pumpkins lead all fruits and vegetables in their vitamin A content. There is more vitamin A in a Sugar Pie than we could possibly ever use. Sugar Pies are very low in calories and carbohydrates, with one of the most complete ranges of nutrients found in all of the fresh-food world. It is no wonder that pumpkins were one of the three basic foods, along with corn and beans, that sustained so many North American Indian civilizations. It's a pity that we only seem to find it in pies loaded with sugar, butter, flour, and cream.

Sugar Pies are the first pumpkins to be harvested. They are available by the third week in September or earlier, and many stores have high-quality inventories as late as Thanksgiving. Sugar Pies keep very well if they are free from bruises and skin scratches and if they retain their stems.

NUTRITIONAL DATA

Weight (g)	Calories	Protein (g)	Carbohydrates (g)	Crude Fiber (g)	Water (g)	Fat (g)	Cholesterol (mg)
100	33	0.9	7.9	1.4	90.4	0.3	—

Vitamin A (IU)	Vitamin C (mg)	Thiamine B-1 (mg)	Riboflavin B-2 (mg)	Niacin (mg)	Vitamin B-6 (mg)	Vitamin B-12 (mcg)	Folic Acid (mcg)	Pantothenic Acid (mg)
33990	5	0.02	0.06	0.4	0.06	—	—	0.4

Sodium (mg)	Potassium (mg)	Calcium (mg)	Phosphorus (mg)	Magnesium (mg)	Iron (mg)	Zinc (mg)	Copper (mg)	Manganese (mg)
5	219	18	36	25	0.74	0.17	0.11	—

Cooked—²⁄₅ cup

JACK-O'-LANTERN PUMPKINS

Only in America. Although pumpkins are grown in other parts of the world, only the United States uses them to celebrate and symbolize Halloween. Jack-o'-Lanterns are oversized hybrid versions of the true pumpkin. They have a thin wall with a huge seed pocket that is mostly empty or filled with airy fibrous seed tissue. When selecting your Jack-o'-Lantern pumpkin, choose one with no bruises or soft spots. It can be green or greenish; green Jack-o'-Lanterns will turn orange and probably last longer. Don't handle the pumpkins by their stems—they break off. A jack-o'-lantern without a stem is like a Christmas tree without a star.

Almost all Jack-o'-Lanterns are used to carve faces. The following is the best technique for carving a face in a way that helps the jack-o'-lantern look best and last longest.

How to carve a jack-o'-lantern:
Cut a star-shaped or many-pointed lid around the stem (a circular lid shrinks and falls into the hole after a couple of days). Cut the lid at an angle so that it's bigger at the skin surface, smaller on the inside.

Remove all fibrous material and seeds from within, scraping the inner walls with a spoon. Scoop out a small space at the bottom for a glass cup to hold a candle. Save the seeds.

Carve the mouth first. The mouth determines the personality and size of the face and should be cut on the front surface of the pumpkin; cutting it under the curve causes the face to collapse.

Cut out the nose and eyes. Laugh lines, wrinkles, and scowls can be carved into the outer rind, taking away only a portion of the yellow meat but not cutting through the pumpkin.

Place a short fat candle in a glass candle cup, light it, and set inside the bottom of the jack-o'-lantern. Then replace the lid and set the jack-o'-lantern outside.

PUMPKIN SEEDS

After cleaning a pumpkin, separate the seeds from the fibrous tissue. Do not wash. Take a cookie sheet and brush it with butter or margarine. Sprinkle the pumpkin seeds generously on the sheet and season with salt, garlic or onion powder, chile powder, black pepper, oregano, or whatever your favorite seasoning. Place in a preheated 375° F. oven for ten to fifteen minutes or until toasted. Remove, pour into a brown paper bag, let cool, and serve in a bowl.

Pumpkin seeds are very high in usable protein and contain valuable amounts of minerals.

NUTRITIONAL DATA

Weight (g)	Calories	Protein (g)	Carbohydrates (g)	Crude Fiber (g)	Water (g)	Fat (g)	Cholesterol (mg)
28	155	8.1	4.2	0.5	1.2	13.1	—

Vitamin A (IU)	Vitamin C (mg)	Thiamine B-1 (mg)	Riboflavin B-2 (mg)	Niacin (mg)	Vitamin B-6 (mg)	Vitamin B-12 (mcg)	Folic Acid (mcg)	Pantothenic Acid (mg)
—	20	0.07	0.05	0.7	—	—	—	—

Sodium (mg)	Potassium (mg)	Calcium (mg)	Phosphorus (mg)	Magnesium (mg)	Iron (mg)	Zinc (mg)	Copper (mg)	Manganese (mg)
—	—	14	320	—	3.14	—	—	—

Raw—1 ounce

Mother Hubbard's Pineapple Squash

¼ hubbard squash
2½ cups diced pineapple
2 tablespoons sugar
½ cup chopped walnuts
1 teaspoon cinnamon
4 tablespoons butter or margarine

Seed, cut up, and boil squash. Drain. Cool. Peel and mash. Cook pineapple with sugar until soft. Crush with fork. Mix pineapple, squash, walnuts, cinnamon, and butter and place in buttered baking dish. Bake in preheated 350° F. oven for 20 minutes. *Serves 4.*

Butternut Boats

2 medium Butternut squashes
½ cup chopped walnuts, hazelnuts, or pecans
4 tablespoons butter or margarine
2 teaspoons cinnamon

Cut squashes in half. Seed and place cut side down in half inch of water in shallow baking dish. Bake in preheated 350° F. oven for 30 minutes or until soft. Cool slightly. Scoop out flesh. Mash with remaining ingredients. Replace in shells and place in baking dish, filling up. Bake at 400° F. for 5 to 10 minutes.

Baked Butternut halves may be served simply baked in their skins, and butter, toppings, spices, and seasonings may be added at the table. *Serves 4.*

Acorn Halves

1 acorn squash
1 tablespoon butter or margarine
1 teaspoon cinnamon

Cut squash in half. Seed and place cut side down in one-quarter inch of water in shallow baking dish. Bake in preheated 350° F. oven for 30 minutes or until soft. Remove from oven. Place cut side up. Dot with butter and sprinkle with cinnamon. Place under preheated broiler for 3 to 5 minutes. *Serves 1 or 2.*

Stuffed Sweet Buttercup Squash

3–4 small Buttercup squashes
2 stalks celery, chopped
1 medium yellow onion, chopped
6–8 sprigs flat-leaf parsley, chopped
2–3 sprigs tarragon or thyme, chopped
4–6 tablespoons corn kernels
¼ cup raisins
4 teaspoons pignola nuts
Light vegetable oil

Remove center gray knobs from blossom ends of squashes, making caps. Also remove stems. Seed. Place squashes, open ends down, and knob caps in half inch of water in baking dish. Bake in preheated 350° F. oven until soft. Sauté remaining ingredients in thin layer of oil. When squash cools enough to handle, fill with sauté. Cap and replace in oven at 350° F. for 10 to 12 minutes. Serve individually.

Stuffed Golden Nugget Squash

4 Golden Nugget squashes
2 stalks celery, chopped
1 green pepper, diced
4 scallions, chopped
1 yellow onion, chopped
2–3 cloves garlic, chopped
6–8 sprigs parsley, chopped
¼ teaspoon salt
¼ teaspoon freshly ground black pepper
1 cup cooked rice
Light vegetable oil

Cut circle around stem of each squash and remove cap. Seed. Place squashes, open ends down, and caps in one-quarter inch of water in baking dish. Cook in preheated 350° F. oven for 20 to 30 minutes or until soft. Sauté remaining ingredients except for rice in light vegetable oil. Mix with rice. Cool squash. Stuff, replace caps, and bake at 350° F. for 10 to 12 minutes. Serve individually.

SPAGHETTI SQUASH AMERICANA

1 large spaghetti squash
4 ounces butter
4 ounces Parmesan cheese, grated
4 ounces Monterey jack or cheddar cheese, shredded
3–4 sprigs basil, chopped
Salt
Freshly ground black pepper

Split and seed squash. Bake cut side down in half inch of water in baking dish in preheated 350° F. oven for 20 to 30 minutes or until soft. Melt butter and mix with cheeses. When squash is cooked, remove spaghettilike strands with a fork. Mix in bowl with cheese mixture. Sprinkle chopped basil on top. Season with salt and pepper to taste. *Serves 4.*

SPAGHETTI SQUASH ITALIANO

1 large spaghetti squash
1 pound ground beef (optional)
Olive oil
1 medium yellow onion, chopped
3 cloves garlic, coarsely chopped
6–8 sprigs flat-leaf parsley, chopped
½ teaspoon dried oregano
¼ teaspoon salt
¼ teaspoon freshly ground black pepper
1 cup tomato sauce

Split and seed squash. Place cut side down in half inch of water in shallow pan. Bake in preheated 350° F. oven for 20 to 30 minutes or until soft to the fork. Brown ground beef in very little olive oil with remaining ingredients except for tomato sauce. Heat sauce over low heat. Add ground beef when browned and crumbly. Simmer over low heat.

When squash is cooked, cool slightly. Remove spaghettilike strands of squash with fork and place in bowl. Top with tomato-meat sauce. *Serves 4.*

TRADITIONAL PUMPKIN PIE

Pastry for 8-inch pie shell
2 eggs
1 cup mashed cooked Sugar Pie pumpkin
1 cup mashed cooked hubbard squash
1 cup half-and-half or evaporated milk
½ cup sugar
½ teaspoon grated fresh ginger
½ teaspoon ground cloves
1 teaspoon cinnamon
¼ teaspoon salt

Preheat oven to 425° F. Line 8-inch pie pan with pastry. Beat eggs and combine with remaining ingredients. Pour into pastry shell. Bake for 15 minutes. Reduce oven temperature to 350° F. and bake for 35 to 45 minutes more. Remove pie when knife blade inserted near center of pie comes out clean. Cool.

PUMPKIN SOUP

4 cups cut-up peeled Sugar Pie pumpkin
4 tablespoons butter or margarine
2 carrots, cut up
2 stalks celery, cut up
1 medium yellow onion, quartered
4 sprigs parsley, chopped
¼ teaspoon chopped fresh thyme
1 bay leaf
½ teaspoon salt
Dash freshly ground black pepper
1 quart chicken stock
½ cup dry white cooking wine

Cook pumpkin in butter in large saucepan for 3 to 4 minutes. Add ¼ cup water and carrots. Cover. Simmer until pumpkin is tender. Simmer remaining ingredients except for wine in chicken stock in separate covered saucepan for 30 minutes. Remove celery, onion, and bay leaf with slotted spoon and discard. Blend pumpkin and carrots in blender until smooth. Stir into broth. Simmer several minutes more. Add wine after removing from heat. Stir. Cover. Let stand for 2 to 3 minutes. *Serves 4.*

WINTER SQUASH

BULBS AND ONIONS

1. SPANISH ONIONS
2. BERMUDA WHITE ONIONS
3. ITALIAN RED ONIONS
4. VIDALIA ONIONS
5. YELLOW GLOBE ONIONS
6. WHITE BOILING ONIONS
7. PEARL ONIONS
8. GIANT ITALIAN GARLIC
9. MEXICAN PURPLE GARLIC
10. SHALLOTS
11. LEEKS
12. GREEN ONIONS

A kitchen without onions is a furnace without fire. There is nothing in the cook's repertoire as invaluable as the onion. Onions were the people's food in Egypt. The Renaissance found great medicinal qualities in the onion. Today onions are produced in great quantities in every agricultural society in the world.

The onion has a kitchen use from its earliest stage as an unformed bulb to its dried and powdered forms with an unlimited life. Onions are eaten raw, boiled, roasted, broiled, sautéed, juiced, breaded, sliced, chopped, minced, fried. They are added to soups, meat loaf, hamburgers, salad dressings, vinegar, marinades, omelettes, quiches, dips, crusts, pies, sandwiches, and salads.

Onions are available year-round in wide, colorful variety and great quantity everywhere and always at a modest price. There are onions so strong that tears run at the first slice and onions so sweet that you can eat them like apples. For all the old wives' tales, there is no way to avoid tears from strong onions. A touch of lemon juice in baking soda will, however, remove their odor from your hands, towels, or cooking utensils.

The outside skins of properly cured onions become papery and dry and, in the trade, are called shells. Old-time stores and modern packers run their onions over long slatted inclines to remove the shells and reveal the shiny coppery onion itself. Look for these hard, shiny onions with that copper color. Avoid onions that are sprouted or have soft bottoms, black mold, or brown outer rings. Your kitchen is best served when you keep a variety of onions on hand to be used to create different tastes and cooking results. Although there are some eighty or more specific varieties with subtle differences, the classifications I will describe include all of the inventory available from most fruit stands and markets.

Onions are nutritionally modest. Their chief recommendation is that they enhance the flavor and the pleasure of the more nutritious foods they accompany.

NUTRITIONAL DATA

Weight (g)	Calories	Protein (g)	Carbohydrates (g)	Crude Fiber (g)	Water (g)	Fat (g)	Cholesterol (mg)
100	38	1.5	8.7	0.6	89.1	0.1	—
100	29	1.2	6.5	0.6	91.8	0.1	—

Vitamin A (IU)	Vitamin C (mg)	Thiamine B-1 (mg)	Riboflavin B-2 (mg)	Niacin (mg)	Vitamin B-6 (mg)	Vitamin B-12 (mcg)	Folic Acid (mcg)	Pantothenic Acid (mg)
40	10	0.03	0.04	0.02	—	—	—	—
40	7	0.03	0.03	0.2	—	—	—	—

Sodium (mg)	Potassium (mg)	Calcium (mg)	Phosphorus (mg)	Magnesium (mg)	Iron (mg)	Zinc (mg)	Copper (mg)	Manganese (mg)
10	157	27	36	12	0.5	—	—	—
7	110	24	29	—	0.4	—	—	—

Raw—1 medium
  Cooked—½ cup

SPANISH ONIONS

The Spanish onion is a large round or slightly oval yellow onion with a deep-coppery outside shell. The Spanish onion is semisweet, mild, and good eaten raw. I like to cut big thick slices of Spanish onions and lay them on coldcut sandwiches, or I break them apart and make a big bowl of onion rings. Chopped Spanish onions are excellent in sautés because they do not overpower the main flavor of the recipe.

For most of the year Spanish onions are sold out of storage. By late spring they are offered for sale in less-than-good condition, often with soft ends, brown middles, and peeled or wet skins. These are signs of old age or poor storage. Always pass on junk unless it is priced low for immediate sale and consumption. By late spring fresh Spanish onions arrive from Chile and Argentina. Mexico also ships some. Texas ships lots of onions by summertime. The best Spanish onions arrive in the fall, around harvest time, from local sources.

Look for hard, deep-copper, shiny onions with very dry papery shells. Hard, big, and shiny is the key to a great Spanish onion.

BERMUDA ONIONS

The Bermuda onion is a large round or oval white onion with a dry white shell and dark-green or gray stripes running vertically from the root to the stem. Unless they have been in storage for a long while, Bermuda white onions are very sweet and can be eaten raw.

Bermuda whites are used in salads or, thickly sliced, for cold meat sandwiches and, though mild, make excellent batter-fried onion rings. Bermuda onions can be prepared with broiled or poached fish, as they don't spoil the delicate taste of seafood.

The fresh Bermuda white onion is more susceptible to black mold than any other onion. This is due to its high sulphur content. Mold occurs when they have been in storage long or in a damp place. When you buy Bermuda onions, make sure they are hard and very white. Soft stem ends often mean an inedible black core. Like all winter onions, Bermudas are sold out of storage. The fresh crops arrive in late summer and early fall. Good, hard, sweet specimens are available through December before their weakness as poor keepers starts to show. As with so many other fruits and vegetables, it's best to buy them at their best.

RED ONIONS

Red onions are always sweet unless they have been stored too long under damp conditions. They are best when they are hard, with deep-maroon dry shells and shiny first layers. Some are flat, some are round, some elongated. Great numbers are imported from Italy during our midsummer. The Italian varieties are round or oval and are often available braided by their long dry leaves. These onion strings are both beautiful and useful. Hanging keeps them exposed to open air, preventing mold, and their braided tresses help them retain their moisture and hardness.

The flat California varieties arrive in the late summer and are always very, very hard because their stems are cured quickly in the dry California sun and little moisture is lost. Red onions are best for salads, either chopped or ringed, and are excellent sliced in sandwiches. Fresh-picked small ones can be eaten whole like a small apple with a minimum of damage to your social life. I like to sit down in the afternoon with a pile of roasted tortilla chips, an avocado, a tomato, a cucumber, and a whole little red onion and chomp away. That particular combination of foods is a wonderful group of contrasting textures, flavors, and spiciness.

The best season for red onions is early summer for the Italian imports, and early fall through December for the California varieties. Avoid red onions that have been peeled to an inner layer, those with double bulbs (they taste too hot to me), or any that are soft at the stem end. I also think big red onions are stronger than small ones, and round ones sweeter than flat. I have no proof, but that's what I buy and eat.

VIDALIA ONIONS

Vidalia onions need be mentioned here only because of their extraordinary sweetness. They grow only in the small town of Vidalia, Georgia, and have a limited supply and distribution. They are, however, the most delicious onion yet produced anywhere, though no one knows the reason that this particular locale would produce such a phenomenon. Vidalia onions are coppery, elongated, and very hard. They are available in early to midsummer and are usually snapped up by aficionados as they arrive at the stands. Vidalia onions have unlimited use, but they are mostly eaten raw and in salads. Many are also pickled to extend their availability through the year. Very few stores carry this variety, but one never knows, do one? If you see them, buy them.

YELLOW GLOBE ONIONS

The yellow globe onion is a medium to smallish round onion with a bright copper-yellow shell and white-yellow first layer. It is the common kitchen onion, and there are many varieties, all closely resembling each other, in this large general category. Yellow globe onions are very flavorful and quite strong. They make excellent sautés and hold their flavor, shape, and texture when steamed, simmered, or fried with other vegetables. Small yellow globe onions can be peeled, parboiled, and placed in with roasts and fowl or chopped and spread on steaks and other broiled or roasted red meats, including fresh calves' or beef liver. They always impart the distinctive onion flavor to whatever they accompany, and that wonderful appetite-stimulating odor of frying onions is best achieved with the common yellow globe onion.

There is a flat variety of yellow onion that is a paler color and tastes slightly milder. I like these onions a lot, especially cooked whole with meats or chicken.

The yellow onion is an excellent keeper, always inexpensive, and readily available everywhere. These kitchen onions don't seem to suffer from storage as much as the big guys. Nevertheless, buy yellow onions that are hard and shiny, with crisp papery shells and no signs of wetness or black mold. Sometimes by late spring they look a little funky, but if they feel hard and still have paper left on their outsides, they'll serve you well.

PEARL ONIONS

There is the small white boiling onion and the pearl onion, which is also white but even smaller. Both possess the same qualities, but you'll find the pearl slightly milder and tastier. These small white varieties are best peeled and cooked whole, either served as a vegetable by themselves or with roasts, in shish kebabs, creamed, braised, or even deep fried. Pearl onions are often avoided because they are difficult to peel and are useless with their skins on. I recommend dropping them into boiling water for ten to fifteen seconds, draining, and cooling them. Cut the root end off, and the skins will slide off easily.

Pearl onions are at their best from late summer through the holiday season into deep winter. Look for hard, shiny onions with dry papery shells and no black mold.

GREEN ONIONS

Unlike the straight, slender scallion, the green onion is a true bulb. After onion seeds are planted or tiny bulbs set, the green-leafed shoots are pulled for thinning. These very white round green-veined onions are about the size of large marbles, with tender white roots and very long tubular green leaves. They are a most excellent vegetable in themselves, especially braised or charcoal-broiled after being brushed with or dipped in olive oil. Their mild, gentle onion flavor goes well with sharp spring salad greens such as dandelions, chicory, arugula, and watercress. Look for green onions, also referred to as spring onions and Japanese onions, in May or early June at farm stands or knowledgeable produce stands where the vendor is aware of the more interesting aspects of seasonal fruits and vegetables.

CREAMY PEARLS

12–16 pearl onions
4 tablespoons butter or margarine
¾ cup light cream or half-and-half
¼ cup instant-blending flour
3–4 sprigs parsley, chopped
¼ teaspoon salt
Dash freshly ground black pepper

Place onions in boiling water. Remove when soft but before the layers begin to separate. Drain. Cover. Melt butter in light cream. Lower heat. Add flour 1 teaspoon at a time, stirring constantly. Cook, stirring, until mixture thickens. Stir in remaining ingredients. Place onions in dark bowl. Pour on cream sauce.
Serves 3–4.

BRAISED PEARLS

12–16 pearl onions
3 ounces sesame or walnut oil
1 small piece ginger, grated
½ teaspoon salt
¼ teaspoon freshly ground black pepper
2 tablespoons tamari sauce

Blanch onions in boiling water. Heat oil, ginger, salt, and pepper in skillet. Place onions in skillet. Brown, turning gently. Add tamari sauce at end.   *Serves 3–4.*

ONION OMELETTE

1 large Spanish onion, diced
2 cloves garlic, chopped
6–8 sprigs flat-leaf parsley, chopped
¼ teaspoon salt

¼ teaspoon freshly ground black pepper
4 tablespoons butter or margarine
6 eggs, beaten

Sauté all ingredients in butter except eggs and half the onions. Beat eggs and pour over sauté. Cover. Heat over medium-low heat until omelette is three-quarters done. Add remaining onions. Fold over. Cook uncovered over medium heat. Flip over. Remove from heat. Cover. Let stand for 1 to 2 minutes. *Serves 4.*

BREADED ONION RINGS

2–3 cloves garlic, finely chopped
6–8 sprigs flat-leaf parsley, finely chopped
2 cups bread crumbs
½ teaspoon dried oregano
½ teaspoon salt
½ teaspoon freshly ground black pepper
Light vegetable oil
2 large Spanish onions, sliced medium-thick
2 cups milk

Combine all ingredients except oil, onions, and milk in flattish bowl. Heat thin layer of oil in skillet. Dip onion slices in milk. Dust with bread-crumb mixture. Fry in hot oil until crisp, controlling heat so oil does not smoke. Drain on paper towels. *Serves 4.*

SAUTÉED ONIONS AND RICE

4 tablespoons butter or margarine
1 large Spanish onion, chopped
2 small yellow onions, chopped
3 scallions, chopped
3–4 sprigs flat-leaf parsley, chopped
2 cups cooked rice
2 tablespoons tamari sauce

Melt butter in skillet. Sauté all ingredients except rice and tamari sauce until just brown. Lower heat and mix in rice. Sprinkle with tamari sauce and simmer. Stir before serving. *Serves 4.*

Onions and Peas

    1 medium yellow onion, diced
    2 tablespoons butter or margarine
    1 pound peas, shelled
    3–4 sprigs flat-leaf parsley, chopped

Soften onion in butter over medium heat. Add peas, parsley, and ¼ cup water. Cover. Lower heat. Cook for 3 to 5 minutes. Serve in bowl. *Serves 4.*

Sweet Red Onion Salad

    1 large red onion
    ½ red pepper, diced
    1 tablespoon finely chopped curly parsley
    Juice of 1 lemon
    2 tablespoons white-wine vinegar
    8–10 peppercorns
    1 bay leaf
    1 teaspoon sugar
    ¼ teaspoon salt
    ¼ teaspoon freshly ground black pepper

Slice onion into very thin rings. Mix and toss with other ingredients. Chill. Remove bayleaf. *Serves 3–4.*

Green-Onion–Avocado Lunch

    1 Hass avocado
    2–3 leaves Boston lettuce
    2–3 tablespoons Mexican salsa
    3–4 green onions

Cut avocado in half. Remove pit. Place halves on lettuce bed. Fill cavities with salsa and serve with green onions. *Serve individually.*

WHOLE ROAST ONIONS

>   4–6 medium yellow globe or small Spanish onions
>   2 teaspoons garlic powder
>   4 teaspoons salt
>   1 teaspoon freshly ground black pepper
>   4 tablespoons butter or margarine, melted

Blanch onions quickly in boiling water. Place two onions per square on aluminum foil squares. Sprinkle with garlic powder, salt, pepper, and melted butter. Wrap. Place in preheated 375° F. oven for 20 to 30 minutes. Unwrap and serve individually.

ITALIAN ONION-TOMATO SALAD

>   2 large red tomatoes
>   2 medium red onions
>   4 cloves garlic, chopped
>   6–8 sprigs flat-leaf parsley, chopped
>   8 basil leaves, chopped
>   1 tablespoon dried oregano
>   ¼ teaspoon salt
>   ½ teaspoon freshly ground black pepper
>   ¼ cup olive oil

De-stem, slice, and fan tomatoes out on plate. Slice one onion thinly and separate into rings. Scatter over tomatoes. Finely chop other onion and combine with remaining ingredients except for oil. Drizzle olive oil onto tomato and onion slices, then top with chopped-onion mixture. Serve with crusty bread and butter.
   Serves 4.

ITALIAN GARNISH

>   1 red onion, finely chopped
>   2–3 cloves garlic, finely chopped
>   6–8 sprigs curly parsley, coarsely chopped
>   8–10 basil leaves, coarsely shredded
>   2 tablespoons olive oil
>   ¼ teaspoon salt
>   ¼ teaspoon freshly ground black pepper

Mix all ingredients. Chill. Serve with salad, tomatoes, antipasto, Italian hero sandwiches, avocado, cheese, or cold vegetables.

GARLIC

It seems that of all the natural foods we utilize throughout the world, garlic is the oldest and most widely used and creates the greatest reactions from its lovers and haters alike. It is both legend and history. It possesses esoteric and practical values equally. Vampires dread it, Mohammed recommended it, the early Olympians relied on it, and Greek and Italian housewives wore it around their necks to protect their children, preserve their fertility, and keep away the common cold. The entire economy of Gilroy, California, is founded on garlic. It is the central focus of the annual Gilroy Garlic Festival, featuring a world of food, all of which contains some form of garlic.

When cooking garlic, be careful not to cook it beyond golden brown. Overcooking increases its acidity and sulphur content, creating bitterness and an unpleasant aftertaste. Garlic should be carefully chopped or sliced or crushed with mortar and pestle. Metal garlic presses or food processors bruise the garlic badly, causing bitterness and undue sharpness. Garlic should be selected for freshness. Avoid heads with small green sprouts, soft cloves, or mold, which indicate dryness or age. Heads that break apart easily are not fresh and have already lost some of the essential oils and moisture that give garlic that special quality.

Garlic is garlic, and all of the varieties cook, taste, and smell basically the same. The differences, however, are genuine. California white-shell garlic seems to peel quite easily, especially if it is allowed to stand several days in your kitchen. Excellent white-shell garlic is grown in northern California, and the finest examples are often braided into strings that retain the cured long garlic stem. These garlic braids retain their freshness for long periods and are both beautiful and handy hanging in the garlic-lover's kitchen.

Mexican purple-shell garlic peels with great difficulty because the cloves are small. But it possesses an extraordinary amount of garlic oil, which prevents it from burning quickly in hot oil. This makes it especially desirable in stir-fry cooking.

Elephant garlic has very, very large wide cloves. It peels easily and has a milder flavor than the other varieties. It is excellent for slicing into very large pieces and serving raw with sliced tomatoes, basil, and Italian dressing.

Imported Italian garlic is what garlic is really all about. No one knows more about garlic than the Italians. It is found in almost every meat, pasta, vegetable, salad, and soup dish in the Italian kitchen. Italian garlic is very white shelled, with hard wide cloves. Italian garlic is shipped loose and braided, both styles the finest examples of Italian garlic production. Look for this import at better fruit stands, where quality and soulful appreciation of fine produce is applied to the business of fresh fruits and vegetables.

In addition to the fresh garlic available year-round from Mexico, California, and Italy, there are excellent specimens shipped in late winter and spring from Chile and Argentina. They are usually large, very white, and easily peeled, with a high oil content and outstanding flavor. Like the Italian varieties, however, they are usually available in the better, more select stores operated by experienced produce-vendors with an appreciation of finer merchandise.

Actually, garlic keeps very well under proper conditions, and with imports available year-round, the American consumer has a constant fresh garlic supply from which to choose. The very best selections, however, are available in the early fall or late summer, when Californian, Italian, and Mexican varieties are all available in abundance. When you buy garlic, make sure it is hard, with crisp papery shells. Look at the cloves, not just the whole head. Some very small heads might have very large cloves, perhaps only six or seven. I always look for this kind first. I never buy garlic in those little boxes with two heads peeping out of their cellophane window. They always look old and weak.

Garlic is not considered a nutritionally powerful food, but it does contain extraordinary amounts of vegetable protein and phosphorus. It is an excellent addition to vegetable-juice combinations and is included in all natural-food diets as a blood-cleanser and a vital aid to liver rehabilitation.

GIANT WHOLE GARLIC CLOVES

    8–10 large cloves elephant garlic
    4 tablespoons butter
    1 teaspoon chopped parsley
    1 teaspoon dried onion flakes
    Dash freshly ground black pepper

Blanch garlic cloves in boiling water and slip off their skins. Melt butter and simmer with remaining ingredients. Add blanched garlic cloves. Cover. Let stand 5 to 8 minutes over low heat. Serve in small bowl as condiment.

GARLIC BREAD

    8 cloves garlic, crushed
    6 ounces butter
    1 teaspoon onion powder
    ½ teaspoon dried oregano
    ½ teaspoon salt
    ½ teaspoon freshly ground black pepper
    1 loaf Italian bread
    2 cloves garlic, chopped

Mix all ingredients except bread and chopped garlic. Cut loaf in half crosswise and then again lengthwise. Spread butter mixture evenly over inside of loaf. Place on aluminum foil and bake in a preheated 350° F. oven for 5 to 7 minutes. Sprinkle with chopped garlic. Broil quickly for 1 to 2 minutes until crisp. Wrap in foil until ready to eat.

SHALLOTS

If garlic is the king of bulbs, then the shallot is most definitely the queen. Of unknown origin, shallots possess much the same qualities as garlic but to a milder, more refined degree. Garlic is the ideal seasoning for robust, earthy Italian and Latin American cooking. Shallots are equally compatible and responsible for the more subtle differences in French cuisine. The more-delicate shallot flavor is unique, neither less garlic nor more onion, simply its own distinct flavor.

Shallots enhance yet do not disturb the delicate balance necessary for the proper preparation of seafood and shellfish. Shallots are essential to butter sauces or light lemon sauté gravies as found in veal and chicken-breast recipes. Almost all lighter recipes do better with less garlic and more shallots. Raw shallots will not burn a salad or overpower a dressing. Indeed, shallots can be prepared as a whole vegetable, being roasted or gently simmered in their skins, then peeled and mixed with green peas or beans or spread on crackers and served with cheeses as hors d'oeuvres.

Select hard, shiny, coppery bulbs. They keep well in the open or under refrigeration, and a bit of green sprouting does little to harm their flavor or freshness. The freshest crops arrive in the fall, as with most onions and bulbs. South American imports are available in the late spring, but only in limited amounts in limited areas. It is a good thing, then, that shallots keep well and that a good supply of quality shallots is almost always available.

Nobody eats shallots for their nutrition, but they are one of those little foods that increase our appetites for those foods they accompany, which are more nutritionally loaded.

NUTRITIONAL DATA

Weight (g)	Calories	Protein (g)	Carbohydrates (g)	Crude Fiber (g)	Water (g)	Fat (g)	Cholesterol (mg)
50	36	1.2	8.4	0.4	34.6	0.1	—

Vitamin A (IU)	Vitamin C (mg)	Thiamine B-1 (mg)	Riboflavin B-2 (mg)	Niacin (mg)	Vitamin B-6 (mg)	Vitamin B-12 (mcg)	Folic Acid (mcg)	Pantothenic Acid (mg)
trace	4	0.03	0.01	0.01	—	—	—	—

Sodium (mg)	Potassium (mg)	Calcium (mg)	Phosphorus (mg)	Magnesium (mg)	Iron (mg)	Zinc (mg)	Copper (mg)	Manganese (mg)
6	167	18	30	—	0.6	—	—	—

Raw—1¾ ounce

LEEKS

*L*eeks are not the most popular vegetable item in the American kitchen. In fact, leeks are often grown for their perennial flowering qualities, producing a giant purple floral bud at the top of a three- or four-foot stalk. As a vegetable, however, the leek is at its best long before the flower stalk even begins to form and shoot.

The best leeks are about fifteen inches long, with six to eight inches of very white leaf topped with bright dark-green leaf tops. The greens are excellent in soup or stew, while the whites can be braised as a vegetable or eaten raw as a salad ingredient. Leek whites are essential to vichyssoise and can add a delicious dimension to steamed or poached whole seafood, shellfish, fish soup, and chowder. Leeks go well sautéed with garlic, oil, and tamari sauce, and they can also make an interesting quiche filling.

When you buy leeks, look for straight specimens with very white whites and very green greens. Examine the root area for excessive trimming. The leaves should be tightly wrapped, with no sprouting. Leeks are best in fall, winter, and spring, although good stores carry an excellent inventory all year-round. When you see broad bright-green leaves and a glistening-white root shaft or bulb, buy a bunch. It is definitely a vegetable worth knowing and knowing what to do with.

The leek is well balanced, if not overly loaded, with nutrition. It is to be recommended for its mineral content, with a good deal of potassium and calcium. Its vegetable-protein content is comparable or superior to that of many more popular and heavily consumed vegetables like string beans, peppers, cabbage, and celery.

NUTRITIONAL DATA

Weight (g)	Calories	Protein (g)	Carbohydrates (g)	Crude Fiber (g)	Water (g)	Fat (g)	Cholesterol (mg)
100	52	2.2	11.2	1.3	85.4	0.3	—

Vitamin A (IU)	Vitamin C (mg)	Thiamine B-1 (mg)	Riboflavin B-2 (mg)	Niacin (mg)	Vitamin B-6 (mg)	Vitamin B-12 (mcg)	Folic Acid (mcg)	Pantothenic Acid (mg)
40	17	0.11	0.06	0.5	—	—	—	—

Sodium (mg)	Potassium (mg)	Calcium (mg)	Phosphorus (mg)	Magnesium (mg)	Iron (mg)	Zinc (mg)	Copper (mg)	Manganese (mg)
5	347	52	50	23	1.1	—	—	—

Raw—3–4 medium

BRAISED LEEKS

    3–4 medium leeks
    3–4 ounces sesame or walnut oil
    2 cloves garlic, chopped
    1 small piece ginger, grated
    ½ teaspoon salt
    ¼ teaspoon freshly ground black pepper
    2 tablespoons tamari sauce

Clip root hairs and remove most of the green leaf from leeks. Cut lengthwise in
half. In large skillet, covered, simmer leeks in ¼ inch boiling water until just soft.
Drain. Heat oil, garlic, ginger, salt, and pepper in skillet. Place leeks in hot skillet.
Brown on one side. Turn and brown other sides. Sprinkle with tamari sauce.
    Serves 4.

EASY VICHYSSOISE

    3 medium white potatoes, peeled and sliced
    4 medium leeks, sliced
    6 cups chicken broth
    1 teaspoon salt
    1 cup light cream or half-and-half
    2 tablespoons chopped chives

Combine potatoes, leeks, and chicken broth in large saucepan. Cover. Simmer
until tender. Process in blender until smooth. Cool. Add salt and cream. Stir and
serve, sprinkled with chopped chives.   *Serves 4.*

POACHED SEAFOOD WITH LEEKS

    2 medium fish fillets or 2 small whole fish
    Juice and grated peel of 1 lemon
    4–6 sprigs parsley, chopped
    6 tablespoons butter or margarine
    2 leeks, leaves separated

Place fish in shallow buttered pan. Sprinkle with lemon peel and parsley. Add lemon juice and dot with butter. Cover all with separated leek leaves. Seal pan with aluminum foil. Bake in preheated 325° F. oven for 25 to 35 minutes.
Serves 2.

BACON-CHEESE BROILED LEEKS

3–4 medium leeks
6 strips bacon
6 ounces cheddar, Monterey jack, or havarti cheese, grated

Split leeks lengthwise and blanch in boiling water. Fry bacon strips. Drain on paper towel. Crumble when cool. Place drippings in bottom of shallow baking pan. Place leeks in pan. Cover with cheese. Bake in preheated 350° F. oven until cheese melts. Place in broiler briefly until cheese just begins to brown. Sprinkle with crumbled bacon.   *Serves 4.*

SPECIAL CROPS

1. LONG GREEN ASPARAGUS
2. THICK WHITE ASPARAGUS
3. 'WILD' ASPARAGUS
4. GLOBE ARTICHOKES
5. THORNY ARTICHOKES
6. TINY 'HEART' ARTICHOKES
7. WHITECAP MUSHROOMS
8. BUTTON MUSHROOMS
9. BOLETUS MUSHROOMS
10. CHANTERELLE
11. OYSTER MUSHROOMS
12. ENOKI 'STRAW' MUSHROOMS

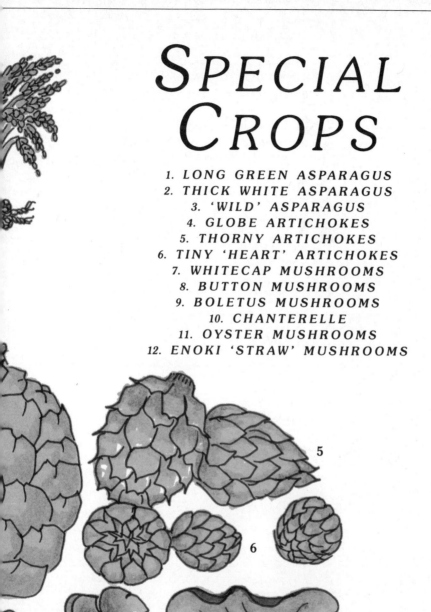

*T*here are three items at every fruit stand that are very individual. They require special cultivation and harvesting, special handling, packaging, and culinary skill. They are delicate, expensive, and wonderfully delicious. These unique items are asparagus, artichokes, and mushrooms.

Asparagus and mushrooms require special soil beds that are constantly refurbished. All mushrooms grow in dark places, and many asparagus as well. Ninety-eight per cent of all artichokes grow in one place in the United States, creating a unique and very specific agri-industry. Asparagus, mushrooms, and artichokes all require special packages specifically designed for their protection and safekeeping in transit. Asparagus requires special tools for cutting and wrapping. Mushroom-pickers wear miners' helmets with battery lamps. Artichoke thorns can tear up your hands if you are not a skilled artichoke picker from Castroville, California.

Their difficult cultivation, their special harvesting and packaging, and their fragile natures have made this trio expensive and the desire of the food elite for centuries. Today they are available for all of us to enjoy. Nevertheless, it still requires special awareness of their qualities and a touch of gastronomic skill to enjoy these beautiful, special foods at their best.

ASPARAGUS

*W*hite, green, purple, thin, fat, long, and short. It's all asparagus, the prize of fresh vegetables, expensive, difficult to harvest, and one of the most delicious and sophisticated of all the earth's foods. Asparagus is a food like no other. It has no known origin. It was probably spread by birds, growing best where the earth freezes. Asparagus requires a special tool for cutting and bunching and is shipped in wooden, and now wax-board, boxes shaped like cut-off pyramids, specially designed for the tapered asparagus shape. The crates contain a special water-retentive pad on the bottom to keep the cut end of each stalk from drying out. No wonder this fragile, difficult plant has such a high price tag.

The most common variety of asparagus is the long green asparagus, about nine inches long and as thick as your finger. White asparagus, grown in Holland by covering the plants with soil as they grow, are considered the best. They actually have the same taste as the green but tend to eat a bit more tender. There is also a very thick white variety with pinkish tips that is quite succulent and very tender. It is often served as a special dish at serious formal affairs and dinners. Purple varieties have a deep color at the tips and at the leaf points and are striated at the whitish end. In the springtime some very long, very thin wild varieties appear at local markets and farm stands, never supermarkets. They are extremely flavorful, very tender, and less expensive.

Select asparagus with tight velvety tips and a minimum of woody white stalk at the bottom. Wrinkled or pithy stalks indicate age and dehydration, and you'll lose a lot in the preparation. If asparagus is fresh, there is no difference in taste or texture between thin stalks and fat stalks. Asparagus of uniform size should, however, be purchased whenever possible to allow for even cooking and consistent doneness. Asparagus is available most of the year, but the off-season price inhibits many customers from buying it.

The proper way to cook asparagus is to tie it in a bunch after cleaning (some people peel or cut off the leaf tips that grow along the stalk, which is not really necessary). After bunching up the spears, stand them upright in a tall pot in one or two inches of water. Cook until the stalks soften, but do not allow them to fall over. Overcooking ruins the delightful texture and robs the asparagus of its springtime freshness. If a tall pot is not one of your possessions, make a very light butter sauté in a wide, flat pan. Cool. Add a touch of water and heat. Lay the washed wet asparagus in, cover, and steam or simmer for three to four minutes. Lift the asparagus out carefully. After cooking, asparagus may be served with butter and lemon sauces; in cheese melts, omelettes, quiches; simmered quickly in shallots and butter; breaded and fried; or marinated in a vinaigrette and served cold. Asparagus makes an excellent soup and may be added raw to salads, where its crisp texture and unique flavor may be truly appreciated.

Asparagus gets fair points in nutrition. It contains better-than-average vegetable protein and large amounts of vitamins A and C. It is digested easily and is beneficial overall to the digestive system.

NUTRITIONAL DATA

Weight (g)	Calories	Protein (g)	Carbohydrates (g)	Crude Fiber (g)	Water (g)	Fat (g)	Cholesterol (mg)
100	20	2.2	3.6	0.7	93.6	0.2	—

Vitamin A (IU)	Vitamin C (mg)	Thiamine B-1 (mg)	Riboflavin B-2 (mg)	Niacin (mg)	Vitamin B-6 (mg)	Vitamin B-12 (mcg)	Folic Acid (mcg)	Pantothenic Acid (mg)
900	26	0.16	0.18	1.4	—	—	—	—

Sodium (mg)	Potassium (mg)	Calcium (mg)	Phosphorus (mg)	Magnesium (mg)	Iron (mg)	Zinc (mg)	Copper (mg)	Manganese (mg)
1	183	21	50	—	0.6	—	—	—

Cooked—⅔ cup pieces

Raw Asparagus Salad

4 ounces peas, shelled
18–20 thin asparagus spears
½ red pepper, diced
1 tablespoon chopped parsley
Juice of 2 lemons
1 tablespoon apple-cider vinegar
¼ teaspoon salt
Dash freshly ground black pepper
2 tablespoons chopped yellow onion

Blanch peas in boiling water. Drain and chill. Wash and trim asparagus. Break or cut into 1½-inch pieces and place in bowl. Combine remaining ingredients except for onion. Toss with asparagus and peas. Top with chopped onion. Serve cold. *Serves 4.*

Simple Asparagus with Many Sauces

18–24 asparagus spears

Trim off white ends of asparagus and place whole spears in ¼ inch water in large flat pan. Cover and simmer over low heat until just soft. Drain water from pan and add any of the following sauces: *Serves 3–4.*

Melted Cheese Sauce

¼ cup milk
2–3 ounces butter
4 ounces cheddar, Monterey jack, Swiss, or
American cheese, grated

Heat milk to boiling. Add butter. When it melts, add part of cheese. Stir. Pour over asparagus spears. Add remaining cheese. Cover. Simmer for 3 to 5 minutes.

GRATED CHEESE AND BUTTER TOPPING

2 ounces butter
4 ounces cheddar, Monterey jack, Swiss,
or American cheese, grated

Dot asparagus with butter and grated cheese. Cover. Simmer for 2 minutes.

GARLIC-BUTTER SAUCE

4 cloves garlic, chopped
¼ teaspoon salt
¼ teaspoon freshly ground black pepper
4 tablespoons butter

Sauté all ingredients in butter. Pour over asparagus spears. Simmer for 3 minutes.

PARSLEY-WALNUT GARNISH

¼ cup chopped walnuts
4 tablespoons butter or margarine
¼ cup finely chopped curly parsley

Steam walnuts in butter in covered pan. Pour over asparagus. Add parsley just before serving.

BAKED CRUSTY-CRUMB ASPARAGUS

18–24 asparagus spears
1 small yellow onion, chopped
3 cloves garlic, chopped
4–5 sprigs flat-leaf parsley, chopped
3–4 sprigs tarragon, savory, or marjoram, chopped
4 tablespoons butter or margarine
2 cups bread crumbs

Blanch asparagus in boiling water and place in buttered shallow baking pan. Sauté onion, garlic, parsley, and tarragon in butter and mix with bread crumbs. Cover asparagus with bread-crumb mixture. Cover pan with aluminum foil. Bake in preheated 350° F. oven for 15 to 20 minutes. Remove foil. Raise oven to 400° F. for 5 minutes. *Serves 3–4.*

EGG-CRUST ASPARAGUS

    18–24 asparagus spears
    3–4 eggs
    2 cups flour
    ½ teaspoon onion powder
    ½ teaspoon garlic powder
    ½ teaspoon salt
    ¼ teaspoon freshly ground black pepper
    Light vegetable oil

Blanch asparagus spears in boiling water and cool. Beat eggs and place in shallow pan or bowl. Mix flour and seasonings. Heat thin layer of oil in frying pan until hot but not smoking. Dip spears in egg and then in flour mixture. Place in hot pan. Turn until brown. Drain on paper towels. *Serves 3–4.*

CREAM OF ASPARAGUS SOUP

    1 small yellow onion, chopped
    1 stalk celery, chopped
    4 tablespoons butter or margarine
    3 tablespoons flour
    3 cups milk
    12–16 asparagus spears, cut up
    4–6 sprigs savory, tarragon, or marjoram, chopped
    ½ teaspoon salt
    ¼ teaspoon freshly ground black pepper

Sauté onion and celery in butter in medium saucepan until they begin to soften. Stir in flour. Add milk and asparagus. Cook over medium heat until mixture thickens and boils, stirring continuously. Add seasonings. Cover. Let stand over very low heat for several minutes before serving. Garnish with chopped nuts, parsley, or scallion greens or sliced mushrooms. *Serves 4.*

ASPARAGUS VINAIGRETTE

    18–20 asparagus spears, cut into 1-inch pieces
    2 shallots, crushed
    ½ cucumber, peeled, seeded and cubed
    ½ small yellow onion, chopped

Juice of 2 lemons
½ cup white-wine vinegar
¼ cup walnut or olive oil
½ teaspoon salt
¼ teaspoon freshly ground black pepper

Blanch asparagus pieces quickly in boiling water. Drain. Mix with remaining ingredients. Cover. Chill for 3 to 4 hours. Serve asparagus with vinaigrette on the side. *Serves 3–4.*

ASPARAGUS IN TOMATO SAUCE

2 cups tomato sauce
16–20 asparagus spears
3–4 ounces Parmesan or Romano cheese, grated

Heat tomato sauce in shallow wide pan. Place asparagus spears in sauce. Cover. Simmer over low heat for 15 to 20 minutes. Add grated cheese. Cover. Let stand for 2 to 3 minutes. *Serves 3–4.*

ASPARAGUS WITH SIMMERED SHALLOTS

6–8 shallots, sliced
2 tablespoons chopped parsley
1 tablespoon chopped tarragon or savory
4 tablespoons butter or margarine
18–24 asparagus spears

Sauté shallots, parsley, and tarragon in butter. Lower heat and add asparagus spears and ¼ cup water. Cover. Simmer for 10 to 12 minutes or until tender. *Serves 3–4.*

ARTICHOKES

Artichokes require table techniques not found in any other food. The leaves, or bracts, are removed one at a time and pulled between the teeth to remove the soft, tender flesh of the underleaf. Once the large green leaves are eaten, the small, pointy, transparent inner leaves and fuzzy choke are cut away and discarded. What remains is a dense, sweet, meaty bottom. This is salted or seasoned to taste and eaten with a knife and fork. For those who take the time to learn the ways of the artichoke—actually the giant bud of a large plant that produces blue or pink thistlelike flowers—there is much pleasure in both the preparation and the eating of this singularly wonderful vegetable.

Fairly new to the American table, it has long been a favorite of the Mediterranean peoples, especially Spaniards and Italians. Artichokes can be boiled, steamed, fried, pickled, stuffed, and baked. They can be chilled and served with many different sauces and dips, from simple mayonnaise to vinaigrettes, béarnaise, and hollandaise. They make a delightful light lunch, or when stuffed Italian style, they can dominate an elaborate full-course dinner.

There are three kinds of artichokes. The most common is the large round globe artichoke. The leaves are rounded, with very small thorn tips or none at all. A second variety is the more oval-shaped thorny artichoke. It has pointed leaves with obvious thorny tips. Both varieties cost the same, cook the same, and taste the same, though I find the globe artichoke prettier and more tender-textured. The third sort are simply small artichokes shaped like toy tops. These little specimens can be opened at the bottom, dechoked, pickled or marinated in either vinegar or olive oil, and served in salads, with antipasto, or as appetizers with bread, cheese, and wine.

Artichokes should always be selected with a bright dusty-green color free from black marks. They should be dense and tightly packed. Open-leaved specimens have a woody taste and lose their soft, sweet leaf flesh. Sometimes in winter or during a cold snap artichokes will show a little bronzing on the outside leaves. This is not bad. I think it actually helps the artichoke flavor.

The peak season for artichokes is springtime, but they are available most of the year. American artichokes are grown in California, mostly in the coastal town of Castroville, where the entire economy is built around the cultivation and distribution of this unique plant. Learn to select and prepare artichokes. They will grace any table and add a wonderful dimension to your kitchen fare.

Artichokes are a high-mineral food with above-average quantities of protein. They also have above-average quantities of sodium, so if you're a low-salt person, don't sprinkle any more on your artichoke.

NUTRITIONAL DATA

Weight (g)	Calories	Protein (g)	Carbohydrates (g)		Crude Fiber (g)	Water (g)	Fat (g)	Cholesterol (mg)
100	44	2.8	9.9		2.4	86.5	0.2	—

Vitamin A (IU)	Vitamin C (mg)	Thiamine B-1 (mg)	Riboflavin B-2 (mg)	Niacin (mg)	Vitamin B-6 (mg)	Vitamin B-12 (mcg)	Folic Acid (mcg)	Pantothenic Acid (mg)
150	8	0.07	0.04	0.7	—	—	—	—

Sodium (mg)	Potassium (mg)	Calcium (mg)	Phosphorus (mg)	Magnesium (mg)	Iron (mg)	Zinc (mg)	Copper (mg)	Manganese (mg)
30	301	51	69	—	1.1	—	—	—

Cooked—1 large

Stuffed Globe Artichokes Italiano

2 small yellow onions, chopped
2 cloves garlic, chopped
2 teaspoons chopped parsley
½ teaspoon dried oregano
¼ teaspoon salt
¼ teaspoon freshly ground black pepper
Olive oil
3 cups bread crumbs
2 ounces freshly grated Parmesan cheese
2 large globe artichokes

Sauté all ingredients except bread crumbs, cheese, and artichokes in olive oil. Mix with bread crumbs and add Parmesan cheese. (Anchovies may be chopped and added *alla siciliano.*)

Cut off pointy, thorny end of artichokes, and snip off thorns from leaf tips with scissors. Spread leaves apart and spoon in bread-crumb mixture, stuffing it evenly between leaves. Place stuffed artichokes in pot and add water to cover them halfway. Cover and cook over medium heat until leaves pull off easily, about 45 minutes. Watch that water does not evaporate completely; add water if necessary. Remove artichokes from pot. Place in loaf pan. Bake in preheated 400° F. oven for 10 to 12 minutes, or broil quickly on low setting. Serve individually.

Steamed Artichokes for Sauce-Dipping

Cut off thorny ends and leaf-tip thorns of artichokes. Place in water to cover them halfway. Cover. Simmer over medium heat until leaves can be pulled off easily. Drain. Serve with any of many dipping sauces like mayonnaise, hollandaise, béarnaise, French or Russian dressing, sauce vinaigrette (my favorite), or a warm lemon, butter, garlic, and parsley sauce. *Serve individually.*

Marinated Artichoke Hearts

8–12 artichoke hearts
1 small red onion, finely chopped
2–3 cloves garlic, finely chopped
1 tablespoon chopped parsley
1 teaspoon grated horseradish
Juice of 1 lemon
2 tablespoons red-wine vinegar
¼ teaspoon dried oregano
1 teaspoon salt
½ teaspoon freshly ground black pepper

This recipe calls for artichoke hearts already prepared in jars or cans, or dechoked hearts from the small thorny artichoke. Fresh hearts must be blanched until the meat of the heart is soft to the fork.

Combine all ingredients in bowl. Cover. Chill. Serve artichokes without marinade.

Artichoke Hearts in Garlic Butter

1 medium yellow onion, finely chopped
2 cloves garlic, finely chopped
2 tablespoons chopped savory or tarragon
¼ teaspoon salt
¼ teaspoon freshly ground black pepper
3–4 ounces butter or margarine
8–10 fresh, jarred, or canned artichoke hearts
¼ cup bread crumbs

Sauté all ingredients except artichoke hearts and bread crumbs in butter. When they begin to brown, drain artichoke hearts and place in pan. Sprinkle with bread crumbs. Cook over medium heat, turning frequently. Serve artichoke hearts and sauté mixture separately.

CULTIVATED MUSHROOMS

Although there is only one commercial variety of mushrooms, there are endless uses for them, and only your imagination limits their versatility. Cultivated mushrooms can be eaten raw, fried, batter-dipped, sautéed, stuffed, sliced, marinated, pickled, or used in soup, sauces, and poultry stuffing. Even the stems can be chopped and used for stuffing or in recipes of their own.

The medium and large white mushrooms are used for their umbrella-shaped caps or sliced whole. One of my all-time favorite things to make is stuffed mushroom caps Sicilian style. I do a whole basket at one time. The small Snowdrop button mushrooms are excellent eaten whole, dipped in cheese dips, or simmered slowly in a light sauce or gravy. The buttons seem to have more flavor, though white mushrooms in general are considered bland food. The United States, even more than France, Italy, or China, produces and consumes an endless year-round supply of artificially cultivated magic vegetables. It is quite a sight to enter a dark, damp, airy cave or cellar and see there in the blackness brilliant white mushrooms emerging from the moist beds of peat, soil, and manure. They are truly magic, and their subtle but deeply pleasing taste and unique texture create outstanding cuisine.

Select firm white mushrooms whose caps are attached to the stem by a delicate veil. Open caps indicate age and loss of delicate flavor. Some varieties have a tan tinge or slight scale, so look for dense, closed-capped specimens of any size.

Mushrooms should never be immersed or held under running water. To clean them it is best to wipe them with a damp soft cloth or brush them with a specially designed mushroom brush. Mushrooms need not be peeled. They refrigerate well in plastic bags or well wrapped in paper or tissue.

Mushrooms have little nutritional value except for above-average protein levels. They have almost no carbohydrates and are excellent weight-loss food.

NUTRITIONAL DATA

Weight (g)	Calories	Protein (g)	Carbohydrates (g)	Crude Fiber (g)	Water (g)	Fat (g)	Cholesterol (mg)
100	28	2.7	4.4	0.8	90.4	0.5	—
70	78	1.7	2.8	0.7	—	7.4	—

Vitamin A (IU)	Vitamin C (mg)	Thiamine B-1 (mg)	Riboflavin B-2 (mg)	Niacin (mg)	Vitamin B-6 (mg)	Vitamin B-12 (mcg)	Folic Acid (mcg)	Pantothenic Acid (mg)
trace	3	0.1	0.46	4.2	—	—	—	—
—	trace	0.05	0.27	2.9	—	—	—	—

Sodium (mg)	Potassium (mg)	Calcium (mg)	Phosphorus (mg)	Magnesium (mg)	Iron (mg)	Zinc (mg)	Copper (mg)	Manganese (mg)
15	414	6	116	13	0.8	—	—	—
—	—	8	81	—	0.7	—	—	—

Raw—10 small
Sautéed—4 medium

WILD MUSHROOMS

Wild mushrooms are gathered, not cultivated and picked. They are not available in most markets and stores, but select produce stands and high-quality metropolitan markets usually carry one or two of the most popular. Their shapes vary, from the large-capped boletus to the amorphous fluted chanterelle.

Boletus mushrooms are usually very large, and though slightly tougher than the large white cultivated mushrooms, have a deeper delicious flavor. They can be used in every way the more familiar cultivated white variety can.

Enoki, or straw, mushrooms are small-capped, very long-stemmed mushrooms. They are a Japanese variety that is excellent in clear broth, simmered in a light gravy, or stir-fried with thin slices of beef and boneless chicken prepared Japanese style.

Oyster mushrooms are large dark leafy mushrooms with a delectable taste and thick texture that is wonderful. They are good sautéed, roasted, or braised in the company of a generous cut of meat. To enjoy oyster mushrooms for themselves make a well-salted chicken broth, and allow time for the oysters to flavor the pot. Oyster mushrooms can be sliced or chopped and served in a thick gravy on duck, goose, or pheasant.

Chanterelles are flowery yellow mushrooms that grow in cool, damp pine forests. They appear in summer and fall and are best prepared in stews with other vegetables and meats.

When selecting wild mushrooms, make certain they are firm, dry, and free from discoloration and dehydration. Gathered mushrooms *must be fresh*.

STUFFED MUSHROOMS ITALIANO

16–20 large mushrooms
6–8 sprigs flat-leaf parsley, finely chopped
1 teaspoon dried oregano
3–4 cloves garlic, finely chopped
1 small yellow onion, finely chopped

½ teaspoon salt
½ teaspoon freshly ground black pepper
Olive oil
2 ounces freshly grated Parmesan cheese
1½ cups bread crumbs

Remove stems from mushrooms. Wipe caps with damp paper towel. Finely chop half of the stems (reserve remaining stems for another use) and sauté with parsley, oregano, garlic, onion, salt, and pepper in olive oil. Mix with cheese and bread crumbs. If mixture appears dry, add just enough water or olive oil to give a moist feel to bread crumbs. Stuff each cap with mixture and place in well-buttered baking pan. Drip olive oil onto stuffing. Cover pan with aluminum foil. Bake in preheated 350° F. oven for 25 minutes. Remove foil. Broil for 2 to 3 minutes to crisp.

SAUTÉED MUSHROOM STEMS

2 stalks celery, chopped
1 small onion, chopped
2 cloves garlic, chopped
2 tablespoons chopped flat-leaf parsley
1 teaspoon chopped savory or thyme
4 tablespoons butter or margarine
Stems of 1 pound mushrooms, sliced or cut up
¼ cup white cooking wine

Sauté all ingredients in butter except mushrooms and wine. When sauté mixture is hot, add mushrooms. Stir. When mushrooms begin to soften, lower heat and add wine. Cover. Let stand for 2 to 3 minutes.

This recipe can be made with slices or pieces of wild chanterelles or oyster mushrooms. *Serves 4.*

BATTER-DIPPED WHOLE MUSHROOMS

1 pound medium mushrooms
½ teaspoon garlic powder
½ teaspoon salt
¼ teaspoon freshly ground black pepper
½ small yellow onion, finely chopped
2 cups prepared pancake batter

Light vegetable oil
Herb mix or garlic salt

Wipe mushrooms with damp paper towel. Chill. Mix seasonings and onion into batter. Chill. Heat 1½ inches oil in saucepan or deep-fryer. When oil is hot but not smoking, dip mushrooms in batter. Deep fry until golden brown. Drain on paper towels. Season with herb mix or garlic salt. *Serves 4.*

CREAM OF MUSHROOM SOUP

8 ounces white, oyster, or chanterelle mushrooms, sliced
1 stalk celery, chopped
1 small yellow onion, chopped
4 tablespoons butter or margarine
3 tablespoons flour
2½ cups milk
2 chicken bouillon cubes
4–6 sprigs parsley, chopped
½ teaspoon salt
¼ teaspoon freshly ground black pepper
½ cup white cooking wine

Sauté mushrooms, celery, and onion in butter in medium saucepan until they begin to soften. Stir in flour. Add milk and bouillon cubes broken down in a little water, stirring constantly. Cook over medium heat until mixture thickens and boils, stirring constantly. Add seasonings. Cover. Let stand over very low heat for several minutes before serving. Add white wine. Stir. *Serves 4.*

MARINATED MUSHROOM BUTTONS

½ pound enoki or button mushrooms
1 small yellow onion, chopped
2 cloves garlic, finely chopped
6–8 coriander leaves, chopped
Juice of 1 lime
½ cup red-wine vinegar
½ teaspoon salt
¼ teaspoon freshly ground black pepper

Combine all ingredients in bowl. Cover. Refrigerate for 2 to 3 hours. Drain mushrooms before serving. *Add to salad or serve as condiment.*

MUSHROOM SIMMER

 4 tablespoons butter or margarine
 ¾ pound mushrooms, cut up
 Juice of 2 lemons
 ¼ cup tamari sauce

Melt butter in covered saucepan. Mix mushroom pieces and lemon juice. Add to butter. Cover. Simmer over low heat for 5 to 8 minutes. Add tamari sauce. Simmer for 2 to 3 minutes.

MUSHROOM-STEM GRAVY

 Stems of 1 pound mushrooms
 6 slices bacon
 4 tablespoons butter
 ¼ cup milk
 ¼ cup instant-blending flour
 ¼ teaspoon onion powder
 ¼ teaspoon garlic powder
 ¼ teaspoon salt
 ¼ teaspoon freshly ground black pepper

Blanch mushroom stems in boiling water and drain. Fry and drain bacon. Combine bacon drippings and butter over low heat. Mix in milk, flour 1 teaspoon at a time, and seasonings, stirring constantly. When mixture thickens, add mushrooms. Cover. Simmer over low heat for 3 to 5 minutes. Serve on meats, string beans, rice, noodles, or chicken.

MUSHROOM RICE

 ½ pound mushrooms, sliced
 1 small piece ginger, grated
 1 small yellow onion, sliced
 2 scallion greens, chopped
 4–6 tablespoons sesame or walnut oil
 2 cups cooked rice

Sauté mushroom slices with ginger, onion, and scallion greens in oil. When mushrooms soften, add rice. Stir-fry until mixture is heated. *Serves 4.*

MUSHROOM-CROUTON POULTRY STUFFING

Poultry giblets, boiled and chopped
6–8 sprigs parsley, chopped
2 cloves garlic, chopped
1 small yellow onion, chopped
2 stalks celery, chopped
½ teaspoon salt
¼ teaspoon freshly ground black pepper
Dash dried oregano
6 tablespoons margarine
¼ pound mushrooms, cut up
2 cups croutons
1 small apple, peeled and cut up

Sauté giblets, parsley, garlic, onion, celery, and seasonings in margarine. When onion and celery soften, add mushrooms. Cook until soft. Mix with croutons and apple pieces. Use to stuff chicken, capon, turkey, or Cornish hens.

MUSHROOM SHISH KEBABS

2 large green peppers, cut in 12 thick slices
12 large cloves garlic
12 white boiling onions
12 medium mushrooms
4 tablespoons margarine
¼ cup vinegar
2 tablespoons chopped parsley
½ teaspoon salt
¼ teaspoon freshly ground black pepper
12 cherry tomatoes

Cut each green pepper into 6 wide strips. Blanch green peppers, garlic, onions, and mushrooms in boiling water. Melt margarine and combine with vinegar, parsley, salt, and pepper. Alternate mushrooms, tomatoes, garlic cloves, onions, and peppers on 12 skewers. Brush with margarine sauce while on barbecue, turning while browning. Serve when browned. *Serves 4.*

MUSHROOM OMELETTE

4 tablespoons butter or margarine
½ pound mushrooms, sliced
3–4 scallions, chopped
1 small yellow onion, chopped
6–8 sprigs parsley, chopped
¼ teaspoon salt
Dash freshly ground black pepper
6 eggs

Sauté all ingredients in butter except for eggs and half the mushrooms. Beat eggs. Lower heat and pour in eggs. Cover. Allow to cook until three-quarters done. Add reserved mushrooms. Fold omelette. Raise heat. When one side of omelette is done, turn over. Finish cooking. Place on platter. Cut in wedges. *Serves 4.*

MUSHROOMS

263

CITRUS FRUITS

1. FLORIDA RED GRAPEFRUIT
2. FLORIDA GOLD GRAPEFRUIT
3. FLORIDA SMALL JUICING GRAPEFRUIT
4. CALIFORNIA RED GRAPEFRUIT
5. FLORIDA NAVEL ORANGE
6. FLORIDA VALENCIA ORANGE
7. CALIFORNIA NAVEL ORANGE
8. CALIFORINA VALENCIA ORANGE
9. BLOOD ORANGE
10. TANGERINE   11. TANGELO
12. MURCOTT 'HONEYBELLE'
13. TEMPLE ORANGE
14. CLEMENTINE
15. GREENSKINNED MANDRIN
16. CALIFORNIA LEMONS
17. KEY LIMES   18. PERSIAN LIMES
19. UGLI FRUIT

The citrus family of fruit is as varied as any group of natural foods. Citrus includes everything from the sugary-sweet Honeybelle orange to the inedible citron and ranges in size from tiny Key limes and kumquats to enormous heavy grapefruits that yield twelve to sixteen ounces of juice each. Citrus provides us with fresh fruit, juice, flavorings, pies, teas, peels, marmalades, and perfumes.

In the United States citrus is grown in Florida, southern California, and some limited areas of Arizona and Texas. Citrus in one form or another is available year-round in excellent quality and abundant quantity. Always select firm, heavy fruit, the thinner-skinned the better.

California citrus is almost always thicker-skinned. California oranges seem sweeter and yield darker juice. California lemons are decidedly superior, but Florida grapefruit is the best in the world, and Florida limes are seedless and very juicy. Fresh citrus juice is excellent for your health and supplies refreshment and energy. Fresh oranges and grapefruits provide important nutrition without contributing useless calories. Any fresh citrus-fruit juice is far superior to any frozen, bottled, or canned variety in every way.

It is too bad that our appetites for convenience have affected our consumption of fresh citrus juice. If we search in our memories, we can recall wonderful breakfasts with glasses of thick fresh-squeezed orange juice, or a hot summer afternoon when mom would call us to the front porch or backyard and pour out a pitcher of homemade lemonade. Boy, that was good. Most of us thaw our orange juice, pour our grapefruit juice, mix our lemonade—and if the tangelos, grapefruits, Temples, and navel oranges don't have a zipper on the skin, you can forget a citrus-fruit cocktail unless there's a can on the shelf or a jarred version in the refrigerator. All citrus juices are adulterated in some way; "100%" pure on a label is a lot of bunk. The processors call reconstituted juice 100% pure. They call dehydrated, concentrated, pasteurized, filtered juices 100% pure. To me 100% pure means cut, squeeze, and drink. All processed citrus juices are crushed, which adds the peel oil to the juice and diminishes natural sweetness. Bottled grapefruit juice is always bitter, insipid, and pulpless. I miss those tiny little fruit cells that float in the juice and add the all-important dimension of texture to citrus-fruit juice.

Our juice bar at Sunfrost serves homemade lemonade with two to three times the amount of natural fresh-squeezed lemon juice as of simple sugar syrup (one cup sugar to three cups water). Packaged lemonade is so bitter that the principal ingredient (if you read your label carefully, and you should) is sugar! We serve fresh-squeezed grapefruit juice to people who suddenly, for the first time, find out grapefruit juice is a sweet, smooth, refreshing experience without the throat-searing, tongue-drying, mouth-puckering taste of that stuff they pass off in cardboard and glass. Canned orange juice and fresh orange juice are like zircons and diamonds. The processors will always say you can't tell the difference. You can. Frozen orange juice is insipid, highly acidic, and flavorless. It tastes like the water you thaw it in.

Nobody wants to peel small oranges anymore. Nobody wants to peel and segment a grapefruit. Too bad, because once you start making fresh, live citrus part of your daily life, your body and your family will feel the benefits of better health, freshness, and sweet variety.

There are some general rules for citrus care and use to give you maximum enjoyment. Citrus should be kept under refrigeration. Exposure to warm air dries the fruit quickly and ferments the sugars. Orange and grapefruit juice should be drunk the instant it is squeezed. Lemon and lime juice keep for weeks in a closed jar in your refrigerator. I always keep a jar of lemon juice and some grated peel, plus wedges of lemon and lime, in the refrigerator.

All citrus should be cut crosswise for best squeezing and segmenting. Lemons and limes should be cut from stem end to blossom end for wedges. Thick skin always peels better than thin skin. Thick skin grates better also. Seedless always means a few seeds. If you compare prices, compare sizes. Even one size can make a big difference in price and quality. If you buy oranges in a cello bag, count them. Ten can look like twelve with no trouble at all.

There's a lot to citrus to know and enjoy. Shop in stores and markets that offer full variety and fresh inventory. Try some of everything every which way. You might like something. Next thing you know you'll have a juice habit.

GRAPEFRUIT

The very best grapefruit grown in the world is grown in Florida. It is also cultivated in California, South America, and Mexico, as well as in the Mediterranean countries and Israel. Texas is a heavy grower and shipper of grapefruit, and the first pink crops were developed there, although the original mutant sport was found in Florida. I find Texas grapefruit inconsistent, not as juicy or sweet as the Florida crop or as meaty and full as the California.

The thin-skinned, shiny, heavy Florida grapefruit is the juiciest, sweetest, most nutritious variety available. Although Indian River Florigold grapefruit is the pick of the crop, many Florida varieties are equally good though not as consistent. A special grade labeled Orchid is far superior to any other variety grown anywhere and is available March to June. Grapefruit got its name from the way it sometimes grows: in large clusters that resemble bunches of grapes on the vine. This is an oddity for such a large, globular fruit, but nevertheless it grows and is harvested in these grapelike bunches.

Grapefruit is at its peak from early winter to June, but excellent varieties start to arrive from Florida by late October. California grapefruit is available from mid-August until October. The difference between Florida grapefruit and California varieties is the thickness and acidity of the peel and the rind. Florida fruit is always thinner-skinned, less pulpy, and more alkaline (or sweeter). Always look for shiny, heavy grapefruit (a touch of green is okay). Pink grapefruit is always sweeter, except that some prime large golden specimens of the Duncan variety have a balance of tart and sweet that is more interesting and delicious than even the sweetest pink-meat grapefruit. Almost all grapefruit sold today is seedless, which means each one seldom has more than eight seeds.

There is no comparison between frozen or prepared grapefruit juice and fresh-squeezed grapefruit juice. All bottled, canned, or frozen grapefruit concentrate contains peel and rind oil, which are bitter, acidic, and leave an unpleasant aftertaste. Fresh-squeezed grapefruit juice is clean, refreshingly sweet, and full of little cells and meat that give the juice dimension and texture as well as that precious natural taste.

Different varieties of grapefruit in different seasons have certain qualities that make them best for eating or juicing. Large red Florida Marsh Rubies have the sweetest taste of all. They are best for segmenting and eating out of the half shell, make excellent juice, and are best from November through June.

Large Florida golden grapefruit, especially the Duncan or Orchid varieties, have that best tart-sweet juicy grapefruit taste. They have very large segments and thin webbing, which makes them excellent for segmenting. They also make the best juice and are at their best from November through June.

Small Florida pink or golden early grapefruit varieties arrive in October and May. They are very juicy but not very sweet. They make excellent juice, however. Later crops in late fall and winter and midsummer are sugary-sweet and have a very high juice yield.

California grapefruit arrives in late summer and early fall. The skin on all sizes and both colors tends to be thicker than that of Florida fruit, and California grapefruit is not heavy with juice. The flavor is very sweet, and California fruit is easy to peel and break up into full segments. The small sizes in both colors make excellent juice. California grapefruit is best for salads.

The best way to prepare a grapefruit for eating is simple. After cutting the fruit in half at the equator, insert the tip of a serrated knife at the center of each section. Cut along the web, then the rind, and back along the web to the center point. Press the blade against the outside edge as you cut the fruit away from the rind and web.

When buying grapefruit, choose the variety in peak season. If you don't have this book with you, ask a clerk or your fruit man, or look for hard, shiny, thin-skinned fruit that feels heavy in your hand. Avoid airy, pebbled, dull fruit or anything with a bloated or protruding stem end.

Grapefruit is an alkaline fruit and therefore a body-sweetener. It helps reduce cholesterol and is a wonderful cleansing agent for the liver and gall bladder. It helps break down fat cells and can greatly assist in any cleansing, detoxification, or weight-loss program. Grapefruit offers a full range of nutrition, including protein and vital minerals, especially potassium. It has significant amounts of vitamins B-1, B-2, and B-6. Red grapefruit has twenty-five times more vitamin A than gold, but otherwise they are almost nutritionally equal. Neither color of any variety contains any sodium.

NUTRITIONAL DATA

Weight (g)	Calories	Protein (g)	Carbohydrates (g)	Crude Fiber (g)	Water (g)	Fat (g)	Cholesterol (mg)
118	39	0.8	9.9	0.2	106.8	0.1	—
123	37	0.7	9.5	0.3	112.4	0.1	—

Vitamin A (IU)	Vitamin C (mg)	Thiamine B-1 (mg)	Riboflavin B-2 (mg)	Niacin (mg)	Vitamin B-6 (mg)	Vitamin B-12 (mcg)	Folic Acid (mcg)	Pantothenic Acid (mg)
12	39	0.04	0.02	0.3	0.05	—	12	0.33
318	47	0.04	0.03	0.2	0.05	—	15	0.35

Sodium (mg)	Potassium (mg)	Calcium (mg)	Phosphorus (mg)	Magnesium (mg)	Iron (mg)	Zinc (mg)	Copper (mg)	Manganese (mg)
—	175	14	9	11	0.07	0.08	0.059	0.015
—	158	13	11	10	0.15	0.09	0.054	0.012

White, raw—½ medium
Pink and red, raw—½ medium

BROILED GRAPEFRUIT

1 grapefruit
2 teaspoons honey
2 teaspoons sugar
1 teaspoon grated orange peel

Cut grapefruit in half and section. Spoon honey, then sugar on each half. Preheat broiler and broil on low rack in oven for 2 minutes. Sprinkle with grated orange peel.

ORANGES

One day a young mother and her small son were shopping at the fruit stand, and the mother asked the boy if he would like her to buy some oranges for orange juice. "No, Mom," he replied, "get the regular kind, in the can." How sad. Price and convenience have redirected the orange from the squeezer to the freezer, and we are the losers. Fresh-squeezed orange juice—not fresh-thawed or reconstituted—is an elixir that can refresh, fortify, and delight the drinker any time of the day, even far beyond its traditional role at breakfast. A good citrus-squeezer in the kitchen will yield years and years of sweet healthful juice that is immeasurably superior to any processed juice or concentrate you will ever find. Cost more? Yes. Taste better? Yes. Better for your health? Yes. Better for your spirit? By far, yes.

Like all fruits, oranges have their seasons, their best times and hard times. But there are always oranges available and always good ones. Always select shiny, heavy oranges without a puffy feel. They should be free from bronzing and should smell like orange blossoms, without a heavy acid or fermented smell.

Oranges come in three categories: navel oranges (known by the navel growth at the blossom end), which are the best peeling and eating sweet oranges; Valencias, which are oval-shaped and have a closed blossom end and are best for squeezing juice; and blood oranges, which have red fruit and juice and sometimes even red skins.

The Florida Valencia juice orange, ounce for ounce, penny for penny, is the class act of juice oranges. It produces the most juice per weight and has the thinnest peel, least citric acid, and least pulp. It is best from November to mid-July, with its sweetest peak and lowest price from mid-April to early June. The best size to buy is either 100 or 125 (this number indicates the number of oranges in a standard orange box; a higher number means a smaller orange). In November and April Valencia oranges sometimes have a tinge of green. This is totally acceptable and indicates freshness and high juice content without loss of sugar.

The California Valencia juice orange has a thicker skin and yields a thicker, richer-colored orange juice than the Florida variety. The California Valencia orange is at its peak, with an extraordinarily high sugar content, in July, August, and September. The best size to buy is 113. California orange juice loves a sprinkle of lime. A combination of fresh Florida and California juice gives a great balance of thick, rich juice with high volume and sweetness.

There is no finer eating orange than the California navel orange. California navels begin to appear in November, lasting all winter and into spring. Always seedless (perhaps with a stray seed or two) and easy to peel, they are excellent snacks and useful in salads and orange desserts. Segmented navel oranges are perfect as companions to roast fowl and ham or simply to eat out of hand, sweet as any candy, only better for your body. Their peak sweetness season is March, April, and May, but they are never tart. Late inventories in June or July tend to be dry at the stem end and should be avoided, as they are expensive and disappointing. Size 56 is the most popular, least puffy, and thinnest-skinned. Look for bright, heavy fruit without spots. Avoid fruit that is at all shriveled or has puffy stem ends.

There are some Florida navel oranges that come to market in late January for several weeks. They have a bronzed light-orange peel, with pale fruit inside. Their thin skin makes them a bit difficult to peel, but they are sweet and very juicy and grow very large.

Jaffa oranges are very close to the California navel orange. They are grown in Israel, are very easy to peel, and are very sweet. Since they are a prime export product of Israel, their quality is always high and consistent. Look for them in late winter and early spring.

Blood oranges can be slightly tart or very sweet and have medium-heavy peels that can be slightly tinged with red or very red. They are mostly grown in Israel and Mediterranean countries. Sicilians love blood oranges best and use them in salads, desserts, as appetizers, and for juice. They are becoming increasingly popular as an orange crop in California and are beginning to appear regularly in our markets and stores, but probably not in supermarkets. Blood oranges can be peeled, halved, sliced, or squeezed. Their red juice deepens the color and appeal of fresh O.J., and their color makes them delightful additives to salads and punch drinks. Look for blood oranges from December to early spring. They are delicious and fun to eat and put a new sparkle and attraction in the fruit bowl on your table.

All navel and Valencia oranges, whether grown in California or Florida, offer the same high-quality well-balanced nutrition. They are high in vitamins A, C, B-1, B-2, and B-6. Oranges provide a full range of all the essential minerals, especially potassium, iron, and copper. Oranges are high in fruit carbohydrates, which are in a highly usable form as energy and do not store up as fat.

NUTRITIONAL DATA

Weight (g)	Calories	Protein (g)	Carbohydrates (g)	Crude Fiber (g)	Water (g)	Fat (g)	Cholesterol (mg)
140	65	1.4	16.3	0.6	121.5	0.1	—
121	59	1.3	14.4	0.6	104.5	0.4	—

Vitamin A (IU)	Vitamin C (mg)	Thiamine B-1 (mg)	Riboflavin B-2 (mg)	Niacin (mg)	Vitamin B-6 (mg)	Vitamin B-12 (mcg)	Folic Acid (mcg)	Pantothenic Acid (mg)
256	80	0.12	0.06	0.4	0.1	—	47	0.35
278	59	0.11	0.05	0.3	0.08	—	47	0.30

Sodium (mg)	Potassium (mg)	Calcium (mg)	Phosphorus (mg)	Magnesium (mg)	Iron (mg)	Zinc (mg)	Copper (mg)	Manganese (mg)
1	250	56	27	15	0.17	0.08	0.078	0.038
—	217	48	21	12	0.11	0.07	0.045	0.028

Navel, raw—1 medium
Valencia, raw—1 medium

THE MANDARINS

In addition to the varieties of sweet oranges, there is a whole group of peeling oranges classified as mandarins. It is supposed that all oranges began as small bitter fruit in the ancient provinces of China. History, time, and man directed the evolution of these fruits into the sweet, succulent varieties we eat today. This group of easy-peeling sweet dessert oranges, called mandarins, includes tangerines, Murcotts, Temples, tangelos, clementines, and the small green-skinned elegant mandarin orange itself.

TANGERINES

Tangerines were named for the city of Tangiers, that mysterious port city of Morocco. Sicily imported them, replanted them, and made them popular in the Western world. Tangerines are flattish oranges, very deep in color, with a soft depression at the stem end. The peeling of a tangerine almost always begins at this point. Tangerines are highly aromatic, and the perfume permeates the room as soon as the skin is broken. Tangerines peel very easily—too easily, in fact, as it causes us "hurry up" people to leave behind the clinging white web. Take the time to remove all the webbing carefully from the fruit, and you will be surprised at the improvement in the tangerine's taste, texture, and sweetness. Tangerines are best from December through February, although they appear earlier and hang around longer. Select shiny tight-skinned fruit, which is almost always best in the medium sizes. Very large fruit is often tasteless; very small is bitter. Avoid puffy skins, soft fruit, and popped stem ends. These are usually signs of tasteless fruit. Late-season crops, generally sold out of storage, can be very sweet if the fruit is still firm and shiny.

Tangerines seem to be the most nutritious of all the mandarins, with a full range of vitamins, especially vitamin A, and vital minerals. They are practically salt-free and are very low in calories, with about-average usable amounts of carbohydrates for sweet fruit.

NUTRITIONAL DATA

Weight (g)	Calories	Protein (g)	Carbohydrates (g)	Crude Fiber (g)	Water (g)	Fat (g)	Cholesterol (mg)
84	37	0.5	9.4	0.3	73.6	0.2	—

Vitamin A (IU)	Vitamin C (mg)	Thiamine B-1 (mg)	Riboflavin B-2 (mg)	Niacin (mg)	Vitamin B-6 (mg)	Vitamin B-12 (mcg)	Folic Acid (mcg)	Pantothenic Acid (mg)
773	26	0.09	0.02	0.1	0.06	—	17	0.17

Sodium (mg)	Potassium (mg)	Calcium (mg)	Phosphorus (mg)	Magnesium (mg)	Iron (mg)	Zinc (mg)	Copper (mg)	Manganese (mg)
1	132	12	8	10	0.09	—	0.024	0.027

Raw—1 medium

TANGELOS

*T*angelos are the largest of the mandarin group. No wonder—they are a cross between a tangerine and a pomelo, a now-forgotten ancestor of the grapefruit. Tangelos have a deep-orange skin shaded with bronze. They grow flattish or round (are the tangerine and pomelo still struggling for the upper hand?) and have a knobby stem end. They have a light-orange flesh and are not seedless. Tangelos produce a high volume of tart, sweet juice with a distinctive flavor. I recommend a mix of tangelo juice with fresh orange or grapefruit juice for yet another delicious citrus-juice experience.

Tangelos are shipped from Florida in December and are available until February. Look for hard shiny specimens that feel heavy in your hand. Soft, puffy, dull tangelos taste that way.

Tangelos have no vitamins except C but have many minerals, especially potassium.

NUTRITIONAL DATA

Weight (g)	Calories	Protein (g)	Carbohydrates (g)	Crude Fiber (g)	Water (g)	Fat (g)	Cholesterol (mg)
170	39	0.5	9.2	—	—	0.1	—

Vitamin A (IU)	Vitamin C (mg)	Thiamine B-1 (mg)	Riboflavin B-2 (mg)	Niacin (mg)	Vitamin B-6 (mg)	Vitamin B-12 (mcg)	Folic Acid (mcg)	Pantothenic Acid (mg)
—	26	—	—	—	—	—	—	—

Sodium (mg)	Potassium (mg)	Calcium (mg)	Phosphorus (mg)	Magnesium (mg)	Iron (mg)	Zinc (mg)	Copper (mg)	Manganese (mg)
2	296	27	20	19	0.2	—	—	—

Raw—1 medium

MURCOTTS

Murcotts are my favorite mandarin oranges. They are also called Honeybelles, and sweet as honey and beautiful they are. The skin is light orange with bronze speckles and is very tight and shiny. Nevertheless, it still peels easily, without a webbing, to reveal the juicy segments within. Murcotts are practically seedless, smaller than tangerines, and extremely sugary. They can be squeezed and the juice added to grapefruit juice for sweetness, to lemonade to reduce the sugar content, and to orange juice to enhance its smoothness. Murcotts are grown only in Florida. They appear at fruit stands and knowledgeable markets in January and, in a good season, last until April.

TEMPLE ORANGES

Temple oranges are larger than tangerines. They are flattish and heavily bronzed and can be slightly rough-skinned. They often have a slight protrusion, or cap, at the blossom end, and the skin emits a distinctive oil fragrance that definitely identifies the Temple.

Their taste is also quite distinctive. They are delicious quartered and eaten out of the skin. Temple oranges peel rather easily, and I always enjoy them after a holiday meal. Temples get you with their flavor, not just their sweetness, and they seem to have a pleasant effect that moderates the full feeling that follows a full meal.

Temples have a high juice content and can be added to fresh orange juice, less for sweetness than to add another delicious aspect to the overall flavor. Select shiny, heavy fruit. Avoid puffy fruit and any with obvious soft stem ends. Temples are shipped from Florida and are at their best from December until February.

CLEMENTINES

Clementines are the candy of oranges. They are a cross between the sweet orange and the original small Chinese mandarin. They are, in turn, small, very orange, and very sweet. Clementines have a tight little structure but are easy to peel, with no webbing and few, if any, seeds. Clementines make excellent snacks with excellent carbohydrate energy.

Most people have no idea of the varieties available in oranges and look upon clementines as some sort of small tangerine. They are not. They are their own thing, with a distinct taste and very tender segments. Select solid, shiny, heavy (for the size) fruit, without puffiness or soft stem ends. Look for clementines in fine produce markets and intelligently stocked fruit stands. You will probably not find them in supermarkets unless they are mislabeled tangerines.

MANDARIN ORANGES

Mandarins are the source of all the other mandarin varieties, with their different colors, shapes, tastes, and quirks. The oldest of all the oranges and probably the least-known, mandarin oranges are sold for big prices drenched in sugar in tiny little cans. They are much better fresh.

I never saw mandarin oranges in any market until several years ago during the late fall. I saw a small box of small emerald-green, shiny, tangerinelike fruits that sparkled with life. The label said "140 mandarin oranges." Green oranges? I peeled one and discovered a brilliant deep-orange morsel inside. It was seedless and netless. I bought five boxes and spent the whole day peeling samples for my customers and selling out. I went back to the market, looking for more. None was available. My customers returned, also looking for more. The truth be that good as they are, mandarin oranges are mostly produced for canning. People don't buy green-skinned fruit, especially deep emerald-green fruit. Only knowledgeable, creative, and innovative fruit vendors like me take the time to sample and then educate and promote the unfamiliar. Nevertheless, the mandarin is a rare and deliciously wonderful fruit of the *Citrus reticulata* family. I urge you to keep up a sharp eye and a sweet tooth in hope that the fresh mandarin will come your way.

PEELING ORANGES WITHOUT A KNIFE

All oranges peel best from the stem end. Remove (peel or bite) a quarter-sized piece of peel from the stem end, and using your thumb, peel downwards a little at a time around the fruit. Work the peel loose. Patience and practice will develop technique and speed.

PEELING ORANGES WITH A KNIFE

Cut a quarter-sized piece of peel off both ends of the fruit. Make six or seven peel-deep cuts from end to end. Remove the peel sections.

ORANGE SLICES ESPAÑA

2 large navel oranges
6–8 mint leaves, chopped
Olive oil
Freshly ground black pepper

Peel and slice oranges. Place in overlapping pattern on plate. Sprinkle with mint leaves and lightly with olive oil. Sprinkle significantly with black pepper. Chill.
Serves 3–4.

ORANGE MARMALADE

12 Florida or California Valencia oranges
2 cups sugar

Squeeze juice of 4 of the oranges. Finely cut up peel of all 12 oranges. Heat 1 cup water to simmer. Gradually add orange peel, juice, and sugar, stirring often. Simmer for 2 to 3 hours. Add more water if mixture begins to get too thick. Cool. Pack in 2 sterilized ½-pint jars. Seal (see "Canning process" entry in Glossary, page 000).

MANDARIN SEGMENTS IN SIMPLE SYRUP

8 mandarin oranges
1 cup sugar or ¼ cup honey
8–10 mint leaves, crushed

Peel and segment 6 of the mandarins. Squeeze remaining 2 mandarins. Melt sugar in 2 cups water. Cool. Mix with mandarin juice. Place segments in bowl with crushed mint. Pour cooled syrup over. Chill for several hours. Serve with ice cream, shortcake, cottage cheese, sour cream, gelatin molds, Boston lettuce, or mixed with grapefruit segments. The syrup makes an excellent basting liquid for duck, with the fruit served along with the carved meat.

LEMONS

Lemons are the sparklers of the fruit world. The peel and its oil, the fruit and its juice, add a delightfully refreshing zing to cocktails, punches, salads, dressings, stuffings, seafood (especially shellfish), melons, sherbets, teas, fruit ades, even seltzer and mineral water. Grated lemon peel mixed with bread crumbs makes an excellent fish breading or stuffing, while the juice of one lemon mixed with hot water and drunk at rising time and bedtime will cleanse the system, stimulate liver activity, sweeten the stomach, stimulate mouth tissue, and brighten your eyes.

Excellent-quality lemons are available year-round at consistent prices, except for the very peak of summer heat, when lemon consumption is at its maximum and supply at its minimum. Although Texas and Florida produce some lemons, the best lemons come from California under the Sunkist label, although Pure Gold is also excellent. All lemons are picked green and ripened to yellow artificially. Lemons cannot be tree-ripened. They become insipid, soft, and easily perishable. Lemons are one of the very few examples of manipulation creating a better, more useful, more economical fresh food.

Look for shiny bright lemons that feel dense and heavy for their size. Avoid puffy skins, dull or brownish coloring, and very large lemons, which are probably mostly peel and little juice.

Lemons are not high in nutrition but provide a broad range of vitamins and minerals. They act as flavor stimulants to increase our use of

more-nutritious foods, and they work in other mysterious ways to benefit
our bodies.

NUTRITIONAL DATA

Weight (g)	Calories	Protein (g)	Carbohydrates (g)	Crude Fiber (g)	Water (g)	Fat (g)	Cholesterol (mg)
58	17	0.6	5.4	0.2	51.6	0.2	—

Vitamin A (IU)	Vitamin C (mg)	Thiamine B-1 (mg)	Riboflavin B-2 (mg)	Niacin (mg)	Vitamin B-6 (mg)	Vitamin B-12 (mcg)	Folic Acid (mcg)	Pantothenic Acid (mg)
17	31	0.02	0.01	0.06	0.05	—	6	0.11

Sodium (mg)	Potassium (mg)	Calcium (mg)	Phosphorus (mg)	Magnesium (mg)	Iron (mg)	Zinc (mg)	Copper (mg)	Manganese (mg)
1	80	15	9	—	0.35	0.04	0.021	—

Raw—1 medium

LEMON DRESSING

Juice and grated peel of 3 lemons
¼ teaspoon finely chopped tarragon
¼ teaspoon finely chopped savory
½ cup olive oil
¼ teaspoon freshly ground black pepper

Combine all ingredients and mix thoroughly. Serve dressing separately from salad.

LIMES

It is amazing how little use the lime gets besides as a wedge or squeeze at the bar. Limes grow in Florida and in the Caribbean islands, where they are used in everything, even in eggs. And why not? Nothing tastes quite like the fresh sharp taste of a sprinkle of lime. It is especially good squeezed on tropical fruits like mango, papaya, pineapple, and watermelon. Sprinkle lime juice on broiled fish after cooking, and it will definitely bring out the best the fish has to offer. It will lighten the syrupy sweetness of a Coca Cola and makes a world of difference between a Cuba Libre and an ordinary rum-and-Coke. Squeeze lime on your salads, add it to any dressing, sprinkle the juice of a wedge or two on lamb chops or a charcoaled steak.

Learn to use limes, limes on everything. Look for bright-green thin-skinned limes called Persians, which are seedless and no less than sixty-three per cent juice. The small round yellow Key limes, which are almost pure juice, are hardly available anywhere anymore except for southern Florida, where they are the key ingredient of Florida's famous Key lime pie. It is, by the way, wonderful.

Always select hard shiny fruit without brown blotches or dryness. Limes tend to be expensive in winter and spring, but mostly they are inexpensive and readily available. Keep your limes in a plastic bag in the refrigerator. They hate the heat, and they hate to be chilled below 38° F.

Limes, like lemons, are not nutritionally impressive. They are well-balanced and do well in the mineral department. Mostly, though, limes make foods taste better and open up new ideas in salad dressings, fish preparations, and refreshing drinks.

NUTRITIONAL DATA

Weight (g)	Calories	Protein (g)	Carbohydrates (g)	Crude Fiber (g)	Water (g)	Fat (g)	Cholesterol (mg)
67	20	0.5	7.1	0.3	59.1	0.1	—

Vitamin A (IU)	Vitamin C (mg)	Thiamine B-1 (mg)	Riboflavin B-2 (mg)	Niacin (mg)	Vitamin B-6 (mg)	Vitamin B-12 (mcg)	Folic Acid (mcg)	Pantothenic Acid (mg)
7	20	0.02	0.01	0.1	—	—	6	0.15

Sodium (mg)	Potassium (mg)	Calcium (mg)	Phosphorus (mg)	Magnesium (mg)	Iron (mg)	Zinc (mg)	Copper (mg)	Manganese (mg)
1	68	22	12	—	0.4	0.07	0.044	—

Raw—1 medium

UGLI FRUIT

No one knows what ugli fruit started out as, but somehow it became the hybridized result of an orange or tangerine and grapefruit—maybe. Whatever the mating, the result is a yellow, pink, bronze, green-skinned fruit with a puffy protuberance at its stem end. In this case, however, ugly is only skin deep. Inside are sweet pink segments separated by a thin tender webbing. Ugli fruit is juicy and flavorful, without any of the tart taste found in grapefruit.

Ugli fruit is harvested in Florida and found in many stores from late winter until mid-spring. In case you don't recognize it, look for the little label that says "Ugli," and pick out the prettiest ones. It is, of course, still a novelty fruit, but it is appearing more often in stores throughout the country. I even saw some stale specimens in the local supermarket. Unfortunately, they were displayed next to some very fine grapefruits, which made them look uglier and a lot less desirable. We sell ugli fruit, peeling many samples to develop a wider appreciation and demand for this delicious, worthwhile fruit.

ORANGES

285

BERRIES

1. WILD STRAWBERRIES
2. CALIFORNIA COMMERCIAL STRAWBERRIES
3. LOCAL FARM STRAWBERRIES
4. BLACK RASPBERRIES
5. RED RASPBERRIES
6. GOLD RASPBERRIES
7. GOOSEBERRIES
8. MULBERRIES
9. BOYSENBERRIES
10. BLACKBERRIES
11. BLUEBERRIES
12. CRANBERRIES

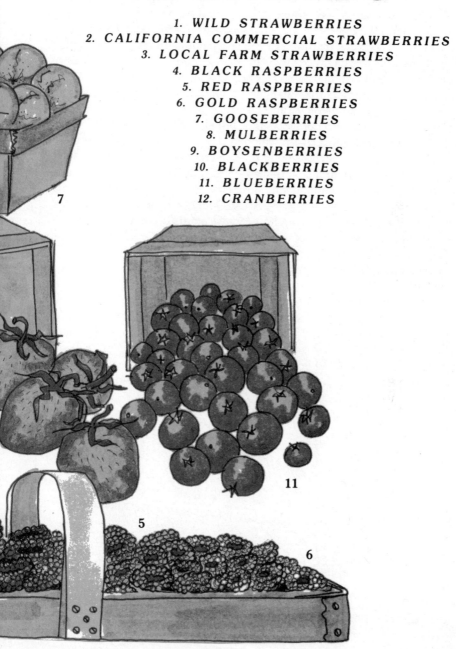

No fruits are more popular nor are any fruits better for you than the beautiful berries that come to us in great abundance from spring through fall. Whether it's the tiny wild strawberry gathered from a meadow on a sunny afternoon or the succulent giant red raspberry that tempts us with its beauty and sweetness, berries are the crown jewels of fruit. They are all rich in vitamin B-1 and B-2, with enormous amounts of vitamin C and considerable quantities of calcium, magnesium, and iron. They provide gentle digestive and cleansing action. Berries are not only beautiful, delicious, and nutritious; they are also extremely versatile. They can be eaten fresh, with cream whipped or not, stewed, sliced in salads, puréed, blended, made into yogurt and ice-cream flavors, served in wine, frozen, canned, or turned into pies, tarts, cobblers, jellies, jams, and preserves. They grow abundantly in almost all areas in season and are truly one of nature's greatest gifts. Always look for bright-colored berries, except for blueberries, which should have a frosty bloom upon them. Avoid wetness, dark stems, and crushed, bruised, or damaged fruit, except if it's for cooking or canning.

I recommend that every home keep a small raspberry or blackberry patch nearby. Just passing it by each day in season will provide you with a constant daily supply of fresh sweet berries you can enjoy out of hand as you go on your way about your day. Make berries part of your breakfast, your summer desserts, and your winter frozen and canned shelves. They'll improve your diet and your menu, provide a better afternoon snack, expand your fresh-food experience, and maybe replace some sugar and carbohydrate starch with more healthful and usable food values.

WILD STRAWBERRIES

*F*or centuries the only strawberries available were the tiny ultrasweet wild ones. Whenever they were first found in a new place, it was always noted that the fields were so full that there was no place to step. Strawberry time became a family affair. And for all the berries' goodness, it took many hours over many days to bring home any real quantity.

Wild strawberries grow best in low grassy alfalfa or clover fields, where they appear in late June—earlier where it's warmer, later in colder climates. Deer love them, and many a fawn has spent most of its early life bedded down in a wild strawberry field, with its delicate senses overwhelmed by the fragrance and taste of this pure sweet fruit.

Wild strawberry plants send out runners at the same time as they produce both blooms and fruit. The runners plant themselves, establish a root and leaf system the first year, then produce flowers and fruit and runners the following two years. This way the wild strawberry field is perpetually maintained. "Strawberry Fields Forever," John Lennon, 1967.

Wild strawberries are the parents of all modern varieties, and they are definitely American. Plants brought to France from Virginia were planted quite by accident next to plants relocated from the western shores of Chile. The unplanned hybrid of two small sweet fruits, one red, one yellow, created the much larger red-hearted gold-seeded strawberry from which all others have been developed.

LOCAL STRAWBERRIES

Local strawberries are the large cultivated descendants of the original wild varieties. They are grown for sale within a short distance of the farm. Since the shipping distance will be short, the fruit is allowed to ripen fully on the plant. These last days of natural sun-ripening increase both the sugar and the vitamin C content enormously. The berries become darker, sweeter, and nutritionally superior. They are also more fragile. Most farm stands and local produce vendors get local strawberries fresh every day in mid-June and early July. The season is short, so buy lots of them. Eat them, freeze them. Make jam, jelly, and preserves. They are wonderful.

The Rosella, or Scarlet, variety is a very popular local farm strawberry, very round, with a tight red heart and deep stem. Look for dark, dry berries with fresh leaves and green stems in quart baskets that show no stains or seepage. Since local strawberries are picked ripe, they stand very little handling. They are usually sold in baskets as they were picked and packed in the field. Fresh high-quality local strawberries have very little waste *if they have been handled gently* and are not three- or four-day-old leftovers. Look for stores that know their berries and have a reputation for quality. If you are not sure, ask your clerk or professional fruit man to tumble a quart and show you the bottom berries.

COMMERCIAL STRAWBERRIES

Make no complaint. Commercial strawberries are far better for all of us than no strawberries. Thanks to modern agricultural packing and shipping techniques, we can have excellent-quality strawberries year-round. Strawberries at Christmas are just one of the many benefits of the marvelous concept called the American food chain.

Whether they come to us from New Zealand, Argentina and Chile, or California, Mexico, and Florida, the varieties of strawberries available are very close relations. They all had the same two wild parents. They tend to be an orange-red rather than maroon-red, elongated, often with a raised stem socket and an open heart. Commercial strawberries have a very firm texture, with definitely larger, tougher seeds. If you select ripe strawberries with fresh green leaves and stems and a bright sparkling color, you'll be more than pleased with the taste you get. Though somewhat denser and slightly less sweet than local varieties, most commercial strawberries are better than good enough.

The calendar for strawberries is twelve months long, although winter-season berries are imported and cost a lot more money than seasonal inventories. During the deep winter months strawberries arrive by plane from the lands on the opposite side of the Equator: Chile, Argentina, and New Zealand. The Chilean and Argentinian specimens suffer a bit from South American carelessness in handling and packing. New Zealand berries arrive in top form due to the country's concentration on building a national industry of fine exported fruits. As spring approaches, Guatemala and especially Mexico ship big luscious berries by truck to our southern areas and by plane to areas further north. Mexican growers have developed a sizeable frozen-strawberry business, catering to American ice-cream, baked-goods, and jelly companies. Mexico exports better fruit than it ships internally because its internal distribution system suffers from bad roads, unreliable trucks, and poor handling and refrigeration.

By April Florida strawberries are abundant, cheap, and usually of high quality and excellent flavor. California strawberries arrive on the heels of Florida's crop, and the early harvests are horticultural marvels. The berries are giants, and they are delicious. California also ships an unusually large berry with a long stem under the Driscoll brand. Four or five berries fill a pint basket. They have a deep, delicious flavor, almost winelike. These Driscolls are not cheap, but neither are they unreasonably expensive. They are simply top of the line. California strawberries keep on coming until local varieties are picked in mid-June. When these supplies finish up, usually by midsummer, there is a short slack season until the early fall, when California ships a second harvest; some of the imports start reaching us by midautumn.

All commercial strawberries are shipped in pint baskets, twelve to a flat. Sometimes merchants refill quart baskets with two pints and lower the price to less than that of two pints. This is a good deal for the storekeeper and the consumer. The merchant sells more, and the consumer gets to eat more. Most stores no longer "plate" strawberries. Plating is the art of arranging the top layer of the fruit in the basket in a pattern so that each berry is presented equally. Unfortunately, most merchants plated their berries to hide the little ones. We still plate berries at Sunfrost. We do not put all the big ones on top. What you see on the top is what you get on the bottom. We empty, refill, and plate every basket so every berry is perfect when you buy it and when you eat it. We sell lots and lots of strawberries.

The whole world loves strawberries, but nobody loves them like Americans. Of the 300,000 acres of world strawberry cultivation, almost half are in the United States. Strawberries are one of the foods America gave the world, and when our strawberry season ends, the world harvest comes here.

Strawberries are top-shelf nutrition, with a complete range of significant food values. Although their protein content is not high, it is above average, and they contain very important amounts of vitamins C, B-1, B-2, and B-6, plus all the essential minerals, especially manganese, potassium, copper, and iron. Strawberries are recommended as essential for cardiac health and offer good nutritional energy that is easy to digest and process.

NUTRITIONAL DATA

Weight (g)	Calories	Protein (g)	Carbohydrates (g)	Crude Fiber (g)	Water (g)	Fat (g)	Cholesterol (mg)
149	45	0.9	10.5	0.8	136.4	0.6	—

Vitamin A (IU)	Vitamin C (mg)	Thiamine B-1 (mg)	Riboflavin B-2 (mg)	Niacin (mg)	Vitamin B-6 (mg)	Vitamin B-12 (mcg)	Folic Acid (mcg)	Pantothenic Acid (mg)
41	85	0.03	0.1	0.3	0.09	—	26	0.51

Sodium (mg)	Potassium (mg)	Calcium (mg)	Phosphorus (mg)	Magnesium (mg)	Iron (mg)	Zinc (mg)	Copper (mg)	Manganese (mg)
2	247	21	28	16	0.57	0.19	0.073	0.432

Raw—1 cup

STRAWBERRY PIE

3 pints strawberries
1 cup sugar
3 tablespoons cornstarch
1 9-inch pie shell, baked
½ pint heavy cream, whipped

Wash, drain, and de-stem strawberries. Crush enough strawberries to make 1 cup. Combine sugar, cornstarch, crushed strawberries, and ½ cup water in saucepan. Cook until mixture thickens, stirring constantly. Cool. Slice, halve, or quarter remaining strawberries and spoon into pie shell. Pour cooked mixture over top. Refrigerate for 3 hours or until strawberries set. Top with chilled whipped cream.

RASPBERRIES

Raspberries are older than Christianity by many centuries. Their delectable essence and pure sweetness are the taste benefits of a highly nutritious food. They are almost completely assimilated as vitamins, minerals, and energy. This high-quality fruit, selling for as high as six dollars a half-pint out of season, is never really inexpensive. Raspberries are not easy to pick, since any broken berries or crushed drupelets make them unacceptable for shipping or packing. Picking is slow and tedious, and many raspberry-growers sell their crops on the bush to "pick your own" customers who fill their pails (and tummies too) for a fixed price. Raspberries come in three colors: red, the most popular; black, the most durable; and golden, which are almost pure sugar and leave not even a trace of the rasp (as in raspberry) associated with the familiar raspberry taste.

Raspberries can be enjoyed plain, mixed with yogurt or ice cream, blended with other fruits or juice, baked in pastries, sprinkled on whipped cream, chilled with heavy cream, or elaborately prepared as a frozen cream-cheese–meringue–almond parfait! Raspberries also make a sauce that wonderfully complements duck and other game fowl.

Local raspberry crops are available twice a year at roadside and farm stands, once in late June and early July, and again in September and early October. Shipments of large, durable, and expensive raspberries leave Oregon, Washington, and Idaho in April and May and are available with expensive price tags at knowledgeable fruit stands and even some supermarkets. Look for dry, firm fruit with excellent form and hollow centers. Although they are never inexpensive, try to enjoy raspberries as often as you can, especially in season. There is no finer fruit.

Raspberries lead all the berries nutritionally. They lead all fresh foods in fully digestable elements and are almost totally assimilated by the body during digestion. They provide completely usable energy, with very high amounts of vitamins A, C, B-1, B-2, and B-6, and are extraordinarily rich in manganese and potassium. Manganese is essential to the fertility processes, and raspberry tea is excellent for building healthy tissue during pregnancy. It is also recommended for relief of menstrual cramps and is an effective antidiarrheal agent. Raspberries, along with avocados and mangos, are the richest sources of copper in fresh foods.

NUTRITIONAL DATA

Weight (g)	Calories	Protein (g)	Carbohydrates (g)	Crude Fiber (g)	Water (g)	Fat (g)	Cholesterol (mg)
123	61	1.1	14.2	3.7	106.5	0.7	—

Vitamin A (IU)	Vitamin C (mg)	Thiamine B-1 (mg)	Riboflavin B-2 (mg)	Niacin (mg)	Vitamin B-6 (mg)	Vitamin B-12 (mcg)	Folic Acid (mcg)	Pantothenic Acid (mg)
160	31	0.04	0.11	1.1	0.07	—	—	0.3

Sodium (mg)	Potassium (mg)	Calcium (mg)	Phosphorus (mg)	Magnesium (mg)	Iron (mg)	Zinc (mg)	Copper (mg)	Manganese (mg)
—	187	27	15	22	0.7	0.57	0.091	1.246

Raw—1 cup

GOOSEBERRIES

Gooseberries are not very popular in the United States. We like our berries sweet, and gooseberries are not. The white gooseberry has a veined smooth, transparent skin, while the red gooseberry has a hairy surface. Gooseberry seeds are enclosed within the berry. The European gooseberry is much larger and easier to prepare in jams, preserves, and cooking sauces. Some New Zealand varieties are available during the winter months, when more elaborate cooking is done at home, while domestic varieties make rather rare appearances in stores and fruit stands in late summer.

The gooseberry plant hosts a bacterial disease that is deadly to the American white pine tree. As a result, the cultivation of gooseberries is prohibited in many American states. With their less-than-sweet taste, their difficulty of preparation, and their danger to the very important white pine, gooseberries will never make it big in the United States. Look for gooseberry preserves and jams, imported mostly from England and Germany.

For all their problems, gooseberries are nutritionally excellent. They possess very high quantities of minerals, especially potassium; protein; and vitamins A, C, B-1, B-2, and B-6.

NUTRITIONAL DATA

Weight (g)	Calories	Protein (g)	Carbohydrates (g)	Crude Fiber (g)	Water (g)	Fat (g)	Cholesterol (mg)
150	67	1.3	15.3	2.9	131.8	0.9	—

Vitamin A (IU)	Vitamin C (mg)	Thiamine B-1 (mg)	Riboflavin B-2 (mg)	Niacin (mg)	Vitamin B-6 (mg)	Vitamin B-12 (mcg)	Folic Acid (mcg)	Pantothenic Acid (mg)
435	42	0.06	0.05	0.5	0.12	—	—	0.43

Sodium (mg)	Potassium (mg)	Calcium (mg)	Phosphorus (mg)	Magnesium (mg)	Iron (mg)	Zinc (mg)	Copper (mg)	Manganese (mg)
1	297	38	40	15	0.47	0.18	0.105	0.216

Raw—1 cup

MULBERRIES

Mulberries come in white, red, and black. None is available in commercial quantities in any store, and only the red mulberry seems to grow well in the United States. The mulberry grows in a weeping-willow–type tree, the thin hanging branches forming an umbrella reaching almost to the ground.

Mulberry trees were originally grown for silkworm production. The white tree, having proven better for silkworm life, is used exclusively for that purpose. The white mulberry is almost pure sugar and is purely tasteless, with no acidity or flavor whatsoever. Silkworms like white mulberry leaves best. I, for one, am willing to leave all of the white variety to these busy silk-makers. The black mulberry tree grows mostly in Europe. The red mulberry tree, which thrives best on our East Coast from Maine to Florida, produces a small berry of many drupelets that retain its stem when picked.

Red mulberries are very sweet and flavorful but do not keep well and stain heavily. Best advice: Eat 'em as you pick 'em.

The mulberry is classified as a berry, and many botanists relate it to the fig, also a berry. Nutritionally this fruit has the same kind of high mineral content as figs and other berries. I have never seen any mulberries that were commercially cultivated. If you find them in their natural state, you will enjoy a refreshing, delightful fruit with rather high amounts of important minerals and better-than-average fruit protein. Happy hunting.

NUTRITIONAL DATA

Weight (g)	Calories	Protein (g)	Carbohydrates (g)	Crude Fiber (g)	Water (g)	Fat (g)	Cholesterol (mg)
140	61	2.0	13.7	1.3	122.8	0.6	—

Vitamin A (IU)	Vitamin C (mg)	Thiamine B-1 (mg)	Riboflavin B-2 (mg)	Niacin (mg)	Vitamin B-6 (mg)	Vitamin B-12 (mcg)	Folic Acid (mcg)	Pantothenic Acid (mg)
35	51	0.04	0.14	0.9	—	—	—	—

Sodium (mg)	Potassium (mg)	Calcium (mg)	Phosphorus (mg)	Magnesium (mg)	Iron (mg)	Zinc (mg)	Copper (mg)	Manganese (mg)
14	271	55	53	25	2.59	—	—	—

Raw—1 cup

BLACKBERRIES AND BOYSENBERRIES

Blackberries and boysenberries are not hollow like raspberries. When picked, they leave their stems behind but keep their caps within. Boysenberries are the result of an experimental blackberry-raspberry cross made by Rudolf Boysen. What he got was a slightly larger, lighter, and slightly sweeter blackberry that looks exactly like the black parent it came from. Blackberries are ripe only when every drupelet turns black. Boysenberries are ripe when each drupelet turns deep purple. Just one light-colored drupelet will make the whole berry bitter, so keep a sharp eye.

Both blackberries and boysenberries appear in late summer and fall after being picked from very thorny bushes with long arching canes. The very best blackberries and boysenberries are cultivated varieties where the canes are cut back and undergrowth is removed. This creates bigger berry clusters and much less thorny material to work around. A small patch of blackberries and boysenberries is fun to have near the house. The fruit ripens daily and gives goodly amounts of sweet fruit for about six weeks.

Blackberries and boysenberries are excellent with cream in cereal, with whipped cream, and on ice cream. Blackberries also make excellent sherbet and fruit ice, while one of the really delicious ice creams is made with boysenberries. Actually, boysenberry ice cream could be my favorite.

Boysenberries and blackberries are available at roadside stands, farm markets, and knowledgeable fruit stands. Most vendors will, however, label their boysenberries blackberries and let it go at that. Blackberries are *really black*. I think it is also correct to say that blackberries are slightly firmer. My best advice is that if you see either, buy them and enjoy them by any name. The extra-large blackberries that thrive in the Adirondack region of New York State are called sheep's tits.

Blackberries and boysenberries are significantly high in nutrients, especially essential minerals and vitamins A and C. They have excellent nonfattening energy and are completely digestible.

NUTRITIONAL DATA

Weight (g)	Calories	Protein (g)	Carbohydrates (g)	Crude Fiber (g)	Water (g)	Fat (g)	Cholesterol (mg)
72	37	0.5	9.2	3.0	61.7	0.3	—

Vitamin A (IU)	Vitamin C (mg)	Thiamine B-1 (mg)	Riboflavin B-2 (mg)	Niacin (mg)	Vitamin B-6 (mg)	Vitamin B-12 (mcg)	Folic Acid (mcg)	Pantothenic Acid (mg)
119	15	0.02	0.03	0.3	0.04	—	—	0.17

Sodium (mg)	Potassium (mg)	Calcium (mg)	Phosphorus (mg)	Magnesium (mg)	Iron (mg)	Zinc (mg)	Copper (mg)	Manganese (mg)
—	141	23	15	14	0.41	0.2	0.101	0.93

Raw—½ cup

BLUEBERRIES

Blueberries come in cardboard pint baskets with or without a cellophane cap. They also come on wild blueberry bushes and can be picked to your heart's content. The difference between the cultivated blueberry in the basket and the wild blueberry on the bush is sweetness and size. Cultivated blueberries are larger and sweeter. Some folks prefer the tarter taste of the wild ones, as well as the pleasure of eating them immediately, almost as they grow. I like my blueberries big and sweet.

Good inventories of cultivated blueberries are available throughout the United States from June through early September from the Carolinas, Michigan, and Massachusetts, as well as the Garden State of New Jersey. Without a doubt the biggest, best, sweetest ones come from the southern part of New Jersey around Hammonton, Vineland, and Whitesbog, where Elizabeth White began offering prizes for outstanding bush specimens in the early 1900s. All commercially cultivated blueberry varieties sold in the United States today are the result of her work. There are several shipping co-ops in New Jersey that package and distribute all over the United States. I think the best label is Blue Buck. Like Florigold is to grapefruit, Sunkist is to oranges, and Chiquita is to bananas, Blue Buck is to blueberries. They only ship fresh, dry, luscious, top-of-the-line merchandise, always reasonably priced and always of consistent, reliable quality.

Near the end of the season Massachusetts picks and ships open containers of a smaller wilder variety that I find especially delicious. These smaller blueberries don't have the meat and size of the Jersey blues, but they have a wonderful flavor, and their small size makes them delicious by the handful.

Huckleberries are a different kettle of fish, or pail of berries, as the case may be. Huckleberries are smaller than blueberries and less sweet. They also have large seeds, or "bones," which makes eating them a bit more difficult. Although blueberries are often called huckleberries, they are not the same, and huckleberries only grow wild. Indians, nature people, and vacationers gather and eat huckleberries, but they never appear in stores.

Blueberries are one of the most delightful summer fruits, with a wide variety of uses. They can be eaten plain or with cream and sugar (and bananas!), with whipped cream, ice cream, cheesecake, or yogurt. They make great jam, jelly, syrup, juice, and dessert topping. And let us not forget our local American hero, the blueberry pie. Does anything say America more or taste better than blueberry pie topped with ice cream? Fresh-baked blueberry muffins and blueberry tarts are other American favorites that are part of our kitchen-table culture. And when dinner's done and the children are quiet and the fire begins to cozy up the house, there's blueberry wine, blueberry cordial, or blueberry brandy to cap off your all-American blueberry day.

When buying blueberries, avoid pint baskets with wrinkled cello caps or stained cardboard baskets. Look for plump uniform deep-blue berries with the frosty dew or dusty, powdery bloom that is the sign of freshness. Fresh blueberries can be frozen right in their cardboard container with the cellophane cover. Just put them in the freezer. Buy a whole flat of twelve and freeze them all. You'll be able to enjoy them during winter as desserts, toppings, and cereal fruit. They'll bring a touch of summer to your winter table, along with their natural nutrition and delicious taste.

Blueberries are an excellent food nutritionally. They provide complete and balanced nutrition, with very high amounts of minerals and usable nonfattening energy.

NUTRITIONAL DATA

Weight (g)	Calories	Protein (g)	Carbohydrates (g)	Crude Fiber (g)	Water (g)	Fat (g)	Cholesterol (mg)
145	82	1	20.5	1.9	122.7	0.6	—

Vitamin A (IU)	Vitamin C (mg)	Thiamine B-1 (mg)	Riboflavin B-2 (mg)	Niacin (mg)	Vitamin B-6 (mg)	Vitamin B-12 (mcg)	Folic Acid (mcg)	Pantothenic Acid (mg)
145	19	0.07	0.07	0.5	0.05	—	9	0.14

Sodium (mg)	Potassium (mg)	Calcium (mg)	Phosphorus (mg)	Magnesium (mg)	Iron (mg)	Zinc (mg)	Copper (mg)	Manganese (mg)
9	129	9	15	7	0.24	0.16	0.088	0.409

Raw—1 cup

BLUEBERRY SHORTCAKE

4 shortcake cups or pound-cake slices
1 pint blueberries
¼ cup sugar
1 tablespoon cornstarch
½ pint heavy cream, whipped

Lightly toast shortcake cups. Cool. Wash and drain blueberries. Cook blueberries, sugar, cornstarch, and ½ cup water in medium saucepan until mixture thickens, stirring constantly. Cool slightly. Spoon into shortcake cups. Refrigerate for several hours. Top with chilled whipped cream.

Ice cream may be substituted for the whipped cream or served along with it for truly sinful pleasure.

ALL-AMERICAN BLUEBERRY PIE

Pastry for a 9-inch pie
2 pints blueberries
¾ cup sugar
¼ cup flour
1 teaspoon maple syrup
1 teaspoon lemon juice
2 tablespoons butter

Preheat oven to 425° F. Line 9-inch pie pan with half the pastry. Combine blueberries, sugar, flour, and maple syrup. Mix gently but thoroughly and place in pie shell. Sprinkle with lemon juice and dot with butter. Cover with remaining pastry. Cut slits for steam to escape. Seal and flute edge. Bake for 40 minutes or until juice bubbles up through top-crust holes. Cool.

CRANBERRIES

Cranberries are more than a bridesmaid to a stuffed turkey. Cranberries are a product of the Northeast, where they were utilized by the American Indians in fresh and dried form as vegetable matter eaten with game. Today most cranberries are grown independently, then processed and shipped to market by the Ocean Spray Cranberry Company.

Cranberry sauce, mostly bought in cans, is easy to prepare fresh, either cooked or raw. A simple recipe of fresh oranges and cranberries is made in a food processor in a few minutes; you can add sugar, honey, or maple syrup to your own taste. Cranberries make excellent relish and bake well in breads. Apple-cranberry sauce or apple-cranberry pie and cranberry chutney are all delicious and different.

Cranberries appear in early fall and are available for most of the winter. Fresh cranberries, however, can be frozen for months in the plastic bags they come in and then used throughout the year as wonderful companions to chicken, duck, turkey, lamb, pork, and sausages. Look for bright, hard dark-red berries, and begin to treat cranberries in a whole new light.

Cranberries are not really considered one of the leading nutritional foods but do provide a complete balance of essential vitamins and minerals. As a health food cranberry juice (not the cheap cranberry-cocktail mix) is an excellent kidney flush, dissolving minerals and excessive acids that can build up and cause distress.

NUTRITIONAL DATA

Weight (g)	Calories	Protein (g)	Carbohydrates (g)	Crude Fiber (g)	Water (g)	Fat (g)	Cholesterol (mg)
95	46	0.4	12.1	1.1	82.2	0.2	—

Vitamin A (IU)	Vitamin C (mg)	Thiamine B-1 (mg)	Riboflavin B-2 (mg)	Niacin (mg)	Vitamin B-6 (mg)	Vitamin B-12 (mcg)	Folic Acid (mcg)	Pantothenic Acid (mg)
44	13	0.03	0.02	0.1	0.06	—	2	0.21

Sodium (mg)	Potassium (mg)	Calcium (mg)	Phosphorus (mg)	Magnesium (mg)	Iron (mg)	Zinc (mg)	Copper (mg)	Manganese (mg)
1	67	7	8	5	0.19	0.12	0.055	0.149

Raw—1 cup

WHOLE CRANBERRY SAUCE

1 pound cranberries
2 cups sugar

Combine cranberries, sugar, and 2 cups water in large saucepan. Heat until mixture boils. Stir occasionally until sugar dissolves. Simmer about 5 minutes or until most cranberries pop. Cover. Chill.

CRANBERRY-ORANGE SAUCE

1 pound cranberries
Grated peels of 2 large California peeling oranges
1 cup orange juice
2 cups sugar

Combine all ingredients in food processor. Grind until almost smooth. Chill overnight.

CRANBERRY STUFFING

1½ cups cranberries, chopped
2 stalks celery, chopped
2 tablespoons finely chopped yellow onion
4 cups stuffing croutons
½ cup orange juice
2 tablespoons grated orange peel
1 teaspoon chopped fresh marjoram or ½ teaspoon
    dried marjoram
¾ teaspoon chopped fresh thyme or ¼ teaspoon dried thyme
½ cup butter or margarine, melted
½ teaspoon salt

Combine all ingredients and toss lightly. This makes stuffing sufficient for 5 to 7 pounds of poultry.

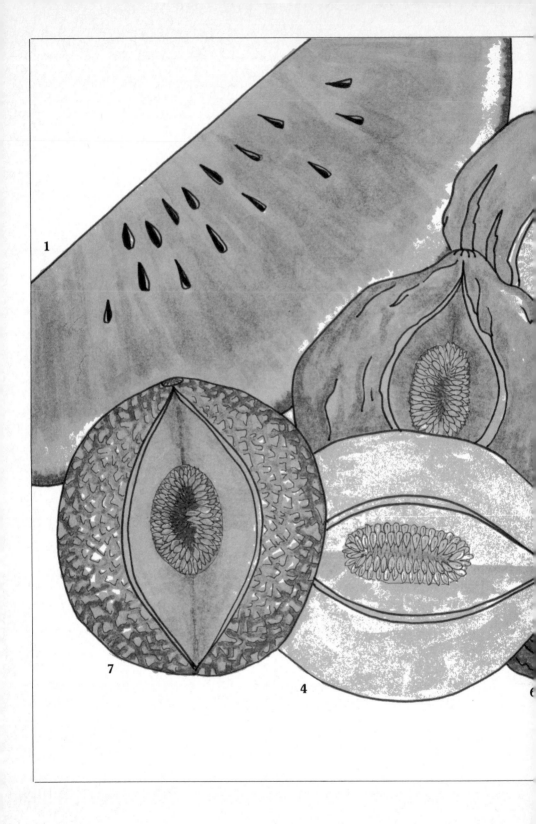

MELONS

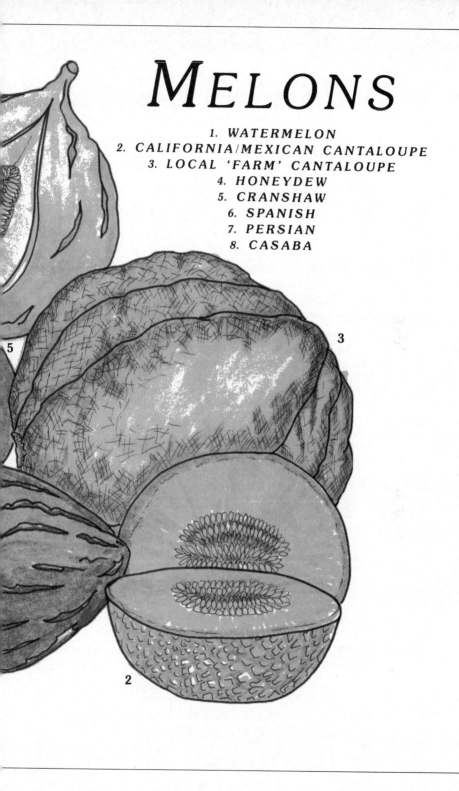

*F*inding a melon in the state of perfect ripeness, when it can fulfill your melon fantasies, is the most difficult shopping task in the fruit and vegetable business, but there are definite signs that will indicate the time when a cantaloupe, honeydew, or watermelon is "right there."

All melons, except watermelons, will perfume the air when ripe. Often the unmistakable sweet musk of a golden cantaloupe, the dewy aphrodisia of a honeydew, or the heady rose-pepper aroma of a Cranshaw is enough to let you know this one is ready. If you have to pick one up and get right down to it for a tiny whiff, forget it. Ripe melon perfume comes to you.

All melons, except watermelons, will show a soft golden blush over much of their surface. This golden glow only happens when ripeness, sweetness, and juiciness are there together. It even shows on dark-green heavily ribbed winter Spanish melons. I don't know how to explain how a dark-green melon glows gold, but it does if it's ready to eat.

All melons, except watermelons, will ripen after picking if they are mature. That doesn't necessarily mean they'll get better. Vine-ripened cantaloupes separate themselves from the vine in the field. If allowed too much time, they will lose their texture and turn mushy. The sugars begin to ferment, and they bruise easily. Hard-rind melons like honeydew and Cranshaw ripen quite nicely after picking, and often that is the only way to enjoy them at all. Any melon that is picked before maturity—before the growth process begins to develop sugar—will always remain hard, shiny, dry, and tasteless. Avoid hard shiny greenish melons, and you'll avoid disappointment. Please don't squeeze the blossom end of any melon. Look for the signs. They are unmistakable, and nobody gets hurt in the selection process.

All melons, except watermelons, fall into three classifications: The first is the true cantaloupe, which has a very green netted rind and an oblong shape. It is grown almost only in Europe and hardly anywhere in the United States. The second is the muskmelon, which we call cantaloupe in the United States and will continue to do. The third is the winter melons, which include the casaba, Cranshaw, Persian, Spanish, and honeydew melons.

All melons, except watermelons, are true melons. Watermelons cannot cross-pollinate with other melons, which, by the way, all melons do promiscuously and continuously, producing many variegated and variformed varieties that appear in small quantities and inconsistently as Canary melons, Christmas melons, Santa Claus melons, and a new variety with widespread commercial possibility, a pink-fleshed honeydew melon that I like very much.

All melons, *including* watermelons, are significantly nutritious and of great benefit to your overall system and body processes. Nutritional details are not available for all specific melon varieties. Cantaloupes and watermelons are superior to the lighter-fleshed winter melons in specific nutritional content. All melons are excellent for dieting, cleansing, and detoxification. They also provide a refreshing, revitalizing feeling when eaten chilled with a splash of lime on a hot summer day.

When shopping for melons, don't rummage through the pile. Look for the golden glow. Don't squeeze the blossom end or the stem end, looking for a soft spot. Follow your nose, not your thumb. Most melons are packed by size, and each crate is usually uniform. That's when melons are sold by the count—eighty-nine cents each, or whatever. Sometimes the sizes are not uniform or the melons are a mixed lot. That's when melons are sold by the pound. We all can agree that God grows 'em all, big and small, but in the case of melons, if they all cost the same, most customers will let God take care of the small ones, and they'll take the big ones, thank you. So would I.

WATERMELONS

*R*emember when we only ate watermelon in July and August? Now watermelons are available almost all year round, and, surprisingly, for a fair price. You can have one whenever you want. Regular shipments of watermelons begin to arrive from Guatemala, Honduras, and Mexico in late winter. They are expensive but delicious. The really great American watermelons begin to appear in May and June from Texas and Florida. These first Southern watermelons weigh between ten and twenty pounds. They are very oval, with a light gray-green skin. They are called Grays. They tend to be a little timid in their flavor, but they have a sweetness that says summer is coming.

By early summer the dark-green striped melons come north and west from Florida, Georgia, and the Carolinas. These are very long and weigh as much as fifty pounds. When ripe, they are always deep red, with a sugary crisp meat that bursts with watermelon delight.

Soon after the big Southern striped melons, the round watermelons arrive. These are called Crimsons. They have fewer seeds than the big stripers and a very large seedless heart section. These Crimsons, weighing about twenty pounds, are the best to buy whole. For my money they have the best flavor and the iciest texture. Three or four people can finish one in an afternoon or two, or once halved, they keep in a refrigerator for at least a day or two.

As summer moves along, the small light-green Sugar Babies and the dark-green Cannonballs arrive. Both are sweet and juicy, with significantly fewer seeds that are also much smaller. A good rule to know is that lighter-skinned melons have lighter meat. This does not mean less sweet or less juicy. However, taste buds seem to react more to a deep-red watermelon, which makes it more appetizing and desirable.

You cannot tell a ripe watermelon without cutting it. You can thump it, twist its stem, see if it floats, and all that, but to really know its ripeness, you must cut it. Most stores sell cut pieces. If the whole melons are the same variety as the cut display pieces, then they are almost sure to be the same color. Trust your vendor. If he deceives you, bring the melon back, or shop somewhere else. Better yet, most produce people will "plug" a whole melon if you ask. That is, they'll cut a small wedge from the melon so you can see the inside.

Watermelons make excellent liquid drinks, summer snacks, and late-night summer treats. If you plan to eat a lot of watermelon, however, avoid other foods for one hour before or after, as all the watermelon liquid dilutes the digestive fluids and makes solid-food digestion difficult. Even the best watermelons love a splash of lime juice or a sprinkle of salt.

It is almost a rule that very popular fresh fruits and vegetables that have widespread acceptance are highly nutritious and good for you. Watermelon is no exception. Basically a water food, it nevertheless delivers significant amounts of vitamins, especially vitamin A, and minerals, especially calcium and potassium. Its high-quality fluid content is an excellent cleanser and detoxifier. Once you've enjoyed the flavor, your body loves the rest. Watermelon drinks made with seeded watermelon, ice, and lime juice are an excellent way to help restore health and internal balance after a cold, infection, or other debilitation.

NUTRITIONAL DATA

Weight (g)	Calories	Protein (g)	Carbohydrates (g)	Crude Fiber (g)	Water (g)	Fat (g)	Cholesterol (mg)
160	50	1	11.5	0.5	146.4	0.7	—

Vitamin A (IU)	Vitamin C (mg)	Thiamine B-1 (mg)	Riboflavin B-2 (mg)	Niacin (mg)	Vitamin B-6 (mg)	Vitamin B-12 (mcg)	Folic Acid (mcg)	Pantothenic Acid (mg)
585	15	0.13	0.03	0.3	0.23	—	3	0.34

Sodium (mg)	Potassium (mg)	Calcium (mg)	Phosphorus (mg)	Magnesium (mg)	Iron (mg)	Zinc (mg)	Copper (mg)	Manganese (mg)
3	186	13	14	17	0.28	0.11	0.051	0.059

Raw—1 cup pieces

CANTALOUPES

*T*here are basically two kinds of cantaloupes, both of which are technically muskmelons, which only Americans refer to as cantaloupes. First, there are the round net-patterned cantaloupes, grown mostly in California, Arizona, and Mexico. These are available mostly year-round but are best from June through September. These "lopes" have a firm deep-orange flesh that is tight-textured, with a compact seed pocket that can easily be removed all in one piece with a tablespoon. The flavor of these "Cal-Mex" muskmelons is very aromatic, with a taste that sometimes reminds me of thick, sweet California orange juice. Texas also grows and ships the same variety of cantaloupe. These melons are, however, weaker, paler, and drier. Texas has a lot of things going for it, for sure, but good lopes is not one of them.

The other variety of cantaloupe or muskmelon is more oval, with a heavily raised netting and deeply grooved rind. These are grown locally all over the United States. They are more fragile, have a juicy, softer texture, and can grow quite large, often up to seven or eight pounds or more. Local cantaloupes have a loose wet seed pocket and larger seeds. These cantaloupes are always picked ripe and do not ship well. In August it is worth a trip to a roadside food or farm market to sample these melons. Local varieties range from a very deep orange flesh with a whitish rind to a lighter meat with tan rind and dark green ribs. Which are the good ones? Your nose knows. So does your professional fruit man. Ask him. The local cantaloupe season begins in early August, and some varieties last until mid-September.

When buying cantaloupes, please do not squeeze the ends. Often soft ends are the result of many thumbs looking for the ripe spot. Choose cantaloupes by their golden glow, their sweet fragrance, and a general allover "give" that only occurs when cantaloupes are ready. Golden cantaloupes that have a slight fragrance but feel hard are ripe and will soften in a day or two if allowed to sit quietly in a warm place out of the sun.

Sometimes when I'm hungry for something, I'll pick out a small golden cantaloupe, cut it in half, scoop out the seeds, and eat both halves out of their skins with a spoon. I always feel satisfied and happier than if I had wolfed down a Milky Way or a couple of crullers and coffee. Cantaloupes taste best eaten at room temperature but still offer plenty of pleasure chilled. They can be cut up in salads or served with smoked meats or cottage cheese. Cantaloupes love lime juice, which gives them an interesting combination of tart-sweet flavors, and they make excellent summer drinks mixed with yogurt, orange juice, or other summer fruit.

Cantaloupes get very high marks for nutrition. They are one of the very best sources of vitamin A and potassium (as are most orange foods) and contribute a full and balanced range of vital nutrients, with very few carbohydrates. Cantaloupes should be eaten year-round when they can be purchased in a high-quality state.

NUTRITIONAL DATA

Weight (g)	Calories	Protein (g)	Carbohydrates (g)	Crude Fiber (g)	Water (g)	Fat (g)	Cholesterol (mg)
160	57	1.4	13.4	0.6	143.6	0.4	—

Vitamin A (IU)	Vitamin C (mg)	Thiamine B-1 (mg)	Riboflavin B-2 (mg)	Niacin (mg)	Vitamin B-6 (mg)	Vitamin B-12 (mcg)	Folic Acid (mcg)	Pantothenic Acid (mg)
5158	68	0.06	0.03	0.9	0.18	—	27	0.21

Sodium (mg)	Potassium (mg)	Calcium (mg)	Phosphorus (mg)	Magnesium (mg)	Iron (mg)	Zinc (mg)	Copper (mg)	Manganese (mg)
14	494	17	27	17	0.34	0.25	0.067	0.075

Raw—1 cup pieces

HONEYDEW MELONS

*T*he honeydew melon is one of the winter melon varieties. And without a doubt it is the prettiest melon of all. Its cool lime-green flesh, bursting with juicy flavor buds and cradled in its thin porcelain skin, is the height of fruit delight. The trouble is finding a honeydew melon that doesn't taste like a raw potato.

A truly ripe honeydew melon will turn its skin from white to a soft gold, and it will feel sticky. There is no need to squeeze either end to feel for a soft spot. The whole melon will give to gentle touch and will exude a sweet honey aroma.

Ripe honeydews are fragile and ship poorly. As a result most are picked green, barely mature, and sold in their hard green state. Some will ripen; most will not. There are rarely if ever any ripe honeydews found in stores or supermarkets before August, although some Mexican fruit may be found around June. I have never seen perfectly ripe honeydews in any supermarket outside of California.

California supplies almost all American-grown honeydews, and most are shipped with labels representing the grower, packer, or shipper. Look for "King of the West" labels. King of the West are the best. They are premium quality, preselected for maturity and ripeness, and unlike the bulk of the honeydew supply, shipped ready to eat or almost so. Pony Boy is also another premium brand and can be relied on for ripeness, taste, and sweet juiciness.

There is now also available a pink-fleshed honeydew that seems to ripen more easily and shows its maturity more clearly. A ripe pink-fleshed honeydew turns *very* golden, the skin "gives in" to a gentle touch, and the fruit releases a strong compelling perfume. Most of these melons carry a bright-orange label that says "Pink Flesh." They are almost always shipped ripe. The taste is not quite as sweet as a perfect honeydew, but its color is appetizing, and it holds great quantities of sweet juice.

I know how many times you have been disappointed by your honeydew melon selection. One time I bought twenty or twenty-five boxes of honeydews. I looked at them in the market for color and ripeness, but unfortunately the one or two boxes I sampled were the only good ones in the whole lot. I bought them in June and finally sold them in September, still unripe, as lottery melons for twenty-five cents each. The sign read, "Lottery Melons, you take a chance with every one," which is the way many people feel about melons every day.

Honeydew melons do not have the nutritional power of orange-fleshed melons and watermelons, but they have a good amount of calcium and potassium and are an excellent weight-loss food. One-quarter of a medium honeydew has half the calories of an apple and one-third the carbohydrates.

NUTRITIONAL DATA

Weight (g)	Calories	Protein (g)	Carbohydrates (g)	Crude Fiber (g)	Water (g)	Fat (g)	Cholesterol (mg)
100	33	0.8	7.7	0.6	—	0.3	—

Vitamin A (IU)	Vitamin C (mg)	Thiamine B-1 (mg)	Riboflavin B-2 (mg)	Niacin (mg)	Vitamin B-6 (mg)	Vitamin B-12 (mcg)	Folic Acid (mcg)	Pantothenic Acid (mg)
40	23	0.04	0.03	0.6	—	—	—	—

Sodium (mg)	Potassium (mg)	Calcium (mg)	Phosphorus (mg)	Magnesium (mg)	Iron (mg)	Zinc (mg)	Copper (mg)	Manganese (mg)
12	251	14	16	—	0.4	—	—	—

Raw—¼ small

CRANSHAW MELONS

Cranshaw melons are the kings of melons. Ripe ones are always golden, sometimes bronzed at the stem end, and exude a sweet thick rose-pepper aroma that tells you they're ready. The skin color and meat color are the same. The taste becomes even more exciting when you sprinkle the melon with lime juice. Cranshaws are usually more expensive than other melons, but remember, they are the best and can be enjoyed totally by themselves for themselves as an afternoon delight, dessert, breakfast, or refreshing cold snack on a hot day.

If a seven- or eight-pound Cranshaw melon is too much for you, ask your fruit-stand man to cut you half. Many stores offer large Cranshaws cut in halves or quarters and wrapped in poly film. This is an excellent way to buy Cranshaws. The meat should be firm, without grainy or glossy streaks or spots, and the center area should glisten with the sugary juice in the seed bed.

For a truly exotic summer fruit salad I recommend peeled Cranshaw squares, mango pieces, shredded coconut, and halved seedless green grapes, chilled and splashed with orange and lime juice.

SPANISH MELONS

Spanish melons, sometimes called Christmas melons even though they are not, are usually found in stores in late December and midwinter. Spanish melons are difficult to select. Their skin remains green and hard, and they have only a slight aroma even at full maturity. The experienced shopper and the knowledgeable greengrocer, however, will be able to detect that certain quality, an overall softness or feel, that indicates ripeness and sweetness. Although the skin doesn't soften, it will lose its slick surface and produce a glow that can be seen and a subtle stickiness that can be detected by touch. The flesh is yellow, very juicy but firm, and eats well by itself or with smoked meats such as prosciutto, Genoa salami, dry sausage, and anchovies. A sprinkle of lime juice is always recommended and creates a delicious contrast between the sweet melon and the citric lime.

A good Spanish melon is rare. Our American hunger for fresh fruit makes us eager melon-buyers in winter, when supplies are at their lowest. We buy melons, especially this variety, with more hope than conviction, and disappointment is often the uninvited table guest. Spare yourself the loss and waste. If the Spanish melons are bricks, pass. They will never become what you want. Buy Hawaiian pineapples, big California navel oranges, and some firm Comice or Bosc pears, and make a sweet, juicy, golden fruit salad that will please you in every way.

PERSIAN MELONS

*I*f Cranshaws are the kings of melons, then Persians are the sultans. They are not as plentiful but are equally delicious and interesting. Ripeness is determined by an overall softness, and the melon's gray-green skin color turns to a gray-gold. The netting lightens and becomes more evident. Persian melons have a deep, deep orange flesh with a firm texture and high juice content. The seed pocket comes out easily in one piece. At its best the Persian melon exudes a perfume that is highly aromatic and unforgettable, something like a night you remember from a summer romance long ago.

Persian melons could be the original of all melons. Historians place their origin somewhere in the Middle East, probably Persia (an excellent guess considering their name). There is some evidence that the contemporary European cantaloupe was first cultivated in China and that the Persian melon was developed from that fruit. Historians, botanists, and taxonomists cannot agree. I say the two kinds of melons are different and could have appeared in either place or both. I know Persian melons have a softer texture that is more sweet than musky. Their skin is thinner and in its ripe state almost transparent. A Persian melon is an exotic and sensual experience in the eating, whereas cantaloupes are good day in, day out. Persian melons make you feel what you taste, make you sit back and say this is not ordinary food. If I were allowed just one of all the fruits I know, I would select the Persian melon. In its perfect vine-ripened state it is the finest fruit I know. Like all rare and special things it is always expensive and always worth it. Look for your Persian-melon experience in late August at really fine fruit markets.

Persian melons eat just as deliciously at room temperature as cold and love lime or lemon juice. They are the best companion for lunch meat salads, antipasto, or cottage cheese and lettuce, but they are also excellent by themselves on a hot summer afternoon when you are lying by the pool or in a tent under the date palms of a desert oasis.

CASABA MELONS

The Persian melon's name comes from its country of origin, Persia (now Iran). The casaba's name comes from Kasaba, Turkey, where this melon was apparently first identified. Unfortunately, the casaba is often a "turkey" as a melon. It is a melon that can give you the farthest extremes of the melon taste spectrum. When it is good, it is very, very good. And when it is bad....

Casabas have the same kind of white rind as honeydews. They have an elongated stem end and deep ridges like Cranshaws. The flesh is white, with a tinge of yellow around the seed bed. The casaba gives off very little fragrance, and often none, unless it is a great vine-ripened specimen. In that state it gives off a definite light floral aroma, and the entire melon has a soft give and takes on a subtle glow. The skin becomes slightly sticky, and you can tell it's ready. Melons must have four qualities to be perfect: great taste, sweetness, good texture, and juiciness. Most melons are more than pleasurable with three of the four qualities. Casabas need all four at their maximum, or you get nothing.

Casabas appear during melon-mania season, near the end of summer and the beginning of fall, when all the late-season melons begin to flood the market. You do not, however, shop for casabas with a list. When you see them, check them out. If you find one that looks and smells like the real thing, gently feel it with your whole hand. If it has a gentle give or stickiness, go for it. Pick up a lemon or lime as you leave the store, and sprinkle either juice over some cold long slices. Crisp, white, glistening casabas, one of the world's best fruits, or....

NUTRITIONAL DATA

Weight (g)	Calories	Protein (g)	Carbohydrates (g)	Crude Fiber (g)	Water (g)	Fat (g)	Cholesterol (mg)
170	45	1.5	10.5	0.9	156.4	0.2	—

Vitamin A (IU)	Vitamin C (mg)	Thiamine B-1 (mg)	Riboflavin B-2 (mg)	Niacin (mg)	Vitamin B-6 (mg)	Vitamin B-12 (mcg)	Folic Acid (mcg)	Pantothenic Acid (mg)
51	27	0.1	0.03	0.7	—	—	—	—

Sodium (mg)	Potassium (mg)	Calcium (mg)	Phosphorus (mg)	Magnesium (mg)	Iron (mg)	Zinc (mg)	Copper (mg)	Manganese (mg)
20	357	9	12	14	0.68	—	—	—

Raw—1 cup pieces

WATERMELON CHUNKS

¼ watermelon (quartered lengthwise)
Juice of 1 lime
Salt

Cut watermelon into 2-inch-thick chunks. Sprinkle with lime juice. Sprinkle lightly with salt.

MELON AND PROSCIUTTO

1 honeydew or Cranshaw melon
16–18 slices imported Italian Prosciutto
8–10 leaves romaine lettuce
4 limes

Cut melon in half and seed. Slice into wedges and peel. Arrange alternating rolled prosciutto slices and melon wedges in wheel-spoke pattern on bed of romaine. Cut limes into wedges and place in center of platter. Serve with a pepper mill; sprinkle with generous amounts of pepper and lime juice before eating.
  Serves 4.

SPANISH WINTER MELON SALAD

1 Spanish melon
3 California navel oranges, peeled and separated into segments
2 kiwis
½ pound Muscat grapes, halved and seeded
4 leaves Boston lettuce
2 limes

Halve, seed, peel, and cut up melon. Remove skins from orange segments. Peel kiwis and cut in wedges. Mix grapes, orange segments, kiwi wedges, and melon pieces. Cut limes into wedges. Serve fruit on individual lettuce beds with lime wedges. *Serves 4.*

Exotic Melon Fantasia

1 Cranshaw melon
1 Persian melon
3 Comice pears, peeled and cubed
1 mango, peeled and cubed
Juice of 2 limes
12–18 strawberries
1 cup orange juice
½ cup grated coconut

Halve, seed, peel, and cut up melons. Combine with pears and mango. Sprinkle with lime juice. Spoon into deep dessert glasses. Chill. Mash strawberries with orange juice. Pour over fruit. Top with grated coconut. *Serves 8.*

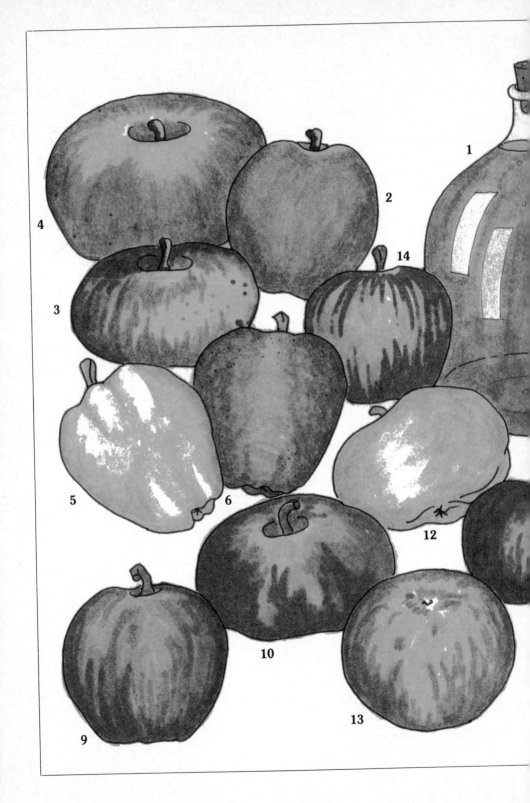

APPLES

1. APPLE CIDER
2. GRANNY SMITH
3. MCINTOSH   4. ROME
5. WESTERN GOLDEN DELICIOUS
6. WESTERN RED DELICIOUS
7. EASTERN RED DELICIOUS
8. EASTERN GOLDEN DELICIOUS
9. WINESAP   10. CORTLAND
11. MACOUN   12. GREENING
13. JONATHAN
14. BALDWIN
15. OPALESCENT
16. LADY APPLES

*T*here is much more to apples than meets the eye. Most of us have grown accustomed to the supermarket varieties of Red Delicious, Golden Delicious, McIntosh, and the relative newcomer from Down Under, Granny Smith. There are actually hundreds and hundreds of wonderful apple varieties grown in the United States. Unfortunately, supermarkets specialize in standardization, and very few of the more interesting varieties reach the consumer through the supermarket chain. At the height of apple season in mid-October the Sunfrost fruit stand carries somewhere between eighteen and twenty varieties at any one time. One must also realize that all apples in the United States are picked from late summer to late fall. Some apples store well, some don't. But all apples do not hold up for long periods after they come out of storage and should be kept under refrigeration or used immediately.

The best apples, the ones that crack and spurt and give off a sweet, tangy autumn essence, are fresh-picked and are available during the September-through-November apple season. It is always worth a trip to the countryside to load up on all kinds of apples—red, green, yellow, russet, golden, striped, flecked, sweet, hard, sour, and all the combinations in between.

If you shop only in supermarkets, you are never going to taste great apples. Supermarkets sell Western Red and Golden Delicious apples, McIntoshes, Romes, Winesaps, and Jonathans. These are all apples that stand up under long storage periods. McIntoshes, Winesaps, and Jonathans tend to soften and turn mealy once out of storage. The Delicious varieties are only really good fresh-picked and during the first weeks of winter. Romes actually improve as baking apples when they have a little age. A green early Rome takes hours to bake and never tastes naturally sweet. If you want really fresh apples, shop the countryside in the fall. Apples at any other time are a substitute for apples.

Apples are the most versatile of all fruits. Aside from their fresh-fruit qualities, they are baked whole, in pies, or in an endless variety of pastries. Apples make sauce, butter, jam, and chutney. They can be dried or pressed and turned into an alcoholic cider, vinegar, or sweet, refreshing, healthful apple juice. For the apple-pie baker it is good to know that three pounds of apples make a flat nine-inch pie, five pounds a high nine-inch pie. If you use firm or hard varieties and mix them, you can keep a high pie from collapsing inside.

There is no finer fruit than the apple grown in the world, and the United States produces the finest and the most colorful varieties. This apple season, break away from the ordinary. Make a visit to the nearest orchard or country fruit stand. Bring home a peck of Macouns, bake a Greening pie or a high-crust Cortland, eat some Winesaps or Baldwins, and just for the fun of it slice up a bag of Golden Delicious, string 'em up to dry, and eat 'em for months. Make apples part of your every day. You'll eat better, feel better, and get better value and nutrition for your money.

There are thousands of apple varieties grown in the United States, and apple horticulture is a study in itself. There are many varieties, however, which crisscross the entire country and are available in season at most roadside stands, farm markets, and worthy fruit and vegetable stands.

More than a fruit, apples qualify as a true food source because of their rich nutritional content. Apples are only ten per cent carbohydrates, all of which are in the form of fructose and glucose, the most easily assimilated high-energy food forms. Apples provide a well-balanced broad range of vitamins and minerals, including calcium, iron, and potassium. The apple is antitoxic and can cleanse the digestive system as well as benefit the kidneys, bladder, and liver.

NUTRITIONAL DATA

Weight (g)	Calories	Protein (g)	Carbohydrates (g)	Crude Fiber (g)	Water (g)	Fat (g)	Cholesterol (mg)
138	81	0.3	21.1	1.1	115.8	0.5	—

Vitamin A (IU)	Vitamin C (mg)	Thiamine B-1 (mg)	Riboflavin B-2 (mg)	Niacin (mg)	Vitamin B-6 (mg)	Vitamin B-12 (mcg)	Folic Acid (mcg)	Pantothenic Acid (mg)
74	8	0.02	0.02	0.1	0.07	—	4	0.08

Sodium (mg)	Potassium (mg)	Calcium (mg)	Phosphorus (mg)	Magnesium (mg)	Iron (mg)	Zinc (mg)	Copper (mg)	Manganese (mg)
1	159	10	10	6	0.25	0.05	0.057	0.062

Raw, with skin—1 medium

GRANNY SMITH APPLES

The Granny Smith apple satisfies our desire for a hard, fresh crunchy apple when there are no hard, fresh, crunchy apples. Granny Smiths are a hybrid derivative of the original green Pippin grown in England and sent abroad on English ships because of their good keeping qualities and high juice content. Today Granny Smiths are grown in Australia, New Zealand, and South Africa and are shipped all over the world, arriving in Europe and the United States in spring and early summer, when domestic apple supplies are either depleted or still slowly coming out of warehouses.

The Granny Smith is a medium- to large-sized apple, oval to round in shape, with a very green, very bright white-dotted skin. The fruit is superhard, white, and juicy, with a delightfully tart taste in early season and a somewhat sweeter flavor later on. The peel is very tough and should be removed. As an imported fruit, the Granny Smith is highly sprayed with an oil formula that, along with its natural tough skin, makes the peel difficult to digest. Peeling a Granny Smith is well worth the effort and actually improves the taste and crunchability. Granny Smith is basically an eating apple. It is too green to bake whole or sauce up but makes a super green-apple pie. It is excellent with crackers, cheese, and mustard or sliced up in a salad.

All supermarkets, all fruit and vegetable markets, and most roadside stands carry Granny Smiths. They are expensive (three to four times the cost of domestic apples in season), so shop carefully. Avoid bruised, yellowed, or black-spotted fruit. Actually, the black spots that seem to show up on Granny are only skin deep. Nevertheless, at something close to a dollar a pound, Granny Smiths should be near-perfect. Bruises are caused by carelessness either in the self-service selection process or in handling by the retailer. Bruised Grannies are unacceptable to me. Yellow Granny Smith apples are tasteless and juiceless. Yellow Grannys are overage or underrefrigerated or both. They are also unacceptable, except for sauce or pies at greatly reduced prices. In the fall the American Greening apple is almost exactly the same as a Granny Smith, although slightly more rustic in taste and less uniform in size and shape.

MCINTOSH APPLES

*T*he McIntosh is America's most famous apple. In New York State, the world's largest producer of McIntosh apples, they are first picked on September 15 and show mostly green with a red blush. Early Macs are excellent apples for an apple-pie mix. They tend to be slightly tart and very crisp. By October the McIntosh is three-quarters red and has sweetened considerably. If kept cold, it will retain most of its crispness. But by November the color is almost all red, the taste is very sweet, and the crispness and crunch seem to soften more than just a little. The late McIntosh makes a thick apple sauce that requires little sugar or sweetener. When you use McIntoshes in pies, it is best to remember that early-season ones will remain firm when baked but that late-season or storage ones will soften and sauce up.

ROME APPLES

*R*ome apples are picked in October. They are usually medium to large in size, with an oval flat-bottomed shape and intense red color all over. Early Rome apples are very hard, with a tart winy taste. They are excellent as part of an apple-pie mix since Rome apple pieces hold their shape and become sweeter in the baking. Later Rome apples, either picked ripe or allowed to ripen off the tree, are the true baking apples. They soften evenly without saucing up. The skin holds the fruit together, and the large size makes them a generous and attractive serving. I do not recommend Romes for eating out of hand. They never taste sweet to me raw, and I always wish they were a little juicier. But as a baked apple they are the best.

WESTERN DELICIOUS
APPLES

The Northwestern Red Delicious apple is America's most popular apple. It has been developed to a high state of color and uniformity, and for most of the selling season it remains crisp and juicy. By March the Red Delicious starts to soften and becomes mealy, especially out of cold storage. The Northwestern Red Delicious is triangular in shape, with the tapered bottom closing with the recognizable five points, or sheep nose, at the blossom end. The Northwestern Red Delicious is always sweet, very juicy in its early stage, and wrapped in a thick peel. The size varies, but most are medium to very large, with the largest—truly magnificent specimens—available around Christmas. Northwestern Red Delicious are almost always eaten out of hand as a snack or dessert apple.

Northwestern Golden Delicious apples are somewhat smaller than their red cousins and actually are a hybrid development from a totally different source. They appear earlier than the red variety and keep longer, have a smooth shiny skin with a slight green tinge in early season, and become a deep, dark yellow by March. Northwestern Golden Delicious are always sweet; make perfect dried apple slices, delicious unsweetened apple sauce, and excellent cheese and wine companions; resist bruising; and keep well for long periods.

All supermarkets carry Northwestern Red and Golden Delicious apples. All supermarket inventories of these apples are subject to a wide range of variables, including shipping, storage, refrigeration, packaging, handling, display, and self-service. So when you shop for these varieties, make sure you select hard, unbruised, heavy-feeling specimens that are true to their color, without dark patches or dull skins.

LOCAL DELICIOUS APPLES

Local Red Delicious apples are less uniform than the Northwestern variety. The color tends to be duller and not completely red, with yellow or orange patches. The size is generally smaller and not as consistent in form, but nevertheless the Eastern Red Delicious is characterized by the sheep-nose blossom end. It is very hard, juicy, and very sweet.

Local Golden Delicious apples appear in September with a green-gold shiny skin and offer a hard bite and semisweet flavor. By mid-November they are deep gold with russet flecks and a very sweet flavor. They are usually sold in random sizes, ranging from small to medium, with some larger sizes available. They are excellent keepers, with proper care lasting well into late winter. Late-season Golden Delicious make great apple sauce and require little or no sweetener to enhance the cooked flavor.

Most supermarkets never carry local Delicious apples. They are not standard-sized and don't have that manufactured shape that supermarkets insist on. To really taste the deliciousness of Red and Golden Delicious apples requires a trip to the nearest farm stand, roadside market, or great fruit stand. The rustic local varieties are many times juicier than the hybridized supermarket versions. The golden varieties are sweet as candy and will keep for months, dehydrating but never turning mealy. If you enjoy the taste of a moist but dried apple slice, you will enjoy a country Golden Delicious six months after it has been picked.

WINESAP APPLES

Winesap apples are one of the oldest American varieties. For the wine, cheese, and apple lover the Winesap is the first choice. It is picked after frost and has a hard, juicy texture, white meat, and a definite wine flavor. The Winesap apple has a rusty-red color with a touch of yellow in its ripe stage. The Winesap is a rugged country apple that holds up well in storage. It has a dull color even fresh-picked, so it doesn't lose any brightness in storage. Winesaps can be found at most roadside stands, farm markets, and good fruit and vegetable stands from midfall through December. After that Winesaps are found in plastic bags, mostly in supermarkets.

Winesap apples make excellent apple pies either as part of an apple mix or by themselves. They cook up into a light, textured sauce with that distinctive Winesap flavor. Sliced and strung, they make delicious dried apples that last all winter or longer. They are excellent cider apples, producing sweeter and sweeter juice as the apple season lengthens.

CORTLAND APPLES

Cortland apples are dark red and are available mostly in medium to large sizes. They have very white meat, good juice, and a semisweet flavor. Cortland apples make great traditional high-crust pies because the pieces do not melt and cause the crust to collapse. Cooking sweetens the flavor considerably, and Cortlands make a sweet, thick apple sauce, but they are delicious eaten out of hand as well. Cortlands are crisper earlier in the season and keep for a few months, although they bruise if handled roughly. Cortland apples also bake whole very well, and their sweet taste and smooth texture take to a cinnamon and chopped-nut topping.

Cortlands are hardly ever found in supermarkets except those near local orchards. Their thin bright-red skin bruises easily, and the bruises spread quickly through the very white meat inside. Peeled and cut up in salads, however, they retain their whiteness and are slow to brown. Cortlands are one of America's great traditional apples. Take a drive out to a local orchard or country market and bite into a big red one on a bright, colorful fall day. The first taste of the peel is sharp and acidic; then the sweet white inside begins to burst with sweet meat and juice. Taste one Cortland, and you'll know why they're still so popular after three hundred years, after thousands of other varieties have come and gone.

MACOUN APPLES

Macoun apples are the finest of all dessert apples. They are a small- to large-sized apple, but the medium size eats best. Macouns first appeared in 1926, when a McIntosh was crossed with a Jersey Black. Today's Macouns have a light gray-maroon to deep vermilion shading with some striping and both white and very dark flecks. The meat is very white, crisp, juicy, and sweet, with no bitter taste to the peel. Fresh-picked Macouns are as good as any apple anywhere and are considered the premium eating apple of the Northeast, where they are mostly grown. Macoun apples, however, are not good keepers and should be purchased and enjoyed in season.

You will not find Macouns in any supermarket. Their season is short, and their sweet, tender skins bruise easily. To enjoy Macouns requires a trip to a country store, farm market, or really knowledgeable fruit stand. Macouns come unsized, so your purchase will usually include some small ones that make excellent pocket snacks for schoolchildren. They are very sweet, and the skins are highly digestible, making them excellent additions to lunch boxes and snack packs as well as afternoon treats. Buy 'em by the peck. Macouns go fast. Make sure they are firm, bright, and free from bruises. Once home, keep them cool and away from direct sunlight.

GREENING APPLES

Greenings are excellent pie apples. They have a tart flavor, with a medium amount of juice that thickens well in baking. Picked after the frost, they are hard and crunchy and make an excellent companion for wine and cheese. Mostly medium to large in size, they have an olive-green skin and yellow-white meat. Many are larger on one side. They also keep very well and resist bruising. Good keepers that don't bruise must have a secret. In Greenings it is the skin, a very tough, thick peel, often slightly bronzed, with a touch of roughness. This protective skin, however, is easy to peel. All the better for preparing Greenings for the pies for which they are so perfectly suited. For those who like a hard tart apple that can be eaten alone or with cheese or baked just as well, the Greening apple is perfect.

JONATHAN APPLES

Jonathan apples were first found on the farm of Philip Rick in Woodstock, New York, where a rock known as Apple Rock now stands in a meadow three hundred yards from Sunfrost Farms, my favorite fruit stand. The original orchard no longer exists, but the acres of apple trees that cover the nearby Hudson Valley abound with this American favorite.

The Jonathan is basically a deep-yellow apple with both bright and dull red stripes. It is very round, with creamy-white meat that is semisweet in early season and sweeter later on. The Jonathan makes excellent eating, mixes well in pies, makes a sweet dark sauce, and is a fine companion for cheese and wine. If ever an apple was successfully created to be all things to all people, it was the Jonathan.

Many supermarkets carry the Jonathan. Most do not. It is an excellent keeper, and storage specimens seem to retain considerable juice and fresh flavor. The Jonathan is not a very hard apple when it is fresh-picked, but neither does it soften to any great extent during storage. There are many reasons why this New York original has remained one of America's five most-eaten varieties. Look for Jonathans in season at roadside stands, farm markets, and fruit and vegetable markets that do more than just buy and sell.

BALDWIN APPLES

Baldwin apples are not available everywhere, but when they appear, they offer a distinctive taste that goes well with wine and cheese. They have a tangy flavor, less than sweet, with a unique aftertaste that is pleasing and invites the appetite for more. Baldwins arrive late in the season—they are picked after frost—and retain a crispness and whiteness well into late November or early December. They have a dull crimson side that on the tree faced the sun, with some yellow flecks and streaks underneath. Baldwins are an interesting addition to an apple-pie mix or applesauce combination, as they sweeten in the cooking but still retain the unique Baldwin flavor.

Supermarkets rarely carry Baldwins. Even farm stands and roadside markets that close right after Halloween rarely carry Baldwins. Shop for them in fruit and vegetable markets that specialize in the widest range of apple varieties possible and country stores that brave the winter. Baldwins keep well, and a half-peck (one-eighth bushel) will do a family quite nicely.

OPALESCENT APPLES

Because of their limited distribution Opalescents are not popular, although there is some distribution to supermarkets because of their timely arrival. They also resist bruising, travel well, and are the first really good local eating apple.

Early crops are mostly green-skinned, with a red blush and large white flecks. They are very hard and super-juicy, with a tart taste. They are the best early eating apple. We sell many boxes of Opalescents in early September. They satisfy the public hunger for a hard, crisp, tart fruit after two or three months of soft, sweet juicy berries, peaches, melons, and plums. Opalescents are the only fresh-picked early apple I know that picks hard, juicy, flavorful, and ripe.

Later in the season Opalescents are much redder and tend to soften a little and sweeten up. These later pickings are great bakers and contribute firm texture and sweet flavor to an apple-pie mix. Look for the white flecks on a large oval but flattish red apple, and you'll find the Opalescent a delightful old-fashioned apple.

LADY APPLES

Lady apples are the oldest apples in history. They were first cultivated by the Etruscans, adored by the Romans, prized by French royalty, and much esteemed by the early American colonists, who included them among the first orchard plantings in their new world.

Lady apples do not appear everywhere but are available from just before Thanksgiving until January. Lady apples are small golden apples with a bright-red side, and their size and bright coloring make them a happy holiday fruit. They are fun to eat—just two or three bites and they're done—with a winy semisweet taste. They should not be peeled; the unusually flavorful peel is part of the whole taste experience. Lady apples can be served cooked whole with roasts, especially pork and ham, along with potatoes, carrots, and other roast vegetables. They can also be dropped into Christmas stockings, hung on Christmas trees, or strung in front of the hearth to dry whole and delicious.

APPLE JUICE AND APPLE CIDER

Apple juice and apple cider are both liquefied apples. The apple juice sold on unrefrigerated grocery shelves comes from apples that are crushed and strained to separate seeds, skins, core particles, and most solid materials. This juice is then pasteurized by boiling. Like ultrapasteurized milk, it will remain for very long periods of time without refrigeration. Most of its vitality is lost in pasteurization and the sugar content is dramatically increased by the heat process. The apple particles that remain, few as they are, darken in the process, and the juice becomes cloudy. This juice definitely has a cooked taste and a sweetness I find unnatural and unrefreshing.

Today apple cider is made by crushing the apples in tubs lined with burlap. The juice is strained and re-strained through finer and finer cloth until the final clear amber-colored juice is attained. It is immediately refrigerated at 36° F. If benzoate of soda is added in the bottling stage, usually one-tenth of one per cent, spoilage is retarded and shelf life under refrigeration is extended. Apple juice to which benzoate is added will never ferment and become alcoholic and bubbly, but it will become vinegar. Apple juice to which nothing is added must be kept under refrigeration at all times and even under refrigeration lasts only ten to fourteen days. This is the best-tasting form of apple juice, full of vitamins and free of all processed interference.

Apple juice that is pressed in the fall and not pasteurized has a lower sugar content, a lighter color, and a thinner density. It lasts longer, can withstand no refrigeration, and often has a sharp, tangy taste. It is early juice, with that traditional autumn flavor that is part of burning leaves, brilliant foliage, pumpkins, and crisp golden days. Apple juice that is pressed as late as the following summer comes from storage apples. It has a significantly higher sugar content, which makes it naturally sweeter. It has a darker color and a thicker texture and must be refrigerated at all times to keep it from fermenting within just a few hours.

I avoid all processed and pasteurized juices. Like so many of our fruits and vegetables—like the apples themselves—apple juice is best in season but available through new storage techniques almost all year round. Know this: All grocery-store unrefrigerated apple juice—in bottles, cans, cartons, and frozen concentrate—is boiled, preservatized, reconstituted, and manipulated. Fresh-pressed apple juice sometimes contains benzoate of soda and will never become anything else but vinegar; fresh-pressed apple juice unadulterated in any way is the real McCoy. Although we often call it apple cider, it is actually simply pure filtered apple juice. The alcoholic cider product is created by the addition of sugar, raisins, or other fermentative additions; temperature control; gas release; and the other precious secrets of the country folk, who can turn a barrel of innocent golden apple juice into a party for half the county.

Look for fresh-pressed apple juice, or cider, at knowledgeable fruit stands, country markets, local orchards, and farm stands. It is one of nature's sweetest gifts, full of nutrition, cleansing properties, and instant energy.

WALDORF SALAD

    2 sweet red or yellow apples, peeled and cubed
    1 semitart red apple, peeled and cubed
    1 tart green apple, peeled and cubed
    1 stalk celery, chopped
    ¼ cup dates, chopped
    ¼ cup walnuts, chopped
    ½ cup mayonnaise
    2 tablespoons sugar
    1 teaspoon lemon juice

Combine all ingredients and mix. Cover. Chill.   *Serves 4.*

POWER BREAKFAST

    1 sweet red or yellow apple
    1 tart green apple
    3–4 strawberries, sliced
    2 tablespoons blueberries or raspberries
    ½ cup apple cider
    2 tablespoons chopped mixed nuts
    1 tablespoon wheat germ or protein powder

Peel, core, and cut up apples. Mix with berries in deep cereal bowl. Pour on apple cider. Sprinkle with nuts and wheat germ.   *Serves 2.*

MOUNTAIN-HIGH APPLE PIE

    8 cups cut-up peeled mixed apples
    1 cup sugar
    3 tablespoons flour
    ¾ teaspoon cinnamon
    ¾ teaspoon nutmeg
    Pastry for 9-inch pie
    1 tablespoon butter

Preheat oven to 425° F. Peel and thickly slice apples. Combine apple slices, sugar, flour, and spices in large bowl. Line 9-inch pie pan with half the pastry. Place apple mixture in pie shell. Dot with butter. Cover with top crust. Cut slits for steam to escape. Seal and flute edge. Bake for 40 to 45 minutes or until juice bubbles up through top crust holes. Cool.

BAKED APPLES

4 large Rome, Cortland, McIntosh, Opalescent, or Winesap apples
¼ cup chopped nuts
¼ cup raisins
2 teaspoons cinnamon
2 tablespoons sugar
4 teaspoons butter

Preheat oven to 375° F. Core apples almost through to bottom. Combine nuts, raisins, and cinnamon. Spoon ½ tablespoon sugar and 1 teaspoon butter into each core cavity. Fill with nut mixture and 1 teaspoon water. Place in baking pan with ¼ inch water. Bake for 1 hour. During baking, baste apples with water from baking pan. Serve warm or cold.

APPLES IN GREEN GARDEN SALADS

Anytime the mood strikes, apples add crunch, flavor, and texture to salads and go with almost any combination of lettuce and dressing.

NO-SUGAR APPLESAUCE

8–10 apples, peeled and quartered
4–6 tablespoons maple syrup
1 tablespoon cinnamon

Combine apple pieces and ½ cup water in large saucepan. Heat until boiling. Cover. Simmer over low heat for 15 minutes or until apples soften. Stir. Add maple syrup and cinnamon. Mix, breaking up but not mashing apples. Cool.
Makes 1 quart.

PEARS

I think pears are elegantly seductive. They are sweet, juicy, wonderfully textured, and highly nutritious. They have the most subtle taste of all orchard fruit and will leave your palate delightfully fresh and clean. But only when they are ripe. Underripe pears are woody. Overripe pears are mealy and grainy. Picked too soon, they will forever remain woody. Picked too late, they are brownish, mealy, and unpleasant. How do you find great pears? At the local fruit stand the professional vendor can tell you which pears are at their seasonal peak and their peak ripeness. In supermarkets you pay your money and take your chances.

Very yellow pears, pears with soft stem ends, bruised pears, and spotted pears are almost always overripe. Very green, very hard, very shiny pears will never be juicy or sweet. Pears must be eaten when they present a subtle golden glow, not an overall yellow color.

All pears must be picked unripe. If pears are allowed to ripen on the tree, they develop little grit cells, or stones, in the flesh. Once the pears are separated from the tree, this process cannot take place, and the pears ripen evenly and smoothly, with a creamy texture. Pears, like lemons, are one of the few fruits to benefit from man's interference. Once picked, pears take approximately twelve days at 65° F. to ripen properly. Since there is no way for the consumer to know how long certain pears have been out of storage, it is important to look for the glow (I call it the bloom of life) and gently feel for a softness to the flesh. Please don't squeeze the pears. Please don't press with your thumb or finger. Simply hold it and feel its softness. If in doubt take the fruit home, and gently set it in a warm, quiet place for a day or two. Without a doubt the pears will ripen and become sweet juicy fruit.

All supermarket fruit is unripe, especially pears. At Sunfrost we ripen our pears according to the variety, the current heat and humidity, and the day of the week. We need more ripe pears for the weekend than for a quiet Wednesday in September. All our pears are displayed on little tissue pads to prevent them from bruising each other. We discourage handling and can provide exactly the ripeness our customers desire. When you go shopping, buy pears that are blemish-free and of good size. If they are not ripe, ripen them at home. But remember that pears ripen from the seeds out, so if you wait for too much softness, the pear will already be past its peak.

Pears can be jellied, stewed, canned, or simmered in a simple syrup for an excellent dessert. Ripe Bartlett, Clapp, or Packham pears are excellent for canning whole and taste wonderful as a January dessert. Cold stewed peeled pears are delightful served with lettuce and cottage cheese, and fresh pears make wonderful companions for wine, bread, and a mixture of sharp cheeses.

Pears are a high-energy food with very little salt and contain vitamins A, B, and C. They are very high in iodine and are excellent for maintenance of the thyroid and kidneys.

NUTRITIONAL DATA

Weight (g)	Calories	Protein (g)	Carbohydrates (g)	Crude Fiber (g)	Water (g)	Fat (g)	Cholesterol (mg)
166	98	0.7	25.1	2.3	139.1	0.7	—

Vitamin A (IU)	Vitamin C (mg)	Thiamine B-1 (mg)	Riboflavin B-2 (mg)	Niacin (mg)	Vitamin B-6 (mg)	Vitamin B-12 (mcg)	Folic Acid (mcg)	Pantothenic Acid (mg)
33	7	0.03	0.07	0.2	0.03	—	12	0.12

Sodium (mg)	Potassium (mg)	Calcium (mg)	Phosphorus (mg)	Magnesium (mg)	Iron (mg)	Zinc (mg)	Copper (mg)	Manganese (mg)
1	208	19	18	9	0.41	0.2	0.188	0.126

Raw—1 medium

PACKHAM PEARS

*P*ackham pears are mostly imported. Although some are grown in California, most of the Packham pears available in America are imported from the Southern Hemisphere countries of Australia, New Zealand, and South Africa. Their picking season begins and coincides with the United States springtime bloom. Domestic supplies from storage are low, and the retail demand for good-quality pears becomes very high. This creates an expensive but popular item.

Packhams come mostly in medium and large premium sizes. They are picked green, but their peel is not shiny. This is a sign of their ripening; Packhams ripen very well and resist bruising. They have a definite pear shape, but the wide bottom is irregular, with a deep blossom end. They are very, very juicy and sweet. They also have a perfume that adds to their exotic flavor. Eat them when they are just turning soft gold all over. A Packham pear at its peak will beg you to take it home. When eaten green and barely soft to the touch, Packhams are still excellent and juicy but less sweet. The firmer texture is interesting and offers a different aspect of the pear experience. You'll find fresh-picked Packhams in your stores in late June, lasting to late summer. Look for them in those fine fruit stands and markets managed by people who know how to buy, care for, and present beautiful produce. Like me.

ANJOU PEARS

*A*njou pears are round greenish pears tapering bluntly to the stem end, mostly of a medium size. They have no waistline and belong to the bergamot group of pears. They are available throughout the United States, shipped from California, Oregon, Washington, and Idaho. Their skin remains green but develops a definite glow when ripe, and they should be eaten when they yield to gentle pressure. Unless you are making pear cider, they must never be squeezed. Although the skin is not tough, it is not as sweet as the meat and has a slight grainy texture, so peeling an Anjou makes it taste better.

Since Anjous never lose their green color, they require a good eye to know their ripeness. Anjous are very popular supermarket pears since they arrive hard as rocks and, by supermarket standards, are not in a bruiseable state. I think all fruit is bruiseable, even coconuts. Nevertheless, you can select supermarket Anjous that are free from damage and take them home. Place them on top of your refrigerator in a low bowl, and watch them for several days. When they begin to glow and feel more like fruit than stone and their lime-green color softens, peel one.

If you shop at a fruit and vegetable market, find out exactly how ripe their Anjous are, and select accordingly. Anjous are a wonderful dessert pear. Their firm texture makes them the best pear for cooking and baking, since they never seem to lose their shape. A good ripe Anjou pear is very juicy, and if you peel one, keep a napkin nearby—or make a real production and cut it in half on a plate, using a small fruit knife and fork. It also loves a splash of lime, a dash of cinnamon, or a dollop of honey.

Anjou pears are available from late fall throughout the winter. Modern oxygen-free storage facilities are now lengthening the season even further, and in the last few years I have seen October-harvest Anjous shipped to retailers as late as April and May of the following year.

BARTLETT PEARS

Bartlett pears are the oldest cultivated pears in the world. They are called the Williams in England, the *bon chrétien* in France, and a very complicated name in Germany. The Bartlett is America's favorite and is grown in every state that has at least two winter months. The Bartlett is a pyriform pear, with a definite waistline and a long stem.

America grows the Bartlett in two colors, yellow and red. The yellow variety ripens very quickly once picked, and while most people wait until it turns a deep yellow with a brown fleck or two, the area around the seeds has already turned brown and the rest mealy and tasteless by then. Yellow Bartletts must be eaten while they are still flecked with green. Yellow Bartletts also bruise easily. Even loud noises hurt them. My best advice is to buy yellow Bartletts while they're green and bring them to their peak at home.

Red Bartlett pears are a development of our Northwestern pear-growers. The red skin is heavier, the pigment resists disease better, and the ripening process is not quite so quick. Red Bartletts are three-quarters red, solid at the cone end and striped below the waist. Red Bartletts are ripe when the yellow area shows slight green and the striped area is still red, not russet or brown.

I like both colors peeled, but that's up to you. Bartletts are excellent canners and dessert fruit but are too fragile for lunch bags, picnic baskets, and carrying around in your pocket. They are an early fall pear, but supplies arrive from all over the country, especially Washington and Oregon, well into winter. Local crops arrive first and are packed loose in mixed sizes. Western pears are later, wrapped, and graded. Nevertheless, I like the local varieties much more than the premium ones. Local pears have a freshness and flavor that cannot be matched by storage pears. Local pears are also less expensive and can be had in pecks and half-pecks for twenty-five to thirty cents a pound.

BUTTER HARDY PEARS

*B*utter Hardy pears belong to an old French pear family called *beurre,* or butter, pears. They are among the oldest and most sensitive cultivated pears in the world, now quite rare but still available.

Butter Hardy pears are like a large Anjou pear that turns yellow with a red blush. They are oval, with a blunt tapered stem end. Butter Hardy pears are very juicy, with a smooth white meat. Like all pears, however, they have a grainy texture when allowed to overripen. Butter Hardy pears are very fine specimens of fruit, and should they come your way, take care in their selection and their handling. Like people, pears, and especially Butter pears, respond to gentleness, and the kinder you care for them, the sweeter they give back. Bruise them, hurt them, chill them, or let them bake in a sunny window—they give up, and everything within loses the will to carry on. Maybe that's why orchardists no longer replant this variety and substitute thick-skinned hard-hearted varieties like Bosc, Anjou, and the red-skinned Clapp. Butter Hardy pears are available in midwinter.

CLAPP PEARS

Almost ten years ago A. R. Mott, fruit-grower par excellence, gave me a gift of thirty sapling fruit trees. Included were four Clapp pear trees, two green, two red. I planted them along the driveway to my house. I pass them many times each day and am constantly aware of the changes that take place from late winter until the last leaf falls in November. These trees have produced abundantly for me with a minimum of care. Some of the annual harvest goes to the store for our customers who appreciate unsprayed local fruit. The rest is eaten fresh, or peeled and jarred for holiday gifts and winter fare. I love to look at the glistening jars of fruit on our shelves, especially the snow-white Clapp pears.

A. R. introduced me to Clapp pears when we began Sunfrost. We handle both the green and red in mixed sizes from late August until the season ends in October. Clapps are a medium-sized pear with very white flesh, very high sugar, and plenty of juice. Sweetness and juiciness always go together in pears. The green Clapp pear has a thinner skin and is one of the few true fruits I enjoy best unpeeled. The red Clapp, like the red Bartlett, has a heavier skin and a slightly firmer texture. It too can be eaten unpeeled.

Clapp pears are hardly ever shipped. They are, however, available at almost all roadside stands, farm markets, and fruit and vegetable markets that make it their business to stay connected directly to growers and orchardists and buy from the source instead of the telephone.

One of my favorite aspects of the fresh-fruit business is to walk the acres of fruit trees that grace the Hudson Valley along our beautiful Hudson River and select fruit for our store while it still hung on the trees. Maybe that's why people who shop at Sunfrost know the food they take home is filled with life. And the Clapp pears that grow along the driveway to my house are daily reminders that the fresh-food business must always remain, for me, fresh first and business last.

BOSC PEARS

*B*osc pears suffer from our impatience. Boscs are the latest pear variety to evolve from the very hard wild pear and have been under cultivation for less than two hundred years. Like all pears, Boscs are picked unripe. They are also picked in their smooth green-colored stage, which is a long way away from their final deep-russet–colored and rough-skinned ripe state. Bosc pears can sit for weeks. Trouble is, we eat them when they are hard and suffer the disappointment of juiceless, sugarless fruit.

The Bosc pear is a member of the conical pear family, long, tapered, and waistless. When properly ripened, it becomes a dark russet color and will respond to gentle pressure. The meat is cream-colored, very juicy, and smooth-textured. Although Boscs come in all sizes, I think the very large ones have the best flavor and sweetness. Large sizes, however, always cost more and are sometimes sold individually. Smaller sizes or mixed lots (called orchard run) are usually inexpensive, and during harvest season local varieties can be had for three or four pounds for a dollar. All Northwestern inventories are sized, graded, and more expensive.

Unless your fruit stand or market makes a special effort to ripen their fruit, I suggest buying green Bosc pears and developing patience and forbearance as your fruit ripens at home. At any stage select clean unbruised fruit without cuts or small holes resulting from stem punctures. These small holes cause rot spots as the pear ripens.

Bosc pears are excellent dessert pears. They can be enjoyed peeled or not and hold up well in lunch pails, tailgate baskets, and fruit bowls. Once ripe, however, they must be eaten.

Bosc pears begin to arrive in late August from local orchards and are available until late fall at roadside stands, farm markets, and good fruit stores. The upper-class Northwestern crops arrive wrapped in tissue paper in their wooden boxes in time for the holidays. Shipments continue until early spring, although the pears get smaller and smaller as the prices get higher.

COMICE PEARS

Comice pears are the best pears in the world. They are so far superior to all other varieties that there can be no quarrel. Even the French and English agree that when it comes to pears, Comice is where it's at.

A Comice pear has a definite pyriform shape, with a short, wide stem end, a waistline, and a very wide blossom end. It is somewhat squattish and irregularly formed. Nevertheless, it projects a presence that echoes its royal qualities. The Comice pear is heavily perfumed with a heady and musky fragrance. The color during peak ripeness is a soft green that glows with a golden aura. It is sometimes slightly bronzed or flecked, and I recommend this sign as an indication of *la crème de la crème*. A ripe Comice will give to gentle pressure, but the skin is so thin and the flesh so wet that anything more than a gentle stroke leaves a mark. It has been said that Comices feel like they are about to burst at any moment. It is little wonder that Comices have limited distribution. They are each carefully picked and immediately wrapped individually and packed in wooden boxes.

Comice pears must be eaten with a napkin. They are almost all juice, and their creamy smooth texture truly melts in your mouth. If you serve them as an after-dinner fruit or dessert, please serve a fruit plate, knife, and fork. The Comice is a complete eating experience. It is never a light snack, an afterthought, or something for your mouth to do while your eyes and ears watch TV.

Comice pears come from California, Oregon, and Washington. I know of no local Eastern varieties. Many are imported from Chile, France, New Zealand, and Argentina, and they are excellent, excellent fruit. They are expensive, not nearly as much so as caviar and champagne but certainly their equals in taste and sensual food pleasure. Comice pears appear in late fall and last throughout the winter and early spring. Look for them at great fruit and vegetable markets, specialty food stores, and great food emporiums like the Los Angeles Farmers' Market, the South Street Seaport in Manhattan, Boston's Haymarket, and of course Sunfrost Farms in Woodstock, New York.

DEVOE PEARS

*D*evoe pears are an old local orchard variety that is picked in October and sold immediately. They have a short storage period and limited distribution. The Devoe is a member of the *calebasse* group of pears, long, narrow, with no waistline and a very pointed stem end. They are green in color when picked and turn slightly lighter with a red blush and a soft golden glow when they ripen. They are best eaten when still firm with a mostly green color.

Devoe pears have an ivory-colored flesh that is sweet and juicy and has a deep, deep pear flavor. They are not commercially available and are found only at roadside markets, farm stands, and fruit stands that feature local fresh-picked produce.

FIORELLE PEARS

*F*iorelle pears are smallish yellow-red pears flecked with deep maroon or rust. They have a smooth shiny skin and a creamy-white flesh and are very sweet and juicy. Fiorelles are a durable fruit and seem to hold up well during shipping, handling, and display. The skin is delicious and yet heavy enough to protect the tender pear flesh.

Fiorelles have been with us for a long time, but it is only recently that they have been cultivated for national distribution. They are grown mostly in California and Oregon and are an excellent snack pear because of their convenient size and sweetness. Fiorelles appear in late fall and are available through December into January. They are especially nice in holiday fruit baskets and fruit bowls.

SECKEL PEARS

Seckel pears are one of the real fun fruits and are excellent children's fruit. They are a true American pear, having been discovered as a mutant sometime around the time of the Revolution. They are always very small and appear in the midfall and early winter in their hard green skins. Buy them and set them up to ripen at home, since they are never sold ripe because of their bruiseability. At the point that the green turns slightly golden and the red blush brightens, start eating. One juicy sweet morsel always leads to another. Children love their sweetness, and one seems just the right size for them. Patience is again a high virtue when it comes to Seckels. Their bright little faces and delectable size make it hard for us to hold back. A green Seckel, however, is nowhere, so I encourage you to wait for the moment, avoid disappointment, and get everything you came for. Seckel pears are available in almost all stores in early fall and continue to be available until January.

HOT HONEY PEARS

8 Bartlett, Clapp, Anjou, or Packham pears
½ cup honey

Peel, halve, and core pears. Pack tightly in sterilized quart jar. Melt honey in 1 cup boiling water. Pour into jar and fill to ½ inch from top. Seal (see "Canning process" entry in Glossary, page 000). Cool. Refrigerate, or store for later use. Serve hot.

This recipe can be used for general canning purposes, substituting sugar for honey; use 1 cup sugar for each cup water.

POACHED PEARS AND COTTAGE CHEESE

4 Bartlett, Clapp, Anjou, or Packham pears
2 tablespoons sugar
½ head Boston lettuce
1–1½ cups cottage cheese
Dash paprika

Peel, halve, and core pears. Place cut side down in ¼ inch water in shallow baking dish. Sprinkle pears with sugar. Cover pan with aluminum foil. Bake in preheated 375° F. oven for 20 to 25 minutes. Chill uncovered. Place pears on lettuce bed with mounds of cottage cheese. Sprinkle with paprika. Serve cold.

CINNAMON BOSC PEARS

4 large Bosc pears
4 teaspoons sugar
4 teaspoons cinnamon

Peel, halve, and core pears. Blanch in boiling water and place in medium saucepan. Melt sugar and cinnamon in 1 cup water. Pour over pears. Cover and marinate 1½ to 2 hours. Heat over low heat until warm. Serve with pork chops, spare ribs, duck, or game fowl.

TINY PEARS ITALIANO

8 small Fiorelle pears or 12 Seckel pears
½ cup orange juice
1 cup sugar
Italian liqueurs or fruit brandies

Blanch pears in boiling water and peel. Steam until soft but not mushy. Place in medium bowl. Melt sugar in orange juice and ½ cup water. Cover. Simmer until mixture is slightly caramelized. Pour over pears. Chill. Serve topped with a dash of choice of liqueur.

PEARS

353

GRAPES

1. *RIBIER*
2. *THOMPSON SEEDLESS*
3. *PERLETTE SEEDLESS*
4. *RED FLAME SEEDLESS*
5. *ALMERIA*
6. *FLAME TOKAY*
7. *EMPEROR*
8. *CONCORD*
9. *MUSCAT*

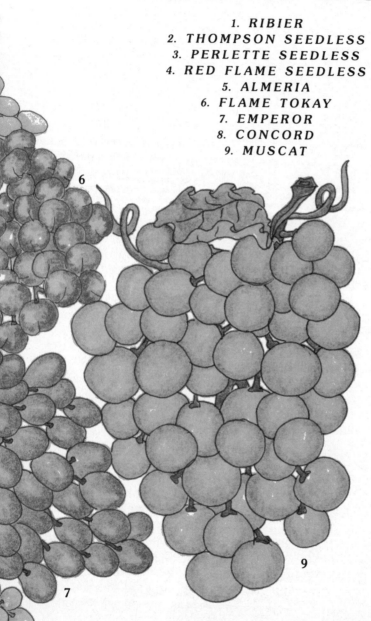

*B*efore there was man, there were grapes. And the best varieties of all grew in America. When the Vikings landed in North America, the wild grapes were so abundant they called it Vinland until Amerigo Vespucci arrived and named it after himself. When the grapevines of Europe were attacked by disease in the late 1800s, native wild rootstock of American grapes was rushed overseas to save the European vineyards. Although Europe may claim the wines, the roots are definitely home-grown.

There are wine grapes that can be eaten as table grapes and table grapes that press up into fine wine. The whole business of wine grapes, however, is best left to the few folks who crush up and bottle their own combinations of grapes for their own use and the hundreds of American wineries large and small that bottle enormous amounts of *vino* for the world. We will discuss table grapes.

Table grapes are selected for their size, appearance, sweetness, and travel ability. Sweet juicy grapes must be picked when they are ripe. Grapes do not ripen off the vine. When you taste a sample of the ones you buy, realize that the taste will not change or sweeten once you are home with the fruit. What you buy is what you get. Select grapes that have a frosty bloom, with fresh stems and firm fruit. Avoid raggedy soft sticky fruit and anything that attracts fruit flies. They are a sign of fermentation and overripe fruit. Wash grapes only when they are to be eaten. Moisture softens the fruit and causes deterioration.

Grapes should be eaten from little bunchlets cut off the stem of the larger bunch. Pulling grapes off one at a time from a large cluster causes the stem ends to bleed, which shrivels and softens the remaining grapes in the bunch. A half-eaten bunch of grapes with its skeletal stems loses its appearance and beauty, which is such an important part of enjoying fruit. Fruit that is well-grown and carefully selected should also be beautifully served and eaten.

All grapes grown east of the Rocky Mountains are the native American varieties like the Southern Scuppernong, the Northern and Midwestern Delaware Pink and Niagara White, and the most famous of all, the deep-purple Concord. These grapes are all slip-skin grapes and contain seeds. All grapes grown west of the Rockies are European stock brought there by the Spanish missionaries. They are adherent-skin grapes, almost always larger than Eastern grapes, and definitely sweeter and easier to eat. More than ninety per cent of all table grapes grown in America are grown in California. California ships grapes from late spring or early summer until late fall and early winter. Naturally, the entire harvest of any single variety is not shipped at once. This would flood the markets, depress prices, and create great slack periods of supply. Grapes are stored for up to six weeks after picking. Grapes do not ripen once picked, but they do age. Grapes held in storage beyond their time develop weak black stems and mold, and they begin to shatter, the grapes dropping from the stems in excessive numbers. Shattered grapes can be sweet and delicious, but they are fragile and difficult to handle and often have a very poor appearance.

The many varieties of California grapes include the black Exotica and Ribier; the green Perlette Seedless and the famous Thompson Seedless; the new Red Flame Seedless; the red Cardinal, Emperor, and Tokay; and the large green Almeria grape, often mistaken for a seedless variety, which it is not. The United States, used to having what it wants when it wants it, imports beautiful table grapes during the domestic off-season. Chile, Argentina, and Mexico ship us wonderfully sweet, delicious grapes from early winter to late spring. Their seedless varieties are excellent although a bit smaller than the Californians. They also ship very large black Exoticas and Ribiers, which I find hard to resist, and so I eat them all day long in season, which must be hurting our profits (they are usually very expensive). Italy exports a very special musky grape called the Muscat, which could be the world's best seeded table grape. Arizona has lately been getting into the table-grape act. They are currently producing excellent varieties of Thompson Seedless that have a beautiful golden color instead of the more familiar dusty green. They are very sweet and highly salable. I think Arizona will become an important source of great American table grapes.

Grapes are eaten fresh. Seedless varieties can also be added to salads and poultry stuffings or prepared as a light sauce for seafood. Dark grapes create the best homemade jams, jellies, and preserves or can be simmered into a syrup for pancakes, ice cream, and other desserts. As for those few individuals with the knowhow and the will, grapes can be purchased in quantity, and nothing sits better on the table than a bottle of your own homemade wine.

Grapes are an excellent food source. Though twenty per cent of the grape is carbohydrate, it is in the form of glucose and fructose, which are easily assimilated. Grapes contain vitamins C, B-1, B-2, and B-6, as well as potassium, iron, calcium, magnesium, and phosphorus. They also maintain kidney function and will reduce an acid condition, helping to restore a proper pH balance.

NUTRITIONAL DATA

Weight (g)	Calories	Protein (g)	Carbohydrates (g)	Crude Fiber (g)	Water (g)	Fat (g)	Cholesterol (mg)
92	58	0.6	15.8	0.7	74.8	0.3	—
160	114	1.1	28.4	0.7	128.9	0.9	—

Vitamin A (IU)	Vitamin C (mg)	Thiamine B-1 (mg)	Riboflavin B-2 (mg)	Niacin (mg)	Vitamin B-6 (mg)	Vitamin B-12 (mcg)	Folic Acid (mcg)	Pantothenic Acid (mg)
92	4	0.09	0.05	0.3	0.1	—	4	0.02
117	17	0.15	0.09	0.5	0.18	—	6	0.04

Sodium (mg)	Potassium (mg)	Calcium (mg)	Phosphorus (mg)	Magnesium (mg)	Iron (mg)	Zinc (mg)	Copper (mg)	Manganese (mg)
2	176	13	9	5	0.27	0.04	0.037	0.661
3	296	17	21	10	0.41	0.09	0.144	0.093

American, raw—1 cup
 European, raw— 1 cup

BLACK RIBIER GRAPES

Black Ribier grapes may not be seedless, but they are so sweet and juicy that who cares? When they are at their best, they crack when you bite them and explode with a combination of juice, sugar, sweet pulp, and delicious skin. Ribiers are very black, very round, and big as large marbles. There is another round black variety, mostly imported from South America, called the Exotica. Exoticas are just like Ribiers, only bigger, harder, and blacker. Most stores sell Exoticas under the Ribier name. Their seasons overlap, and if the truth be known, I'd bet most produce managers, storekeepers, and clerks can't tell the difference and don't even know there is one. Ribiers are always round and often come in small sizes. Exoticas often have a seam or cleft at the bottom and never seem to come in little sizes.

Ribiers first appear in late fall from California and last into midwinter. South American varieties arrive in late spring, followed by the Exotica. Look for hard, clean, bright fruit still covered with a powdery frostlike substance. This is an indication of freshness, ripened maturity, and proper handling. Avoid scraggly bunches, wet spots, or bunches with weak dark stems. I like my black grapes chilled.

Ribiers and Exoticas are both large enough to be peeled. If that seems silly, I recommend you take the time to see what I mean. They can also be halved and seeded, which makes them available for salads, fruit cups, and stuffing. But peeled or not, seeded or whole, these high-quality black grapes are excellent fruit for pleasure, nutrition, and sweet variety.

SEEDLESS GRAPES

Seedless grapes are America's favorite table grapes. They completely satisfy American fruit-lovers' requirements for sweetness, availability, and convenience. Seedless grapes' sugar content is twenty per cent or more.

They have a long season, from the early spring imports to the last California varieties in October, and are available in stores everywhere. They have no peels, no seeds, and no waste.

There are three popular varieties of seedless grapes: Perlettes, Red Flames, and Thompsons. There are two seasons. One is the import season. It begins with South American and Mexican varieties in late winter and ends in late spring. The domestic season begins with the California seedless varieties in June and ends in late fall. Both seasons begin with Perlettes.

Perlettes are small to medium-sized. They are greenish and offer little sweetness until late in their season. At their peak they are supersweet and very juicy. Perlettes are turgid little things and pop when you bite them. It is unfortunate that they are rushed to market before their time. Customers are disappointed and blame the storekeeper, the clerks, and the checkout clerks for their sour grapes. The person who knows when to buy and when to wait looks for Perlettes when they are yellow and slightly flecked. They are always sweeter, juicier, and cheaper when they are riper.

The long pale-green sugar lumps we call Thompson Seedless grapes arrive from South America in late winter and spring and from California (in enormous quantities) from July until the end of October. High-quality Thompson Seedless grapes have a soft frosty dust cover called the bloom of life. This dust is missing from early crops, and they are always less sweet and much greener in color. Ripe sweet juicy Thompsons are yellowish and firm. Their taste is the perfect blend of sweet juicy meat and tart skin. It is this combination that makes our taste buds excited as we eat one after another until they're gone. Later in the season Thompsons develop a tinge of brown or rust. They become a little softer to the bite, but they are pure sugar to the taste. Near the end of their season Thompsons drop a lot of loose grapes. Good fruit men collect these loose seedless grapes in small baskets and sell them at greatly reduced prices. If they are clean and pretty and of good quality, buy them. They are excellent food and excellent value. Late in both seasons some very small seedless grapes arrive. They are the last to be picked, and tiny as they are, they are wonderfully delicious and very sweet. I like to eat 'em by the handful or the mouthful, as the case may be.

I must say I now prefer Arizona seedless grapes to California ones. Arizona grapes grow in a rich limestone soil that requires no chemical fertilization. The grapes are always heavily covered with the bloom of life, and many growers ship grapes that are U.S.D.A.-certified chemical- and pesticide-free. I cannot buy enough of these varieties to satisfy the demand. Without a doubt these chemical-free limestone-grown Arizona seedless grapes are the best America produces.

Red Flame Seedless grapes were created by grafting and hybridizing the Flame Tokay grape, which is not seedless, with a round seedless variety like the Perlette. There is now a very acceptable round Red Seedless grape that is juicy, crisp, and very sweet. It is smaller than the Thompson, more like its Perlette parent. Red Seedless grapes, unlike other seedless varieties, are not sour when their season begins. They are simply bland. Many people believe that this is their natural state and pass. On the contrary; at their peak Red Seedless grapes are crisp, meaty, and sweet and have the touch of wine that characterizes their Tokay parent. Look for imported Red Seedless grapes in May and June and the California varieties from August through October.

ALMERIA GRAPES

Almeria grapes are green and probably Spanish in origin. They are large, long, crisp, and juicy. Almerias have seeds and a definite dry-wine flavor. They are excellent dessert grapes, and their wine-grape ancestry makes them perfectly suited to the last course of large meals and holiday feasting.

Almerias arrive from South America in November and last until January. California Almerias arrive in late August and last until midfall. Look for large oval fruit. They must be firm and tinged with a golden glow and soft dusting. Dull, dark-stemmed soft fruit often has a fermented taste, and the skins become tough and unpleasant. Seeds, unfamiliarity, and no sugary flavor keep the Almeria from becoming very popular. Nevertheless, it is an excellent grape, full-bodied and very satisfying. Almerias also add a graceful and rich touch when they crown a fruit bowl, their long fruits cascading over the bright apples, oranges, pears, and pomegranates.

RED GRAPES

*T*here are three varieties of red grapes that are beautiful, bountiful, and delicious. None, however, comes anywhere near the popularity of the green seedless grape. Blame it on seeds and sugar. Seeded red grapes are not sugary-sweet. They are, however, crunchy, juicy, and full of grape or sweet wine flavor.

The sweetest of the red-grape trio is the Emperor. It is heavily cultivated in Latin America and especially in Mexico, where it seems to be everywhere in season. Americans like their grapes of any color to be crunchy. Our southern neighbors don't mind a bit of softness in their grape. In the case of the Emperor, it becomes quite sugary as it softens. Emperor grapes don't shatter but stubbornly cling to their stems to the very end. It is hard to snap off bunchlets of Emperors. They must be cut or carefully separated at the main stem. Emperor grapes arrive from South America in late winter and are available until springtime. California Emperors are an early-fall arrival and last until midwinter. They are evenly colored maroon, with dark stems and an oval shape. Although most stores sell Emperors when they are firm, they retain their sweetness well past their peak. Emperors grow in very large long loose bunches and seem to be the least expensive of all the grapes.

Cardinal grapes are pink-red and marbled with streaks of green and yellow near their stem ends. The bunches are roundish, tight, and clumpish. The stems are bright green, and little bunchlets can be snapped off easily. Cardinal grapes are very firm and crack when you bite them. They are very round, but many have a definite cleft at the bottom end. Cardinals have a clean juicy taste but only produce their sweetness when they are ripe. Cardinals fall off their stems easily once ripe. The combination of easily broken stems followed by many loose grapes makes the Cardinal a difficult grape to merchandise. It never appears in supermarkets, where the Emperor is preferred. It is found in really good fruit and vegetable markets and specialty stores. The Cardinal is a cracking-good high-juice wine-flavored grape that I like very much. Very few are shipped from South America; most Cardinals appear in midsummer and early fall from California. Cardinals are excellent companions to the sugary green seedless grapes. Eating them together improves them both.

The Flame Tokay grape is the prettiest of all the grapes. It is long, elegant, and bright red, with a definite velvety coat. It has a very firm texture, lots of juice, and plenty of sweetness. It also has large seeds, something Americans hate in their fruit. More's the pity, because the Tokay has the most grape flavor of all the American grapes. Tokays grow in large long heavy bunches. They have thin skins and dark green stems that can be snapped or cut to make bunchlets. Although they are sweet, they are not sugary. I think they are an excellent after-dinner grape. They are not shipped heavily from South America, which concentrates on the durable Emperor. California ships them in late fall, and they can be had at many fruit stands as late as January.

CONCORD GRAPES

The Concord grape is small and deep blue, with large seeds. It is America's true grape. It has been cultivated since early Colonial times, and before that by our Native Americans. I think Concord grapes used to be more popular when people sat on the porch and popped Concords. It was a busy experience. The grapes were eaten by chewing the skins while sucking the juice from the meat and spitting out the pulp and seeds. Today the living room and the TV have replaced the front porch and swing, and seedless grapes don't need no pits spit.

Today folks buy the Concord for the fun and fall tradition, but mostly they are used for jelly, Welch's juice, and wine. Concords appear in early fall, mostly in the Northeast, and are available for about three weeks. They are sold in two-quart baskets with handles. Concords grow in small stubby bunches, with green stems. They break off easily in little clusters and at their peak are full of juice and sugar. If only they didn't have pits!

MUSCAT GRAPES

Muscat grapes are very large green-yellow grapes. They are definitely round, with a slight touch of rust on their skins. They have a distinct musky wine flavor and are excellent with cheese and bread. Muscat grapes are also pressed for muscatel wine, but at their peak they are excellent fruity table grapes with a unique and enjoyable flavor. Their taste conjures up monastic stone wine cellars full of oak kegs brooding in beeswax candlelight. Sometimes their aroma precedes their flavor, and the combination is heady. I imagine Muscats overflowing from large silver bowls surrounded by large orange flowers and portraits of medieval ancestors hanging in the dining halls of a Renaissance prince. Even today Muscat grapes are the pride of Italy, and almost all American inventory is imported from that country. The heavy bunches of Muscat fruit are carefully packed, each one hand-wrapped in soft netted paper, cradled in excelsior, and packed in wooden boxes with colorful Italian labels and tissue, twelve pounds each one. Last year my son gave me a gift of a particularly musky, beautiful, enormous bunch of Muscats that had just arrived. I ate them with some sharp hard Provolone and hot espresso.

Look for large slightly bronzed dusty-green grapes that look Italian. It is very difficult to find bad Muscat grapes. Their season is short. They are expensive—only great ones are exported—and they never go unsold. Look for them from early October until Thanksgiving at high-level fruit and vegetable stands and ethnic markets that cater to the classic taste of Italian Americans.

GRAPES

365

SUMMER FRUIT

17

25

23

26

24

*I*n the good ol' summertime there are picnics and barbecues, beach days and lazy lakeside afternoons. We love the sun and seek the shade, and everything seems full of the best nature can create. Chief among the joys of summer are its fruits. As our kitchens slow down and our appetites lighten up, it's so easy to reach for a peach, sit on the porch chewing cherries, or pick a fat, cold plum and taste its sweet juice. Summer fruit salads and fruit bowls sparkle with textures and juiciness that only come during those eight glorious weeks from the Fourth of July until Labor Day. The days are past when we went out back and gathered our favorites from our own trees. But the fruits are still here at our fruit stands, farm stands, roadside markets, and even some supermarkets. It's an abundance found nowhere else in the world.

Unfortunately, the largest distribution of fresh fruits in the United States is through the supermarket chain. Supermarkets do not sell ripe fruit. Ripe juicy fruit cannot survive self-service. Every thumb, every pinch, poke, and press, will leave a bruise, and nobody seems to want the bruised ones. They get pushed aside and tumble-bumped, and before you know it, the bruised fruit becomes garbage. No business can stand that kind of loss. That's why supermarket fruit is sold green. Sometimes their fruit is packaged in poly-wrapped trays. The intent is admirable, but these airless tray packs cook the fruit and make mealy out of juicy. Much of the summer fruit sold in supermarkets is eaten green, and the true pleasure is lost. Some of it rots before it ripens, because of storage and refrigeration. Sometimes things work out and it tastes great.

Good fruit and vegetable stores ripen the summer fruit they sell so it *all* tastes great. Sunfrost keeps a constantly rotating system of green, almost ripe, and ripe. Like all markets that really care about what they sell, we serve each customer personally and select items our customers want exactly the way they want them: plums for today, peaches for tomorrow, nectarines for the weekend, and apricots to eat on the way home. Or a couple of each every day.

All summer fruit has definite signs of ripeness, sweetness, and juiciness. Peaches glow; plums get deep, dark, and sensuous; cherries get very bright and dark and feel like they're about to burst (some actually do). Apricots develop a velvet down, and nectarines emit a perfume and a dewy softness from their skins. Yellow turns to gold; red turns maroon, vermilion, and ruby; blue turns purple; and shiny hard becomes semi-gloss soft. Summer fruits are different from all others. They are the drupes, cleft fruit that carries a hard stone within its sweet meat, which protects the single seed within. All other American fruit has seeds and can be had almost all year round. Summer fruit is summer only.

Learn your summer fruit. Look for the blush and the bloom that tell you the fruit is at its best. Try new varieties, and create new ways to serve and enjoy the fruits of summer. Take the time to shop at fruit stands, roadside markets, and stores that take the time to treat summer fruit like the special gift it is. Your food pleasures will multiply, your health will benefit, and your life will be richer.

APRICOTS

Apricots began in China. Some say they were brought to Europe by Alexander the Great, then disappeared with the Roman Empire and returned with the Crusaders. Or did they arrive with the Moors when they conquered Spain? One thing is for sure. The Franciscan friars brought apricots to California, and that's where almost all our fresh apricots come from today. Idaho also grows excellent apricots, even though California claims the best in the world. I have eaten apricots in Morocco, Japan, Australia, and Spain. I thought the Moroccan apricots were best, but that was straight off the tree. I have no complaints about California apricots. The trouble is that they are so very fragile that they must be picked and shipped while still hard. If they are placed under refrigeration for too long a period or go through many temperature changes, they will not ripen properly. Instead they become dry and woody.

A really well-ripened apricot is gold all over. There can be no traces of green. Very often a small bruise or soft spot is an indication of an excellent piece of fruit. To be sure, however, look for deep-yellow unwrinkled, firm, unbruised fruit. Ask for a sample even if you have to pay the price. Great apricots are worth the money. Poor ones are not. A perfectly ripened apricot is a great fruit pleasure and a rare one.

Apricots are at their peak in early summer from California. They are the first summer fruits to make their appearance. There are later supplies from Oregon and Idaho that are often excellent and some winter fruit from Australia and New Zealand whose taste has not pleased me yet.

Apricots make the finest jams, fruit toppings, and pastry fillings. They are excellent jarred whole or de-stoned and dried in halves. They are at their peak from early May to September, shipped from California and sometimes Oregon and Idaho. They should be wrinkle-free, unbruised, and firm without any green color. Allow hard fruit to sit quietly for several days in a warm place until soft to the touch.

Although they are deliciously sweet, apricots have a low sugar content. They are, however, extremely high in potassium, phosphorus, and other minerals. They are high in protein for fruit and have one of the highest levels of vitamin A found in any fruit.

NUTRITIONAL DATA

Weight (g)	Calories	Protein (g)	Carbohydrates (g)	Crude Fiber (g)	Water (g)	Fat (g)	Cholesterol (mg)
106	51	1.5	11.8	0.6	91.5	0.4	—

Vitamin A (IU)	Vitamin C (mg)	Thiamine B-1 (mg)	Riboflavin B-2 (mg)	Niacin (mg)	Vitamin B-6 (mg)	Vitamin B-12 (mcg)	Folic Acid (mcg)	Pantothenic Acid (mg)
2769	11	0.03	0.04	0.6	0.06	—	9	0.25

Sodium (mg)	Potassium (mg)	Calcium (mg)	Phosphorus (mg)	Magnesium (mg)	Iron (mg)	Zinc (mg)	Copper (mg)	Manganese (mg)
1	313	15	21	8	0.58	0.28	0.094	0.084

Raw—3 medium

APRICOT CRUSH

1½ pounds apricots
½ cup orange juice
½ cup sugar

*B*lanch apricots in boiling water. Peel and de-stone. Place in blender with ½ cup water. Process until almost smooth. Combine with orange juice and sugar in medium saucepan. Simmer over low heat, stirring frequently, until mixture thickens. Cool. Serve as ice-cream, shortcake, yam, or sweet-potato topping; as ham, fowl, or pork glaze; mixed into cold fruit salads; or as cake filling.

CINNAMON 'COTS

12–16 apricots
1 cinnamon stick
½ cup sugar
2 tablespoons honey
2 teaspoons cinnamon

*B*lanch apricots in boiling water. Peel and de-stone without halving fruit. Place apricots in sterilized quart jar with cinnamon stick. Melt sugar, honey, and cinnamon in 1 cup water. Pour over fruit, leaving ½-inch space at top of jar. Seal per canning instructions (see "Canning process" entry in Glossary, page 000). Serve as dessert or as cold side dish with pork or poultry.

PEACHES

*P*eaches may have begun in China, had their fling with the Persians and the Romans, and dined at the table of Louis XIV, but it was the American Indian who took the peach from early settlers and planted it from coast to coast. It was their favorite fruit, and they made the American peach the best in the world.

Peaches grow best in areas that have two months of cold winter. Georgia claims to be the peach state, and California grows big beautiful red-hearted peaches. But I think the best peaches are grown in the Delaware Valley, throughout Pennsylvania, and northward to Massachusetts. Any peach, however, tastes best tree-ripened and eaten where it grows. Great peaches are warm, syrupy, and sensuous. They are summer in a skin.

The season begins with Florida's first crop. It is not great. Thankfully, Georgia picks and ships before Florida finishes, and Georgia peaches are outstanding (remember the "Georgia Peach," Tyrus Raymond Cobb?). They arrive in early June. By late June California starts shipping, and the seasons move east and north until the last yellow freestone peaches are picked in New England in September.

Peaches are the least expensive of all summer fruit and can be enjoyed in many ways: fresh fruit, salads, pies, juices, jellies and preserves, fruit shakes, ice cream, brandied, canned, puréed topping, or frozen for winter use.

It is always best to purchase ripe peaches, but they must be firm, with good color and a minimum of green skin. Peaches will ripen at home if left to sit quietly in a warm, safe place for two days or so. Sometimes peaches have a small brown spot or two. This is almost always a sign of a great peach. I am not recommending you look for spotted peaches, but sometimes a small skin blemish is a sign of a ripe, juicy, sweet peach inside. Very ripe peaches make excellent fruit shakes blended with yogurt, apple cider, or orange juice. In fact, bruised or spotted peaches are always better for drinks, pies, and stewing.

There are hundreds of varieties of peaches, but basically there are two categories: clingstone and freestone. The clingstone peaches are the early arrivals, and the freestones come later. Both kinds include both red and yellow peaches. There is also an outstanding early white clingstone peach.

Peaches have a significant amount of vitamin A and a wide range of minerals, especially potassium. Although they offer a sweet flavor, the sugar content is less than ten per cent.

CLINGSTONE PEACHES

Clingstone peaches arrive in June and last until August. The first pickings are usually small but, if picked ripe, are very delicious and will sharpen your taste for those peaches yet to come. Most clingstone peaches are either all yellow or yellow with a large red area and red meat around the stone. The stone cannot be removed easily, but clingstones tend to be juicier, sweeter, and softer-textured than freestones. Clingstones are picked earlier than freestones, and most varieties are round rather than oblong. There are many more red than yellow varieties. There is also a white variety. It used to be called the Belle of Georgia, but today it bears any number of names from any number of local orchards. It is a fragile, tender fruit, does not ship well, and bruises easily. The white peach is sold at roadside markets, farm stands, and fruit markets that deal in farm-fresh or orchard-fresh produce. It has pale, pale green flesh, with a white skin and a lovely deep-pink blush. The center of the white peach is always red. The meat is soft and melts in your mouth. White peaches must be eaten with a large napkin. There is no way to describe the white peach adequately. It is so delicate that, once ripe, it can only be eaten, as it bears no peeling, slicing, or cutting up. Look for white peaches in late July, and prepare yourself for a very peachy experience.

Clingstones are excellent dessert peaches, and although they are less convenient to prepare because of the stones, most make excellent canning and jam peaches. Their soft texture makes them less desirable for pies, but they make delicious fruit shakes and ice cream.

FREESTONE PEACHES

*F*reestone peaches range from the red-skinned Havens and Harknesses to the longish golden, yellow-skinned Elberta peach. They can be split in half by hand, and the stone is easily removed. Freestone peaches are the most popular but actually tend to be a little less juicy and have a firmer texture than clingstones. Nevertheless, they are very fragrant and sweet-tasting.

Freestone peaches are excellent for canning and baking pies. They come in larger sizes than clingstones and are easier to cut up, peel, and prepare. Most freestone varieties are later crops and reach their peak in mid-August. California produces a large red-yellow freestone peach called Percoco that is highly desirable for brandying, canning, and serving sliced in chilled new wine.

Look for bright all-yellow peaches or those with deep-red blushes. Avoid all fruit that is shriveled, wrinkled, or has a dull look. Many peaches are over-refrigerated. Unlike tomatoes, which become watery once chilled, peaches become dry and mealy, losing their warm, syrupy moisture to the cold refrigerated air. If you're not sure about the peaches you see, ask the clerk for a sample. Most stores don't mind, especially if you respond to good taste with a good purchase. Farm stands and good fruit stands offer bushels, half-bushels, pecks, and baskets at discount prices late in the season. These are excellent for canning and jamming, which is wonderful family work and provides a summer taste in midwinter.

NUTRITIONAL DATA

Weight (g)	Calories	Protein (g)	Carbohydrates (g)	Crude Fiber (g)	Water (g)	Fat (g)	Cholesterol (mg)
87	37	0.6	9.7	0.6	76.3	0.1	—

Vitamin A (IU)	Vitamin C (mg)	Thiamine B-1 (mg)	Riboflavin B-2 (mg)	Niacin (mg)	Vitamin B-6 (mg)	Vitamin B-12 (mcg)	Folic Acid (mcg)	Pantothenic Acid (mg)
465	6	0.02	0.04	0.9	0.02	—	3	0.15

Sodium (mg)	Potassium (mg)	Calcium (mg)	Phosphorus (mg)	Magnesium (mg)	Iron (mg)	Zinc (mg)	Copper (mg)	Manganese (mg)
—	171	5	11	6	0.1	0.12	0.059	0.041

Raw—1 medium

PEACH CRUSH

2 pounds peaches
¾ cup pineapple juice
½ cup sugar

*B*lanch peaches in boiling water. Peel and de-stone. Cut up into blender. Add 1 cup water and blend until almost smooth. Combine with pineapple juice and sugar in medium saucepan. Simmer over low heat, stirring frequently, until mixture thickens. Cool. Serve as dessert topping, cake filling, or glaze for baked ham, poultry, or pork. Mix with yogurt, cottage cheese, sour cream, or cream cheese as a delightful summer snack or dessert.

PLUMS

*T*here are more kinds of plums than could possibly be detailed. Let it be clear, however, that plums and prunes are quite different. Fresh prunes are never wrinkled and are always blue or purple. Plums are red, maroon, black, pink, green, and yellow. All plums have skins that are shiny when unripe and that change to a dull matte color as they ripen and sweeten. The skin and pit of any plum variety contain more acid than the meat and are never as sweet. A ripe plum should be firm but yielding to gentle pressure. Do not, however, squeeze plums.

Plums travel as far as 3,500 miles to reach some markets. They are delicate, bruiseable things that need to be handled carefully after their journey. In order for plums to reach peak flavor, they must be picked mature if not downright ripe. Plums are almost always in some state of ripeness when they reach the store. The softening and sweetening takes place after picking, once the tree has started the ball rolling. Plums left to ripen on the tree until they are soft develop dark-brown areas that are not good-tasting. Plums picked before maturity wrinkle and will never be good.

Look for dark, firm, fragrant fruit without spots or brown or bruised areas. Plums break down quickly in hot humid weather, so keep a sharp eye during the dog days of summer.

Supermarket and self-service plums can never be presented in their juicy ready-to-eat state. They can only be sold hard or not at all. You must be patient and ripen this fruit at home for two to three days. Roadside markets and fruit and vegetable markets that control their inventory from source to sale always have plums at their peak. When buying plums, make sure the store clerk packs your plums on top of the bag next to the tomatoes, not on the bottom next to the potatoes.

There are also plums that appear in stores in midwinter from New Zealand, Chile, and Argentina. So far I have found them dry, tasteless, and bitter. Plums are the second most cultivated true fruit in the world after apples and are grown practically everywhere except Antarctica. In the United States almost all plums are grown in California. Some are also grown in Michigan and New York State. In season the California growers keep a constant stream of ever-changing tastes, colors, and sizes for almost four months in what has to be an amazing display of horticultural mastery. From the very first red Beauts to the large red thick-meated Presidents in September, the plums of California are one of the great summer fruits.

Plums have three seasons: late spring–early summer, midsummer, and late summer–early fall. Local varieties usually arrive in the third season, except for the lovely little Mirabelles, which make a July appearance. Since there are endless varieties of plums, we'll discuss some of them according to the plum time they appear.

Though nutritional content may vary by variety, plums generally do not offer high numbers. They are very low in calories and offer a wide range of nutritional elements, with good ratings for minerals, especially potassium.

NUTRITIONAL DATA

Weight (g)	Calories	Protein (g)	Carbohydrates (g)	Crude Fiber (g)	Water (g)	Fat (g)	Cholesterol (mg)
66	36	0.5	8.6	0.4	56.2	0.4	—

Vitamin A (IU)	Vitamin C (mg)	Thiamine B-1 (mg)	Riboflavin B-2 (mg)	Niacin (mg)	Vitamin B-6 (mg)	Vitamin B-12 (mcg)	Folic Acid (mcg)	Pantothenic Acid (mg)
213	6	0.03	0.06	0.3	0.05	—	1	0.12

Sodium (mg)	Potassium (mg)	Calcium (mg)	Phosphorus (mg)	Magnesium (mg)	Iron (mg)	Zinc (mg)	Copper (mg)	Manganese (mg)
—	113	2	7	4	0.07	0.06	0.028	0.032

Raw—1 medium

EARLY-SUMMER PLUMS

By early June and sometimes by Memorial Day California ships a small round red plum with yellow meat called a Beauty. It is not a full-flavored plum but sweetens and ripens nicely to a juicy treat. Star Rosas and Santa Rosas are also early plums. They have dark-red skins and pink meat. Santa Rosas have good plum flavor, sweetness, and juice. Look for slightly pointed round plums with a definite cleft. There are also some darker plums in the "osa" group, the Burmosa and the Formosa, which I find a little drier-textured but sweet and flavorful. Plums during this period are at their most expensive.

MIDSUMMER PLUMS

During the second week of July several darker, larger varieties with lower prices and higher flavor appear. The Tragedy is a very dark long sweet plum. The Burbank is a very red plum with firm, meaty sweet orange flesh. Wicksons are yellow, pointed, and deeply cleft. They are large, with deep-yellow meat, a small pointed pit, and lots of juice and flavor. Yellow plums don't turn a lot of heads, but I recommend you begin to enjoy them. They are some of our best. Duarte is an old variety. It has dark, dark skin and a pointed shape. The tip of a Duarte splits at its peak of ripeness, which makes it a difficult plum to sell, especially in self-service markets. There is also a small yellow or pink plum that is mostly a local-source plum. I know it as a Mirabelle. It is a favorite all over Europe as a flavorful but bitter plum that sweetens in the cooking. Perhaps our growers have misnamed their plum, but it is identical to the European variety in every way except that it is sweet, juicy, and flavorful. It is also a fun plum. Mirabelles are thin-skinned, and they eat like candy. A pound can disappear quickly, so buy two. We always sell them two or three pounds for the money. Mirabelles ripen quickly once picked and bruise easily, which makes them noncommercial. They are a farm-stand or great fruit-stand item. The very best midsummer plums are the Eldorados and Friars. They are medium to large plums, and the larger the better. Eldorados have a deep, dark-red skin and golden meat. Friars have a black skin and beautiful rose-red flesh. They are juicy and sweet, with a mouthful of flavor. By early August Friars are especially huge, and three can weigh over a pound.

LATE-SUMMER PLUMS

*T*he Santa Rosa comes around a second time, this time with lots more flavor to go with its sugar and juice. It is also darker and larger. Tragedy does a second appearance as well, but I find it soft and not attractive. The Laroda plum is a red-skinned fruit with bright-red flesh. It comes in all sizes, but bigger is best. The Laroda plum is always sweet from skin to stone, with hardly any trace of acidity. It is so sweet it must be eaten when it is still firm. Too soft is too ripe is too bad. The late-summer Kelsey plums arrive in August. They are large, pointed, and cleft and almost always arrive wearing their California green skins. If allowed to sit quietly in a warm place (not in the sun or humidity), they will turn golden-yellow from skin to stone and fill themselves with syrupy sweet lovely juice. Kelseys have a smooth, firm texture to go with their sweet flavor. Kelseys also have a very tiny stone, and the flesh is firm enough for slicing. They are quite a summer fruit. They used to be called green gages, but nobody likes green fruit, so Kelsey it is. The best plum is the Queen Anne. She has a deep, dark, almost black skin. The flesh is purple under the skin, becoming golden-rose near the stone. Queen Anne plums are usually very large and remain quite firm even when fully ripe. They are very heavy plums because of their dense flesh and high juice content. The stone is very small and can be removed without any cling. Queen Annes resist bruising, may be sliced, and in the ultimate act of sensualizing the fresh-fruit experience, may be peeled. I recommend you take the time. Do two, one for each of you.

The last plum always, for me, is the President. It is a longish medium-sized red plum with orange meat. Ripened perfectly, it is a delicious, juicy, meaty plum. Unfortunately, Presidents are often held back by shippers to lengthen the season and raise the price. Too often this results in dry, mealy fruit that is worthless. Plums finish up like they started, with some small red varieties with yellow flesh. My best advice is to quit when Queen Anne leaves, and start eating apples. The season is so good for so long, why spoil it with anything less than the best?

PRUNES

Prunes are wrinkle-free in their natural state. Their skin is always blue or deep purple. The flesh is greenish but browns up slightly at the stone, which is always freestone and easily removed. Most prunes are grown in California, but the sweetest, thin-skinned varieties come from Idaho. Large prunes called damsons are grown in the Northeast, but they tend to be softer in texture and ripen quickly once picked, which inhibits shipping outside of their local markets.

Almost all prunes are grown for the processors who produce juice and dried and canned versions of the original fruit. Properly sun-dried prunes should be moist and have a bright red-brown sheen that indicates freshness and sweetness. Dark, dull dried prunes are tough and tasteless and need to be stewed to be good again. Prunes, however, are excellent eaten as fresh fruit. Select dark firm fruit that is slightly soft at the stem end. Really soft is really too ripe.

Prunes eaten in their natural state are much lower in nutrients than dried prunes. Dried prunes are relatively high in protein, with a high content of vitamin A, B-2, B-6, niacin, potassium, phosphorus, iron, and copper. They are also high in calories and carbohydrates, which makes them an excellent exercise food but not good for weight loss.

SOFT WHOLE PRUNES

2 pounds blue freestone prunes or deep-purple damson plums
¼ cup sugar

Wash and de-stem fruit. Stones may be removed if desired before cooking, but fruit tends to collapse. Place prunes in medium saucepan with 2 cups water and sugar. Cover. Steam for 10 to 15 minutes or until soft. Chill. Serve whole, with cream, sour cream, or cottage cheese.

NECTARINES

Nectarines are not a cross between anything. Neither are they a hybrid, a sport, or a mutant. Most authorities classify the nectarine as a fuzzless peach. I disagree with that. There are peaches being grown today that are nearly fuzzless, if not totally so, and they have neither the taste, texture, skin, nor perfume of a nectarine. Whatever I read and all that I feel makes me place the origin of the nectarine somewhere between the origins of the apricot and peach in ancient China around 1,000 B.C. For almost three thousand years it has remained a fruit without the individual identity it deserves.

Nectarines come in two colors, red and yellow. Red nectarines are available in many sizes but are mostly large. In the numbers game, a 60 to 70 size is large, 70 to 80 is medium, and over that is small. A 60 to 70 tastes best. Red nectarines are mostly all of the King variety: Sunking, Flame King, and so forth. Whether partially or almost completely red-skinned, all red nectarines have red meat surrounding the red stone. Red-skinned nectarines also have a slightly more acidic peel but in their ripe state are very sweet and juicy and the peel not bothersome. The red skins arrive in the stores and markets in May and continue to be available until mid-July. As the seasons overlap, many stores have both colors in midsummer.

Yellow nectarines are more oval than round. Both red and yellow are usually deeply cleft. Yellow nectarines have a golden color, with a pink center near the stone. Care must be taken to select firm yellow nectarines with tight moist skins. Wrinkled leathery skins mean mealy meat. Yellow nectarines are always better in mid-August than late August or September. Refrigeration for long periods robs them of their juice and flavor.

There is also a small green-red nectarine with a hundred names, one from each of the local areas where it grows. These local orchard-run (packed like they're picked) nectarines ripen to a soft lime green, never yellow, and red, with plenty of flavor, plenty of juice, and lots of sugar. These little nectarines have thin sweet skins and a red stone. They are available in mid-August at roadside markets, farm stands, and country-oriented fruit and vegetable markets but never in supermarkets. During our winter Chile, Argentina, South Africa, and Israel export nectarines to the United States. They are beautiful but often underripe or over-refrigerated. They are always expensive.

All nectarines have a strong fragrance. It is sharp and sweet and rises above the wonderful perfume of the golden melons, the peaches, plums, grapes, pineapples, and bananas, as they breathe their own life as summer fruit. My father always described his fruit as the nectar of the gods. A really good nectarine will tell you why.

Nectarines are almost twice as nutritious as peaches. Nectarines have relatively large amounts of protein and B vitamins. They contain a very large amount of vitamin C and phosphorus. For lots of reasons nectarines can stand on their own.

NUTRITIONAL DATA

Weight (g)	Calories	Protein (g)	Carbohydrates (g)	Crude Fiber (g)	Water (g)	Fat (g)	Cholesterol (mg)
136	67	1.3	16.0	0.5	117.3	0.6	—

Vitamin A (IU)	Vitamin C (mg)	Thiamine B-1 (mg)	Riboflavin B-2 (mg)	Niacin (mg)	Vitamin B-6 (mg)	Vitamin B-12 (mcg)	Folic Acid (mcg)	Pantothenic Acid (mg)
1001	7	0.02	0.06	1.3	0.03	—	5	0.22

Sodium (mg)	Potassium (mg)	Calcium (mg)	Phosphorus (mg)	Magnesium (mg)	Iron (mg)	Zinc (mg)	Copper (mg)	Manganese (mg)
—	288	6	22	11	0.21	0.12	0.099	0.06

Raw—1 medium

CHERRIES

Cherries are everyone's favorite summer fruit. They last from early June until mid-July. In cooler areas cherries are still picked in August, but in most of the country the best cherries are finished by the middle of summer. Cherries must be picked ripe, and the last few days before harvest are critical. During this period cherries become sweeter and heavier as they become juicier. Picked too soon, cherries are pale and tasteless. Picked too late, cherries are too soft and watery. The best time to pick cherries is just before the birds eat them. Birds have an uncanny sense of timing, and once they decide it's time, they can wipe out an orchard in a few days. The best time to buy cherries is when they are dark, bright, and hard. If you're not sure, ask any bird.

Cherries should be clean and dry. Sticky fruit is overripe and probably beginning to break down. Overripe cherries lose their red color and begin to brown up. They begin to leak, and the fermentation causes damage to the other cherries.

Most people take it for granted that cherries can be sampled. After all, how much could one cherry cost? A large fancy Bing cherry at $2.49 or $1.99 a pound weighs out at about 5¢ to 8¢. If 25 to 30 people eat just one cherry each, it costs somewhere around $1.75 to $2.00. That is more than the owner's profit on a box of cherries. If you want to sample when buying cherries, take a small handful of six or seven cherries and pay for them. It's a real classy act, and the store-owner will show his appreciation with better service, better merchandise, and lots of little extras and turn-ons. You're the kind of customer he wants.

For all their delicious flavor, difficult harvesting, intense cultivation, and high price, cherries have very little to offer nutritionally. They are high in vitamin A, with some B-1 and B-2. They have some reputation as a kidney and liver cleanser, based on a pound-a-day consumption. That's a lot of pits.

NUTRITIONAL DATA

Weight (g)	Calories	Protein (g)	Carbohydrates (g)	Crude Fiber (g)	Water (g)	Fat (g)	Cholesterol (mg)
68	49	0.8	11.3	0.3	54.9	0.7	—

Vitamin A (IU)	Vitamin C (mg)	Thiamine B-1 (mg)	Riboflavin B-2 (mg)	Niacin (mg)	Vitamin B-6 (mg)	Vitamin B-12 (mcg)	Folic Acid (mcg)	Pantothenic Acid (mg)
146	5	0.03	0.04	0.3	0.02	—	3	0.09

Sodium (mg)	Potassium (mg)	Calcium (mg)	Phosphorus (mg)	Magnesium (mg)	Iron (mg)	Zinc (mg)	Copper (mg)	Manganese (mg)
—	152	10	13	8	0.26	0.04	0.065	0.063

Sweet, raw—10

SWEET CHERRIES

Except for local crops, which are not shipped at all, sweet cherries are grown in California and the Northwestern states. Sweet cherries are packed in twenty-pound boxes, from which they are sold. Sweet cherries are graded by size, determined by how many cherries fit side by side in a row across the width of a standard cherry box. The smallest size I know of is a 14 row, fourteen cherries across. The largest is an 8½, which is eight cherries in a row with a little space. The alternate rows are staggered to accommodate the half space above. Sweet cherries are packed upside down. The top cover goes on before the bottom of the box. The cherries are placed in by hand, row after row, for two layers, and then the box is filled with loose cherries. This used to be done for all sizes. Today only 10-row or larger cherries are plated. Most sweet cherries are either 11- or 12-row sizes. Larger sizes are considered fancy or premium fruit and demand fancy premium prices.

The two most common varieties of sweet cherries are the Lamberts and Bings. Lamberts are the smaller of the two and are considered a tender cherry, or a heart cherry. Their skin color ranges from deep pink to red. The flesh is tender and watery, rose-yellow to blackish-red, with deep red juice. Lamberts are the first cherries to arrive from California in early June and the first from Washington, Oregon, and Idaho in late June. Lamberts are never large and mostly are packed as 12- or 13-row sizes. Lamberts are very juicy, sweet, and tender-skinned.

Bing cherries are the later sweet cherry. Bings are bigger than Lamberts and are usually packed as 10½- to 11½-row sizes. All premium-sized cherries are Bing cherries. Bings are a hard cherry, or white-heart cherry. They are heart-shaped, cleft, and have a colorless juice. Their skins are deep red to black; the flesh is firm and rose-red, with a white radiant center. Bings crack when you bite them. They last longer once picked, travel better, and keep better. There are some very big black heart-shaped Bings called Oxhearts. They look like small plums and can be eaten in several bites instead of one chomp. They arrive from California about the third week in June and in mid-July from the Northwest. Without a doubt big black hard Bing cherries are the best. I prefer the early California crop to the later Northwestern varieties. Maybe it's because I've eaten so many my taste buds demand something else.

There is a third variety of sweet cherry that has almost disappeared from our markets. It is the Royal Ann, or Napoleon, cherry. I love it, seek it, and sell it at Sunfrost. It is a large heart-shaped fruit, amber to yellow with a red blush. The flesh is firm and juicy, and it is a white-heart cherry. The Royal Ann has more cherry flavor than pure sugar sweetness, and its skin is tender and tastes acid-free. You will never find Royal Anns in supermarkets. Look for them in late June at farm stands, roadside stands, and beautiful fruit and vegetable markets that have a sophisticated knowledge of all the wonderful varieties of fresh fruit.

Sweet cherries are all picked ripe, and like all ripe fruit they are very fragile. They are also susceptible to damage from humidity and often degenerate in a warm retail environment after being refrigerated. The cherry-growers and scientists have developed cherries, especially Bing cherries, with thicker and darker skins. Along with careful cultivation of hardy strains, the process includes systemic hormone injections, which darken the fruit earlier. A darker cherry is usually a softer, sweeter cherry. Cherries are also sprayed with insecticides, antifungal oils, and moisture seal. All this manipulation has resulted in a thick-skinned hard-meated chemical-coated product. Wash your cherries thoroughly before eating them. Chew your cherries thoroughly before swallowing them. I love cherries and eat them all day long in season, often unwashed. Well, nobody's perfect.

SOUR CHERRIES

Sour cherries are not important fruit-stand items. Sour cherries are, however, important processed-fruit items. More sour cherries than sweet cherries are grown in the United States. Michigan probably leads all states in production of sour cherries with the densest population of cherry trees in the world, and it claims to be the cherry capital of the world. Sour cherries are harvested and sold in July. They are smallish cherries with glossy, almost transparent skins. Sour cherries come in endless colors and color combinations. Some have dark juice, some are colorless. Some sour cherries taste sour, while others are acidic, tart, and even semisweet. You can have yellow tart fruit with a dark-red blush; a pink sour fruit with a yellow splash; or a red semisweet fruit with no blush. Whatever the color, the flesh is firm and juicy. Avoid overripe fruit. The flesh is brown and the texture mushy.

Sour cherries, with enough sugar, honey, or maple syrup, produce outstanding pies, cobblers, and tarts. They also create syrup, jam, jelly, wine, and brandy. Sour cherries are easily pitted because of their thin skins, a factor not overlooked by the efficient American fruit-processors. Look for fresh sour cherries for home baking and canning during late July at farm stands and roadside markets. Some country fruit stands may have them for a few days. They are very perishable as well. If you must have sour cherries and you don't mind processed fruit, look for them wherever fine cans are sold.

TROPICAL FRUIT

1. YELLOW BANANAS
2. RED BANANAS
3. CUICADITAS   4. MANGO
5. COCOHUSK & COCONUT
6. SOUTH AMERICAN PAPAYA
7. PACIFIC PAPAYA
8. CHERIMOYA
9. HAWIIAN PINEAPPLE
10. SPANISH CAYENNE PINEAPPLE
11. CALIFORNIA AVOCADOS
12. FLORIDA AVOCADOS

*T*he tropical fruits in any fruit stand or produce department are always the most unusual and colorful of all the produce. They are also some of the most nutritious. Yet most people never think to 'try them. Why? Because coconuts are impossible to open. Pineapples are hit-or-miss and easier to peel when they come in a can. Papayas have those weird black seeds. Mangos...what's a mango? And how do you eat one? A cherimoya looks like nothing that ever grew in Mom's backyard. And a passionfruit sounds like something you share with another consenting adult. But once you get past the unfamiliar, once you learn how to select, peel, slice, serve, and eat the fruits of Eden, all your ideas about and appetites for tropical fruit will change.

Tropical fruits have long seasons. Long before peaches and plums are picked, mangos are ready to satisfy your fruit hunger. In deep winter there are smooth, flowery papayas. Fresh-grated coconut will find its way into so many of your recipes, and fresh coco chunks do everything candy does except make pimples, fat cells, and bad teeth. Pineapples arrive every day by plane from Hawaii and boat from Honduras, Costa Rica, and Puerto Rico. And if you haven't yet discovered all there is to avocados, turn yourself loose. Tropical fruits are bursting with pleasure, nutrition, and appetite excitement. They make the best fruit shakes you can drink. Tropical fruit combined with some orange juice and ice in a blender can create summer fun drinks, breakfast drinks, health drinks, and party drinks, all of which are sweet, aromatic, and sensuously pleasing. There is no salad like a tropical fruit salad splashed with lime, crushed nuts, grated coconut, and orange juice. It is natural ambrosia without the marshmallows.

Until recently all tropical fruit except for bananas was being raised from seed. Seedling reproduction is inconsistent, and many tropical plants produced inferior, poor-tasting, low-quality fruit. About twenty years ago the hybridists got into the act, and the improvement in the color, size, flavor, and quality of our domestic tropical-fruit production is tremendous and increasing every season. I think Florida mangos are better than Mexican or Guatemalan. California avocados have better flavor than any I ever ate in all of Central America, which is where they began. Many people prefer the sweet floral flavor of the Hawaiian papaya to that of the Caribbean papaya. I don't. I'm glad for both. I think the tropical-fruit game is only just beginning for the American consumer. If your local stores don't carry mangos, papayas, coconuts, and great pineapples along with their bananas, ask them to do so. The fruit is here, part of the earth's bounty. It is plentiful and beautiful. The more of it you learn to select and properly prepare, the better life you will enjoy. Shop at stores that make it their business to carry tropical fruits. If they don't, they are definitely out of time.

BANANAS

Bananas don't grow on trees. Bananas grow on very tall plants that produce one long stalk, at the end of which is a large red heavy flower and ten to fourteen hands of bananas. Once the bananas are harvested, the plant dies, and new plants develop from the underground system of rhizomes. Bananas do not have true seeds. They have tiny partial seeds that do not regenerate but serve only to release the hormone that turns green bananas ripe and yellow. All bananas are harvested and shipped green. They are stored in temperature-controlled rooms and cannot tolerate temperatures below 50° F. Bananas cannot be refrigerated. Green bananas that become chilled will turn a muddy brown and never ripen. Bananas are ripened at regional distribution centers throughout the United States at a temperature of around 65° F. Humidity is also a factor.

Bananas should be purchased when they are yellow, with a touch of green at either end. Bananas should be eaten when the peel becomes speckled. Speckled bananas are not overripe or rotting. They are perfect. When the fruit is speckled, it is at the stage when it has the lowest starch, the highest sugar, and the lowest tannin content (tannin is the fresh-fruit element that tastes bitter and puckery). Black-spotted bananas or bruised, soft, or muddy-green bananas should never be purchased. There is no excuse for any fruit stand or supermarket to carry anything but beautiful bananas. Bananas are one of the few fruits that ripen off the plant just as well or better as on the plant. If supermarket managers or fruit-stand owners can't supply perfect bananas, they ought to look for other work. Bananas are elementary.

There are three kinds of bananas: yellow, red, and Chicadita. The yellow banana was introduced to North America by the United Fruit Company, now called United Brands. It is largely because of the effort of this unusual and controversial company that we enjoy a constant high-quality supply of this tropical fruit. Almost all yellow bananas are of the

Valery variety, large, straight, and very bright yellow. Whey buying yellow bananas, select them for home ripening. Three average bananas weigh about a pound and are the least expensive of all tropical fruits and possibly of all fruits.

Red bananas are like yellow bananas, only different. Red bananas are shorter and fatter than yellow bananas. They have thicker red skins, and the fruit is darker and creamier. Red bananas have a higher oil content as well, and a thicker texture. The taste of a red banana as compared to that of a yellow banana is like comparing cream to milk. Red bananas are rich, aromatic, smooth, and sweet. They are delectable and satisfying and they make excellent fruit salad and blender drinks. They hate the cold and cannot be refrigerated.

Red bananas have limited distribution. They are usually available only in metropolitan areas in ethnic markets. They are also found in fruit and vegetable markets that understand and properly merchandise the beautiful fruits of the tropical world.

Chicaditas are small dull-yellow bananas that are also called Ladyfinger bananas. I prefer the name Chicaditas. They are incredibly sweet, and their small size makes them fun to eat. I first ate them in Mexico, but they are everywhere throughout Central and South America and the Caribbean islands. They are also found in Tex-Mex markets; the Spanish markets of Miami, New York, Boston, and Philadelphia; the Mexican *fruterias* of southern California; and once in a while at Sunfrost in Woodstock, New York.

Ask for the red bananas and tiny Chicaditas at your friendly fruit stand or produce department. They'll probably look at you with a blank stare. Don't be discouraged. Ask again. And one day a fragrant bunch of deep-red fruit or elegant little yellow dwarf bananas will arrive from Nicaragua or Costa Rica. Maybe.

Bananas are nutritionally loaded. They are very high in plant protein; vitamins A, B-1, B-2, B-6; and niacin. They also provide some of the largest amounts of zinc, manganese, potassium, phosphorus, iron, and copper of any fruit.

NUTRITIONAL DATA

Weight (g)	Calories	Protein (g)	Carbohydrates (g)	Crude Fiber (g)	Water (g)	Fat (g)	Cholesterol (mg)
114	105	1.2	26.7	0.6	84.7	0.6	—

Vitamin A (IU)	Vitamin C (mg)	Thiamine B-1 (mg)	Riboflavin B-2 (mg)	Niacin (mg)	Vitamin B-6 (mg)	Vitamin B-12 (mcg)	Folic Acid (mcg)	Pantothenic Acid (mg)
92	10	0.05	0.11	0.6	0.66	—	22	0.3

Sodium (mg)	Potassium (mg)	Calcium (mg)	Phosphorus (mg)	Magnesium (mg)	Iron (mg)	Zinc (mg)	Copper (mg)	Manganese (mg)
1	451	7	22	33	0.35	0.19	0.119	0.173

Raw—1 medium

MANGOS

Mangos have been called the fruit of the gods. You get no argument from me. Buddha rested and meditated in his mango grove. I have rested in the shade of a giant mango tree in the scorching sun of an El Salvador afternoon and eaten its fruit. Mango trees are enormous. The leaves are thick and shiny green and create a cool oasis wherever their tree stands. The mangos hang from long stems like giant upside-down lollipops.

Most mangos in the world are grown and eaten in India. Mangos are that country's national pastime. I ate mangos in India. I also ate some chicken that near killed me, and I left India before I got back to mangos. I love Mexican mangos. I prefer the big green and parrot-red ones that are harvested in May to the kidney-shaped green or gold ones you get in March and April. So far I like Florida mangos best. Florida provides a high-quality consistently good crop of mangos from May until September. It grows two varieties, Hayden and Tommy Atkins. Both are medium to very large (about two pounds), with green skins that turn yellow at the blossom end and bright red at the stem end. Haydens have bright-orange meat, and Tommy Atkinses have bright-yellow meat. Both are sweet, juicy, and free from the fibers that spoil the succulence of stringy mangos. There are many varieties imported from Mexico, South America, and the Caribbean, as well as some other Florida varieties.

Mangos come in all shapes, sizes, and colors. They may be round or oval. They are pear-shaped, peach-shaped, and heart-shaped. Some are kidney-shaped, or long and thin, or S-shaped. Some are flat. I like the Florida giant egg-shaped ones best. Some mangos weigh four ounces, some four pounds. The best ones weigh from one to two pounds. Mango colors range from dull green to multicolored parrot hues of yellow, red, orange, rust, and gold. To make things easy look for large oval red, green, and yellow fruits that smell like romantic tropical flowers.

Mangos are becoming more and more popular. Their distribution improves yearly, and I am amazed at the high-level consistent quality of the Florida fruit. The trouble with mangos is that hardly any North Americans know how to eat them. The unskilled mango-eater will, without a doubt, create a mess, ruin the fruit and maybe a shirt or blouse, and never come back for more. A mango cannot simply be sliced and eaten, or peeled and eaten, or even quartered. All these techniques mangle the mango and the mango-eater. All mangos have a long thick seed husk in the center, approximately the same shape as the mango. This seed husk is connected to the center mango meat by a thousand tenacious fibers. The way to enjoy a mango neatly is to deal with the seed husk first. Here is what to do:

Place the mango on its long narrow side. Cut all the mango meat away from the right side of the seed husk in one piece. Then cut away all the meat from the left side, also in one piece. This will leave you with a half-inch center slice containing the seed husk and two mango fillet halves. Hold one half in your hand, and score the meat like tic-tac-toe. Hold the scored half in both hands, and pop it inside out. The mango will present itself in neat little squares. Eat them right off the skin. Do the same with the other half. The husk slice can be peeled and eaten around the edges, but the main meat is in the outer halves.

Mangos are beginning to appear in all kinds of fruit and vegetable stores and supermarkets. In case the storekeeper doesn't know, mature but unripe mangos cannot tolerate temperatures below 50° F. Chilled mangos never ripen. Ripe mangos should not be refrigerated below 45° F. unless they are being chilled for immediate use. When selecting mangos, look for the gold, yellow, green, and red ones that smell wonderful. If you still can't tell, pick one up. Please don't squeeze it. Touch it gently. Does it give? Yes? Eat it. Maybe with a dash of lime or a sprinkle of chile.

Besides being one of the world's great hand fruits, mangos are a basic chutney ingredient. They will liven any fruit salad or cereal and go well with cottage cheese and yogurt. Mangos also make excellent blender drinks, ice cream, and one of my favorite desserts, mango mousse.

To make the mango even more worthwhile, it is a fruit that delivers comprehensive nutritional coverage. Mangos have the largest amount of vitamin A of any fruit. They have very large amounts of vitamins C, B-1, B-2, and B-6, as well as some of the highest ratings of essential minerals. There is every reason in the world to learn about and enjoy mangos.

NUTRITIONAL DATA

Weight (g)	Calories	Protein (g)	Carbohydrates (g)	Crude Fiber (g)	Water (g)	Fat (g)	Cholesterol (mg)
207	135	1.1	35.2	1.7	169.1	0.6	—

Vitamin A (IU)	Vitamin C (mg)	Thiamine B-1 (mg)	Riboflavin B-2 (mg)	Niacin (mg)	Vitamin B-6 (mg)	Vitamin B-12 (mcg)	Folic Acid (mcg)	Pantothenic Acid (mg)
8060	57	0.12	0.12	1.2	0.28	—	—	0.33

Sodium (mg)	Potassium (mg)	Calcium (mg)	Phosphorus (mg)	Magnesium (mg)	Iron (mg)	Zinc (mg)	Copper (mg)	Manganese (mg)
4	322	21	22	18	0.26	0.07	0.228	0.056

Raw—1 medium

MANGO CHUTNEY

    8 mangos, peeled and cut up
    1 pound green apples, peeled and cubed
    1 cup celery, finely chopped
    1 pound onions, chopped
    1 cup seedless raisins, chopped
    1 quart apple-cider vinegar
    4 cups brown sugar
    ¼ cup maple syrup
    ½ teaspoon grated ginger
    1 teaspoon salt
    Pinch cayenne pepper
    Dash freshly ground black pepper

Place mangos, apples, celery, onions, and raisins in pan and simmer in own liquid until just soft. Add remaining ingredients. Stir and simmer until mixture thickens and excess liquid evaporates. Spoon into sterilized pint jars. Cover and refrigerate, or seal and follow canning procedure (see "Canning process" entry in Glossary, page 000).

MANGO MOUSSE

    1 cup milk
    1 envelope unflavored gelatin
    1 cup crushed peeled mango
    ⅔ cup sugar
    2 eggs, separated
    Pinch salt
    1 teaspoon vanilla
    1 cup heavy cream, chilled

Mix milk and gelatin and let stand for 10 to 12 minutes. Cook mango and half the sugar, adding 1 to 4 tablespoons water until fruit is smooth. Add to milk mixture. Beat in egg yolks and salt. Cook over medium heat until mixture thickens, stirring constantly. When mixture clings to spoon, add vanilla. Cool. Whip egg whites until fluffy and add remaining sugar slowly, beating until stiff peaks form. Fold into mango mixture. Whip chilled cream. Fold mango mixture into whipped cream. Spoon into individual dishes or dessert cups. Chill for 2 to 3 hours.
    Serves 2.

TROPICAL FRUIT SALAD

1 mango, peeled
1 Hawaiian papaya, peeled
¼ pineapple
4 strawberries, sliced
1 flecked ripe banana, sliced
½ cup orange juice
2 tablespoons coconut cream
2 tablespoons crushed mixed nuts

Cut mango, papaya, and pineapple into medium-sized pieces. Mix. Add sliced strawberries and banana. Divide among individual bowls. Pour orange juice over fruit in each bowl. Top with coconut cream and nuts. *Serves 2.*

COCONUTS

Nobody gets to see the coco in America, only the nut. Actually, it's a drupe, a hairy shell that surrounds the seed inside. The heavy fibrous green-gold husk is removed before shipping. Coconuts are grown in all the world's tropical and subtropical areas. Most of the ones that become part of our American food chain are grown and shipped from Central America. The big forty-pound burlap sacks are printed in Spanish with pictures of coco palms, native women, and Indian men.

What we eat is the meat of the seed inside the shell. When picked ripe, coconut meat is solid, pure white, and very sweet. The seed liquid is cloudy, aromatic, and pleasing to the taste and body. The meat of an unripe coconut is gelatinous and more nut-flavored than sugary. The liquid is clear and light and tastes both delicious and cleansing. The meat, eaten with a spoon, is fed to babies, used as healing food, and much preferred by natives to the solid ripe meat. Both the ripe and unripe liquids are either drunk straight from the shell or used in tropical drinks, fruit shakes, and cocktails. Sometimes the liquid is replaced in the shell by elaborate alcoholic concoctions for fun-seeking tourists or straight rum for the local boys.

Coconut milk is the juice that is extracted from the meat during grating or grinding. It is an important food commodity for millions of people throughout the tropical world. It is used in place of oil, fat, and butter. Coconut milk is used in tropical baking, cooking, and liquid refreshment. It is sweet, high in protein, and free from cholesterol. Copra is the dried kernel of the coco nut and is the stuff pirate and conquistador legends were made of. It was, and still is, prized the world over. Pressed, it yields coconut oil, which is used for perfumes, cosmetics, cooking, and medicine. The refined oil becomes coconut butter, which can be used as a low-cholesterol butter substitute, as well as for perfume, hair lotion, skin cream, cosmetics, cakes, and candy.

Coconuts are sold in many supermarkets and at all great fruit stands. They are harvested year-round but shipped north during the colder months of the year. Heat causes rapid fermentation of the juice of the dehusked ripe coconut. It is very difficult to find a fresh sweet coconut during hot weather. The trip from the jungle is never easy or short, and to further sit in warehouses, storerooms, and display shelves during hot, humid days is too much to ask even of the incredibly resourceful coconut.

Hardly anyone in the United States knows how to open a coconut. Most folks run for the hammer first and ask questions later. Actually, it is not so difficult. All coconuts have a face at one end: two eyes and a mouth. The eyes are hard, and one should be punctured with a sharp tool. The mouth is soft and can be opened with a small knife. Pour out the liquid. Bounce the shell all around on a hard surface, over and over. This loosens the inside meat. Then bounce the shell very hard, and it will crack open. Remove the loosened meat with a flat-bladed table knife.

Select coconuts that have liquid inside. Shake before you buy. Avoid coconuts with wet spots, cracks, soft eyes, or dry insides. If possible select coconuts with fibrous light-colored hairs. Unfortunately, a coconut may have all the good signs but taste fermented. It was probably exposed to heat or sun. Return it. In selecting fruits and vegetables, the only item that can be squeezed safely is the coconut. Squeezing can do it no damage. Neither can it tell you anything about the real fruit inside.

Coconuts are an endangered species. Fifteen years ago an airborne bacterial disease called the Miami blight destroyed all the south Florida coco palms. The bacteria moved westward and southward, where it has practically destroyed the coco crops of the Yucatan peninsula south to Belize. Widespread destruction caused by the blight, which hits the trees slowly over two to three years, is all over Jamaica, the West Indies, and the Caribbean islands. Signs of the blight have lately been reported in Africa. There is a process of antibiotic infusion of the trees that inhibits the bacteria but does not destroy it. It is available but is costly, fussy, and gives only symptomatic relief. The disease remains. It is possible that the Miami blight could eliminate one of the most important basic food sources of millions and millions of people whose basic survival is already strained by expanding population and decreasing food supplies. At the very least it will leave vast stretches of treeless areas along beautiful coastlines, on mountainsides, and on tropical islands. They will build hotels where the trees were and tell the tourists who come there how beautiful it used to be when the cocos grew.

Coconuts have limited but important nutritional value. Their high-protein, high-sugar, no-fat, no-cholesterol composition makes them a high-energy body-building food that does not create fatty cells and tissue. Both the liquid and well-chewed or grated meat are an excellent digestive aid and cleanser.

NUTRITIONAL DATA

Weight (g)	Calories	Protein (g)	Carbohydrates (g)	Crude Fiber (g)	Water (g)	Fat (g)	Cholesterol (mg)
48	174	1.6	6.8	—	24.4	16.3	—

Vitamin A (IU)	Vitamin C (mg)	Thiamine B-1 (mg)	Riboflavin B-2 (mg)	Niacin (mg)	Vitamin B-6 (mg)	Vitamin B-12 (mcg)	Folic Acid (mcg)	Pantothenic Acid (mg)
—	2	0.02	0.01	0.3	—	—	—	—

Sodium (mg)	Potassium (mg)	Calcium (mg)	Phosphorus (mg)	Magnesium (mg)	Iron (mg)	Zinc (mg)	Copper (mg)	Manganese (mg)
8	373	10	47	—	0.95	—	—	—

Fresh, raw—½ cup shredded

PAPAYAS

*P*apaya. They call it *mando* in Brazil, *lichasa* in Puerto Rico, paw paw in the Caribbean, and *melon zapote* in Mexico, except in the Yucatan, where it is called papaya. *Papaya* is the Cuban term for the female sex organ, so the Cubans call it *fruta bomba,* the hand-grenade fruit. Sex and violence are everywhere. So are papayas. They are eaten in all the tropical and subtropical parts of the world. There are two kinds of papayas. They are different only in size and color. The larger papaya is the South American papaya. It weighs as much as twenty-five pounds. It is always picked ripe, with a green skin that turns yellow or orange when fully ripe. The soft mellow flesh is either yellow, salmon, pink, orange, or rose. The smaller papaya, usually about a pound and a quarter, is the Pacific papaya. Almost all imported papayas are of the Pacific type, from Hawaii. They are picked ripe, with a green skin that turns yellow or gold. The flesh is always yellow. I think Hawaiian papayas have more fragrance, but neither variety is very aromatic. Tropical papayas have a thinner skin but only when ripe. They are much larger but more fragile and cannot tolerate shipping very well. Florida is beginning to develop a sizable papaya industry, but its enormous papaya-loving Hispanic population consumes the entire output.

The flesh of all papayas has a taste unfamiliar to most Americans. It has the look and texture of a cantaloupe, so your mind tells your taste to expect cantaloupe or something like it. Only you don't get melon, you get papaya. Perhaps it's a bit more mushy or jungle-y, something you'd expect to find in a hot, humid, beautiful place, which is exactly where papayas grow best. Papaya fruit is incredibly smooth. Ripe papaya actually melts in your mouth. There is no food so soothing to a dry throat or a heat-weary soul than peeled chilled papaya splashed with lime. Papaya is my favorite tropical fruit.

All papayas have seeds, hundreds of small black shiny oval seeds that look like large caviar. They are nestled in the large central cavity of the fruit and are easily removed. They are all loose in the tropical papaya and slightly attached by their fibers in the Pacific variety. Papaya fruit, and especially the seeds, contain papain, which is an element similar to pepsin, the essential ingredient in digestive juices. Papaya juice is one of the best meat-tenderizers, and papain is found in all packaged tenderizing products.

The healing power of papaya is legendary and factual. Papaya fruit will eliminate gastric indigestion, reduce gas, sooth irritation and inflammation. Papaya as an important part of your diet will improve your skin, nails, and hair; keep eyes clear; cleanse infection; help restore proper intestinal action; and detoxify the entire body. Papaya is a wonderful addition to fruit salad, makes delicious juice or shakes, cuts and serves like melon, and becomes even more delightful with a sprinkle of lime juice.

Look for papayas from Hawaii in fancy city fruit markets, fruit and vegetable markets that do more than buy and sell, some modern supermarkets, and all Latin American and Caribbean ethnic markets. If purchased green, papayas need five to seven days to ripen fully. Half-green fruit needs two to three days. Fruit that has been chilled before it is fully ripe will never ripen. Papayas hate the cold. They are jungle food. Avoid black spots, especially those with ringed discolorations. There are almost no South American papayas available outside of New York and Miami, and those go directly to ethnic retailers, who *must* have them. Papaya is part of the ancestral culture of millions of Hispanic Americans.

Papaya is one of the world's most nutritional fresh foods. It contributes more than its share of vitamins A, C, B-1, and B-2. It is high in niacin, calcium, magnesium, and all the other minerals our bodies need to thrive and survive.

NUTRITIONAL DATA

Weight (g)	Calories	Protein (g)	Carbohydrates (g)	Crude Fiber (g)	Water (g)	Fat (g)	Cholesterol (mg)
304	117	1.9	29.8	2.4	270	0.4	—

Vitamin A (IU)	Vitamin C (mg)	Thiamine B-1 (mg)	Riboflavin B-2 (mg)	Niacin (mg)	Vitamin B-6 (mg)	Vitamin B-12 (mcg)	Folic Acid (mcg)	Pantothenic Acid (mg)
6122	188	0.08	0.1	1	0.06	—	—	0.66

Sodium (mg)	Potassium (mg)	Calcium (mg)	Phosphorus (mg)	Magnesium (mg)	Iron (mg)	Zinc (mg)	Copper (mg)	Manganese (mg)
8	780	72	16	31	0.3	0.22	0.049	0.033

Raw—1 medium

CHERIMOYAS

Cherimoya fruit must ripen on the tree. Once ripe, it is fragile and deteriorates quickly. It cannot be shipped and is almost never found in American stores and markets. But Americans are great travelers, especially to the Caribbean and Latin American worlds. I am sure the *turistos Americanos* have encountered this strange-looking and luscious fruit many times and loved it. I have no doubt the cherimoya, despite its fragility, will join the expanding varieties of tropical fruits that are becoming part of the American food chain. We Americans want what we like, and we get it.

The cherimoya is a heart-shaped fruit with either dull-green or purple-pink skin. Some green ones look artichoke-shaped. The skin is somewhat scaly and irregular. Cherimoyas range in size from something like a large apple to a small cantaloupe. They have heavy ungainly stems. The flesh is pure white and contains several flat black seeds. The texture is like pudding and the taste is very sweet, with a thick fruit flavor. The fruit is either halved and spooned or peeled and eaten out of hand. It is never eaten casually. The fullness and ripeness is sometimes too much for people. For now it is still a jungle fruit. You Tarzan, me cherimoya.

Cherimoya has the highest protein content of all the fresh foods except for lima beans.

NUTRITIONAL DATA

Weight (g)	Calories	Protein (g)	Carbohydrates (g)	Crude Fiber (g)	Water (g)	Fat (g)	Cholesterol (mg)
547	515	7.1	131.3	12	402.1	2.2	—

Vitamin A (IU)	Vitamin C (mg)	Thiamine B-1 (mg)	Riboflavin B-2 (mg)	Niacin (mg)	Vitamin B-6 (mg)	Vitamin B-12 (mcg)	Folic Acid (mcg)	Pantothenic Acid (mg)
55	49	0.55	0.6	7.1	—	—	—	—

Sodium (mg)	Potassium (mg)	Calcium (mg)	Phosphorus (mg)	Magnesium (mg)	Iron (mg)	Zinc (mg)	Copper (mg)	Manganese (mg)
—	—	126	219	—	2.74	—	—	—

Raw—1 medium

PINEAPPLES

*P*ineapples have nothing to do with pines or apples. They are grown on tropical plantations from ground-level plants that shoot a three-foot stalk, atop which perches the pineapple, fruit and crown. Pineapples look complicated but are basically simple. There are two kinds sold at American fruit stands, produce markets, and supermarkets, the Red Spanish and the Cayenne.

The skin of the Red Spanish ripens to a deep red-orange color. It has spiky serrated leaves and whitish-yellow meat. It is usually more acid than sweet and is the smaller of the two most common American pineapples. The Red Spanish is grown in Puerto Rico, Costa Rica, Honduras, and Mexico and imported to American markets. Pineapples are shipped green but, with luck, mature and require time and patience for their ripening. It is mostly Red Spanish pineapples that are sold in stores and markets. Look for these with a touch of orange and red, at least at the base. An all–orange-red Red Spanish without bruises or soft spots will taste delicious. This is the way they are eaten where they are grown.

The Cayenne is greenish to orange. It has short or long smooth-edged leaves and deep-yellow meat. It is sweeter, softer, and juicier than the Red Spanish. The Cayenne often grows quite large and heavy. Cayenne pineapples are the famous Hawaiian export fruit. They are marketed chiefly by Dole and Del Monte as "Jet Fresh Pines." They are picked ripe, carry string labels, and are air-shipped by jet to mainland markets. They are two to three times as expensive as the Red Spanish, which is shipped green as freighter fruit from Central America and the Caribbean islands. They are also much better. A ripe "jet fresh" Hawaiian Cayenne is sweet, soft, juicy, and tantalizingly yellow. Dole and Del Monte also have plantations in Honduras and Costa Rica that are now producing Central American versions of the Pacific Cayenne. These varieties are picked green and shipped by freighter as well. These large Cayennes also wear string tags with the company name, but they are not "jet fresh" or plant-ripened. So read your labels carefully.

All pineapples must be picked mature or they will never ripen. Avoid pineapples that look metallic or what fruit men call dead green. The pineapple you select must have some color and some aroma as well.

Leaves that pull out easily are not a sign of anything. If your pineapple needs further ripening at home, do not refrigerate it. Stand it upside down in a warm, quiet place. This causes the sugary juice to flow towards the crown instead of sitting at the base, fermenting. Wait for the time it takes to color up and smell like the perfume of a Tahitian night.

I think pineapples should always be cut in quarters or sixths for best serving. Remove the core from each wedge. Cut inside the skin along the fruit to remove the peel. Cut up the wedges and mix them in a salad bowl, or serve the individual boats with wedges of lime. Pineapples can always be halved, scooped out, and refilled with small pineapple pieces and kirsch, ice cream, and other fruits, but hardly ever with chopped liver.

Pineapples have impressive nutritional credentials. They have one of the highest mineral contents of all the fruits. They contain significant amounts of papain. Pineapples are excellent digestive aids and cleansers, help clear bronchial tissue, and eliminate mucous waste. Please remember, however, that we're talking about real leaves-and-peel pineapples, not the canned kind.

NUTRITIONAL DATA

Weight (g)	Calories	Protein (g)	Carbohydrates (g)	Crude Fiber (g)	Water (g)	Fat (g)	Cholesterol (mg)
155	77	0.6	19.2	0.8	135.1	0.7	—

Vitamin A (IU)	Vitamin C (mg)	Thiamine B-1 (mg)	Riboflavin B-2 (mg)	Niacin (mg)	Vitamin B-6 (mg)	Vitamin B-12 (mcg)	Folic Acid (mcg)	Pantothenic Acid (mg)
35	24	0.14	0.06	0.7	0.14	—	16	0.25

Sodium (mg)	Potassium (mg)	Calcium (mg)	Phosphorus (mg)	Magnesium (mg)	Iron (mg)	Zinc (mg)	Copper (mg)	Manganese (mg)
1	175	11	11	21	0.57	0.12	0.171	2.556

Raw—1 cup pieces

PINEAPPLE CRUSH

½ **cup sugar**
½ **pineapple, cut up and juice saved**
½ **cup orange juice**

Melt sugar in ½ cup water in saucepan. Add pineapple and its juice. Simmer over low heat, adding orange juice. When mixture thickens, place in bowl and refrigerate. Serve cold as dessert topping, or spoon over baked or roasted ham, duck, or pork.

AVOCADOS

If I am ever in a situation where there is only one food source, I hope it's the avocado. *The avocado is the world's most perfect food.* It is rich in oil and protein and is an excellent meat substitute. Avocados grow abundantly, can only be picked green, ship easily, and are becoming better and better as prices seem to decline.

There are two kinds of avocado, the Guatemalan and the Mexican. In the United States California grows the Guatemalan avocado. It is marketed by growers' cooperatives (one ships under the label Calavo) by their varietal names. The two most popular are Fuerte and Hass. These varieties have rather thick pebbled skin that turn black when the fruit is ripe. California avocados have a higher oil content than Florida ones, and their flesh is smooth and very clean. When ripe, the flesh is a pale yellow-gold. The seed is easily removed. The Hass and Fuerte avocados range in size from medium pears to very large round pears. They used to be referred to commonly as alligator pears. Hass avocados are available from April to November and the Fuertes from November to March. So you can relax. You're covered in avocados.

The United States also imports avocados from Mexico, which grows the Guatemalan varieties, both Hass and Fuerte. Their quality is high and the price is right, but for some reason—probably because the oil content is higher—I prefer the California-grown fruit to the Mexican. The Mexican variety itself is now grown in Florida.

Florida grows the larger smooth-skinned Mexican avocado and a West Indian variety. They have a much higher water content, less oil, and a softer texture. They are picked green but the skins do not turn black as the fruit ripens. If a Florida avocado skin turns black, throw it away. Florida also markets its avocados by varietal name. One source claims the Lula variety is the most popular, but I prefer any of the Booth series and the Taylor variety. Florida avocados, however, are not retailed nearly as often as the California Hass or Fuerte. They are always much larger and less expensive than their California cousins. I don't like them as much, except as cradles for seafood salads. They seem to complement each other better. Grapefruit sections also go better with their Florida friend.

Avocados are definitely one of the fruits you do not squeeze. Look at them. If they are unwrinkled, unbruised, unscarred, and unripe, buy them. Ripen your avocados at home. Keep a constant supply ripening at different stages. It is almost impossible to buy ready-to-eat avocados. Sunfrost is the only fruit stand in America I know that sells perfect ready-to-eat avocados every day, every season, every variety. Avocados are destroyed by self-service, and most fruit men panic when they get a load of ripe ones. They refrigerate them and kill them altogether. Ripe avocados cannot be refrigerated for more than a day. A sure sign of ripeness is a loose stem. If the stem slips out of its little socket, your avocado has surrendered.

All avocados make great salads, dips, and soups and are great food by themselves. Halve the avocado, remove the seed, and fill the hole with fruits or seafood salads, French dressing, or an especially *picante* salsa. This Mexican combination is a wonderful contrast between the soft, mellow avocado meat and the sharp salsa fire. Guacamole has to be one of the world's most delicious light meals. Serve it with chips and little cucumbers. There are also tiny avocados called cocktail avocados appearing now. They are larger than olives, California-grown, and a delight to eat.

I cannot say enough about avocado nutrition. It is the only food I know that has a perfectly balanced pH. It is neither acid nor alkaline. It is neutral, the only state for the world's most perfect food. Avocados are not fattening. The salad dressing, mayonnaise, and tuna fish are. Avocados are, however, the highest-protein fresh fruit of all next to their tropical friend the cherimoya. California avocados are nutritionally richer than Florida fruit, but both offer complete and plentiful vitamins, minerals, and highly usable carbohydrates. Avocados are easily digestible and as a regular part of your daily diet will improve hair and skin quality and soothe the digestive tract. They are an excellent food for children during their growth years due to their high B-vitamin content. How many avocados can you eat?

NUTRITIONAL DATA

Weight (g)	Calories	Protein (g)	Carbohydrates (g)	Crude Fiber (g)	Water (g)	Fat (g)	Cholesterol (mg)
173	306	3.6	12	3.7	125.5	30	—
304	339	4.8	27.1	6.4	242.4	27	—

Vitamin A (IU)	Vitamin C (mg)	Thiamine B-1 (mg)	Riboflavin B-2 (mg)	Niacin (mg)	Vitamin B-6 (mg)	Vitamin B-12 (mcg)	Folic Acid (mcg)	Pantothenic Acid (mg)
1059	14	0.19	0.21	3.3	0.48	—	113	1.68
1860	24	0.33	0.37	5.8	0.85	—	162	2.95

Sodium (mg)	Potassium (mg)	Calcium (mg)	Phosphorus (mg)	Magnesium (mg)	Iron (mg)	Zinc (mg)	Copper (mg)	Manganese (mg)
21	1097	19	73	70	2.04	0.73	0.46	0.422
14	1484	33	119	104	1.6	1.28	0.763	0.517

California, raw—1 medium
Florida, raw—1 medium

Avocado Lunch for Two

2 avocados
4 lettuce leaves
1 cup fresh tomato sauce
1 small red onion, finely chopped
2 cloves garlic, finely chopped
3–4 sprigs parsley or coriander, finely chopped
2 tablespoons finely chopped red pepper
1 small jalapeño pepper, finely chopped
Dash Tabasco sauce
¼ teaspoon salt
¼ teaspoon freshly ground black pepper
4 Kirby cucumbers, sliced
2 plum tomatoes, sliced
6 ounces roasted tortilla chips

*H*alve avocados. Hold half with seed in one hand and strike seed with knife edge to embed blade. Twist knife to remove seed. Place 2 lettuce leaves on each of 2 plates. Place 2 avocado halves on each lettuce bed. Mix tomato sauce, onion, garlic, parsley, red and jalapeño peppers, Tabasco, salt, and pepper. Place in avocado holes. Dress each plate with sliced cukes, tomatoes, and mound of chips.

Great Guacamole

4 lettuce leaves
4 Kirby cucumbers, sliced
2 plum tomatoes, halved
6 ounces roasted tortilla chips
2 avocados, halved
4 red radishes, coarsely chopped
1 small jalapeño pepper, finely chopped
1 small red onion, finely chopped
4 cloves garlic, crushed
4 sprigs parsley or coriander, finely chopped
½ cup fresh tomato sauce
½ teaspoon salt
½ teaspoon freshly ground black pepper
Dash Tabasco sauce

Arrange 2 plates with lettuce leaves, cukes, tomatoes, and chips. Scoop out avocado halves into bowl (save shells). Mash avocado and mix in remaining ingredients. Replace mixture in shells. Place on plates.

AVOCADO-TOMATO SALAD

1 large or 2 medium avocados
2 tomatoes
1 small Spanish onion, thinly sliced
1 clove garlic, chopped
½ teaspoon salt
½ teaspoon freshly ground black pepper
Juice of 2 limes
¼ cup olive oil

Halve avocados. Remove seeds and cut into wedges. Place in bowl. Cut tomatoes into wedges. Mix with avocados, onion slices, garlic, salt, and pepper. Sprinkle with lime juice and olive oil. Serve with hot bread or toasted tortillas. *Serves 4.*

AVOCADO-CHEESE TOSTADAS

2 large avocados
2 cloves garlic, chopped
½ small Spanish or white onion, chopped
2 tablespoons chopped red or green pepper
½ teaspoon salt
¼ teaspoon freshly ground black pepper
10–12 tortillas
6 ounces cheddar, Monterey jack, Swiss,
   or America cheese, grated
Alfalfa sprouts

Halve avocados and scoop out meat. Mash in bowl. Combine garlic, onion, red pepper, salt, and pepper. Toast tortillas and spread with mashed avocado. Sprinkle garlic mixture on avocado. Press in with fork. Sprinkle with grated cheese and place under broiler until cheese melts and begins to bubble. Be careful not to burn. Remove and top with alfalfa sprouts. Salad dressing may be added to taste. *Serves 3–4.*

Avocado Stuff

2 avocados, halved
Seafood salad (made from 1 can tuna fish, crab meat, or lobster) or
egg salad (made from 2 hard-boiled eggs)
Lettuce
Caviar (if egg salad is used)
Crackers or chips

Stuff avocado halves with salad. Make lettuce beds on 2 plates. Place 2 avocado halves on each plate. Garnish egg salad with caviar. Serve with crackers.
Serves 4.

Avocado Italiano

2 large avocados, halved
4 lettuce leaves
2 tomatoes, sliced
4 scallions
4–6 cucumber sticks
1 small onion, chopped
4 cloves garlic, chopped
8 sprigs flat-leaf parsley, chopped
12–14 basil leaves, chopped
¾ cup olive oil
½ teaspoon salt
½ teaspoon freshly ground black pepper

Arrange each halved avocado on plate of lettuce, tomatoes, scallions, and cucumber. Combine remaining ingredients in blender. Blend until smooth. Pour into avocado holes. Serve with crusty Italian bread. *Serves 4.*

EXOTIC FRUIT

1

1. PERSIMMONS
2. POMEGRANATES
3. BREBA FIGS
4. WHITE KADOTA FIGS
5. BLACK MISSION FIGS
6. PASSION FRUIT
7. KUMQUATS
8. PRICKLY PEAR
9. KIWI FRUIT

7

4

5

*I*s the pomegranate really the apple Eve gave Adam? Was the persimmon the world's first fruit? Do kumquats only come candied in Chinese restaurants, or do they have a life of their own? Was the passionfruit the fruit of Christ? And are prickly pears so precious they protect themselves with the thorns of the cactus they grow upon? The answer to these and other questions are lost in time and in legend, but the exotic fruits with strange shapes and fragile temperaments are here, more than ever before, for all of us to enjoy, adding their singular qualities to our daily lives. We all shun the unfamiliar, especially in our food habits. We must realize, however, that there must be good reasons why these exotic fruits, these mysterious fruits, have survived and remained with us for centuries and centuries. It is because they are great. Each exotic fruit is totally unique in its form, its texture, and its taste.

It is estimated that a new food takes two or three hundred years to be accepted after it is first introduced. That may have been true centuries ago when superstition, fear, and ignorance kept a safe distance between the old ways and the new thing. Today the rate of introductions of new foods, especially fruit, and their acceptance is increasing at an astonishing rate. New foods like the kiwi come complete with history, nutritional data, recipes, endorsements, and mouthwatering photographs of how good they look to eat. Most of the fruits we call exotic were introduced long ago but are still slowly working their way from the frontier of fresh food to the mainstream of American fruit favorites. They have never had the benefit—as have the kiwi, star fruit, banana, and broccoli—of a major agribusiness promotion. Most exotic fruits have had to make it on their own, relying on us to learn about their unusual shapes, colors, tastes, and textures by, if you will allow the pun, word of mouth.

This book will attempt to move the exotic-fruit ball along. Everyone should eat figs. They are healthy, digestible, fun, and inexpensive. Pomegranates are both mysterious and delightful. Children love the whole idea of all those little fruits in that shiny 'apple.' Kumquats are one of the world's most delightful after-dinner fruits. Their sharp sweet citrus taste is a refreshing way to end any meal. I think slices of kiwi are one of the prettiest things you can eat. Persimmons are the most sensuous. (A mango and a persimmon back to back by candlelight are probably more than most of us can handle.) And prickly pears are the most bothersome. People who enjoy prickly pears, however, soon forget the effort they require while their pleasure lingers on.

Most fruit stands and hardly any supermarkets carry consistently ripe top-quality exotic fruit. Kiwis, once again, have made it to the chrome-and-glass marketplace, even though they suffer badly from self-service and hand-to-hand combat. Other exotic fruits make brief appearances from time to time. The Los Angeles Farmers' Market does a wonderful job, as do many of the specialty fruit stands in major cities, like Balducci's and Dean & DeLuca in Manhattan. Sunfrost Farms in Woodstock is the fruit stand I know best. It always features several exotic fruits, and by samples, salesmanship, and display we have made exotic fruits important items in our store. Our customers have expanded their awareness and enjoyment of these exotic jewels in the crown of food called fresh fruits and vegetables. In the meantime, sensuous as they may be, please don't squeeze the persimmons.

PERSIMMONS

The persimmon is the national fruit of Japan. It is probably, however, Chinese in origin. At any rate its deep orange-red color, graceful leaf cap, and basic simplicity definitely give the persimmon an Oriental personality. Most varietal persimmons bear Oriental names such as Fuji, Hatia, Hachiya, and Mandarin. Persimmons have an unfair reputation for a mouth-puckering astringent taste, which keeps people from really appreciating the persimmon for the elegant, sensuous fruit it is. The California growers have done an excellent job in developing hardy, top-grade delicious fruit as good as that cultivated in the Orient for three thousand years. All commercial persimmons are shipped from California in two varieties with several significant differences. One variety is the American persimmon, the other the Japanese (Oriental) persimmon.

The American persimmon is fertilized artificially. The male pollen is applied mechanically to the blossoms, and the resultant fruit grows without seeds. Since it has no seeds, it cannot reach full ripeness on the tree and must be picked before cold weather kills it on the tree. The American persimmon will turn bright orange before it is ripe, and it becomes its sweet, soft, juicy, succulent self only when the skin becomes very thin and the fruit soft. The red-orange color never darkens or wrinkles. American persimmons are sold under the names Hachiya and Hatia.

Japanese persimmons have seeds. They are pollinated the old-fashioned way, by bees and insects. The Japanese variety stays on the tree much longer than the American, even longer than the tree's own leaves. The deep-gold fruit hangs from leafless branches well past the frost until almost fully ripe. Japanese persimmons can fall to the ground and continue to ripen if the fall doesn't kill them. They are picked and shipped during the last days of ripening. Japanese persimmons turn very dark and develop wrinkled skins when they are ready to eat. They are not as beautiful as the American artificially inseminated variety, and they have four softish seeds. But they are a touch sweeter, smoother, and muskier. California Japanese varieties are sold by the names Fuji and Mandarin.

All persimmons grow on trees of the ebony family, and all persimmons are the last soft-skinned fruits to be harvested. They all taste the same, but every writer describes their taste differently. I think the persimmon has the texture of a very ripe warm mango, the taste of apricot-papaya custard, and the sweetness of a tree sap like maple syrup. Persimmons are highly sensuous. I like making productions out of eating beautiful food. May I suggest quietly peeling two warm, ripe persimmons, chilling some soft white wine, and enjoying both with someone you love.

Persimmons are available at fine fruit and vegetable markets. They are also found in supermarkets, where they take a terrible beating. They are found only from October to January. Select firm bright-orange fruit, and ripen them carefully at home. If you must touch them to determine ripeness, please be gentle, and please don't squeeze.

Persimmons have very high nutritional numbers, no less than one would expect for a fruit that could be five or six thousand years old.

NUTRITIONAL DATA

Weight (g)	Calories	Protein (g)	Carbohydrates (g)	Crude Fiber (g)	Water (g)	Fat (g)	Cholesterol (mg)
25	32	0.2	8.4	0.4	16.1	0.1	—
168	118	1	31.2	2.5	134.9	0.3	—

Vitamin A (IU)	Vitamin C (mg)	Thiamine B-1 (mg)	Riboflavin B-2 (mg)	Niacin (mg)	Vitamin B-6 (mg)	Vitamin B-12 (mcg)	Folic Acid (mcg)	Pantothenic Acid (mg)
—	17	—	—	—	—	—	—	—
3640	13	0.05	0.03	0.2	—	—	13	—

Sodium (mg)	Potassium (mg)	Calcium (mg)	Phosphorus (mg)	Magnesium (mg)	Iron (mg)	Zinc (mg)	Copper (mg)	Manganese (mg)
—	78	7	7	—	0.63	—	—	—
3	270	13	28	15	0.26	0.18	0.19	0.596

American—1 medium
Japanese—1 medium

POMEGRANATES

*E*ver since man has recorded his thoughts, the pomegranate has been a central character in myth and legend. Solomon walked his pomegranate groves and composed songs to the virtues of this complicated ruby fruit. Moses led his people into the land of wheat and barley, vines and fig trees, and pomegranates. Aphrodite planted pomegranates to add to her bag of lover's tricks. Zeus created pomegranates from the blood of the murdered Agdistis. Persephone, goddess of spring, ate six pomegranate seeds while in the Underworld and created six months of earth life and six months of winterkill. (The story varies geographically. Californians say she ate only one or two seeds. Eskimos say she ate the whole thing.) The nightingale sang Romeo's love song to Juliet from the branches of the pomegranate tree. Some women of Athens ate pomegranates to bear many children, while others ate them to not have children. Adam and Eve ate this fruit of the forbidden tree and changed Paradise into nine to five, car payments, and atom bombs. Whatever its power, the pomegranate is unique, a singular fruit with a botanical classification all its own: *Punica granatum,* or as it is more affectionately and delightfully known, the Chinese apple.

The pomegranate is an exotic fruit most popular in the hot dry desert climates of the Mideast, where it is heavily cultivated. All commercial pomegranates sold in the United States come from California. They are bright red, with a leathery skin and a small crown at the blossom end with the dried remains of the pomegranate flower. Inside the skin are hundreds of brilliant ruby-red kernels of semisweet almost-transparent fruit wrapped around a small white seed. The pomegranate is actually hundreds of tiny whole fruits. These fruits are separated into cluster cells by a yellow-white membrane. Do not eat the membrane. It tastes awful. Eat the little fruits. They taste great. They are slightly acid or tart but are wonderfully refreshing and leave a clean, brisk feeling in your mouth and throat.

Preparing a pomegranate requires patience. Some people cut it apart, spritzing indelible sweet red juice everywhere. The wedges are picked apart, peel from membrane, membrane from fruit. It's fun. Children love it. But sloppy pomegranate work means more work for Mother, and the juice stains don't come out. I recommend one person carefully separate the peel and membrane from the fruit and put all the kernels in a bowl. These fruity kernels love a splash of lime or orange juice, a touch of sugar, or a light pouring of cream and can be eaten with a spoon easily, neatly, and most enjoyably.

Pomegranates are hardy souls, having survived Hell, the Jews' journey out of Egypt, and the serpent himself. Their leathery skins protect the more delicate fruits within. They must, however, be refrigerated. Warm dry air makes them woody and dry. Select only bright-red pomegranates. If they appear flat or squarish, it is okay. They have a tendency to be naturally six-sided, which is weird for a round fruit. Pomegranates are available from September through December. They are colorful, delightful fruit and perfect for fall and holiday feasting. The kernels may also be sprinkled on green and fruit salads and added to blender drinks.

Pomegranates are eaten chiefly as a pleasure food, but they have considerable amounts of minerals, especially potassium, and some B vitamins.

NUTRITIONAL DATA

Weight (g)	Calories	Protein (g)	Carbohydrates (g)	Crude Fiber (g)	Water (g)	Fat (g)	Cholesterol (mg)
154	104	1.5	26.4	0.3	124.7	0.5	—

Vitamin A (IU)	Vitamin C (mg)	Thiamine B-1 (mg)	Riboflavin B-2 (mg)	Niacin (mg)	Vitamin B-6 (mg)	Vitamin B-12 (mcg)	Folic Acid (mcg)	Pantothenic Acid (mg)
—	9	0.05	0.05	0.5	0.16	—	—	0.92

Sodium (mg)	Potassium (mg)	Calcium (mg)	Phosphorus (mg)	Magnesium (mg)	Iron (mg)	Zinc (mg)	Copper (mg)	Manganese (mg)
5	339	5	12	—	0.46	—	—	—

Raw—1 medium

FIGS

*F*igs are actually thousands of tiny fruits held within a thick, soft sack. It is the same packaging concept found in the pomegranate. The tiny fruits, each with a stem and a seed surrounded by sweet flesh, are always pink or crimson. The skin color can be white, green, purple, red, or any combination. Figs are round, flat, oval, wine-baggish, or very long. The three most popular fresh figs are the sacklike Breba, the flat green Kadota, and the round black Mission. Most figs are bigger than golf balls but not much. All figs have a navel at the bottom, or blossom end, that at the peak of ripeness secretes a sweet sap through a tiny opening. This fig sap is a sure sign that the fruit is sweet, succulent, and ready to eat. It also means the fig is in a highly delicate, fragile, bruiseable state, and once damaged, figs ferment and degenerate quickly. Gently is the only way to handle figs. At Sunfrost we cradle them in waxed paper and wrap them snugly in small paper bags.

Figs grow extensively from India westward to Europe, with heavy cultivation in Iran, Turkey, Greece, and Sicily. Some accounts report that the Greeks sent the fig eastward to Egypt and beyond. Others argue that figs travelled the other way. My first fig came from Sicily and grew in my grandfather's backyard. I remember my very young summers, climbing the peach tree that grew next to the fig tree (you had better not get caught in the fig tree) and watching the figs grow bigger and bigger until they turned dark purple, almost black. Grandma always got the harvest, but not before I managed to pluck a few off. I ate them slowly, trying to taste every one of the little fruits inside. They were wonderful. Spending time with the figs, the peaches, the tomatoes, and the grapes in my family's garden was an experience that has remained one of the sweetest memories of my life's many. But it was only years later that I could really appreciate how sacred a place the garden must have been for my immigrant grandparents. They came to America from the fruit-rich world of Sicily and the Mediterranean to the factory city of Newark, New Jersey. Their garden was their homeland, and every fall my grandfather Bartolo would drag out the old carpets, linoleum, blankets, and rope from the cellar. He'd carefully tie up all the fig-tree branches and wrap and tie the whole thing, root to tip, in layers against the winter. He always finished by putting his five-gallon garden pail on the top. You could always tell an Italian house. They all had these carpeted and roped things in their yards, with a pail on top. In our house only Grandpa touched the fig tree. But every spring when he decided winter was over, we'd all help him unwrap the tree and put the stuff away. What a joy. Spring. Summer. And figs again.

All American figs come from California. They are shipped in single-layer wooden or cardboard boxes, with each fig cradled in its own plastic nest. They are available early in the summer and late in the season. There are no bad figs, only overripe ones or damaged ones, which can be stewed, cooked, or used in baking. I prefer fresh figs and recommend you buy high-quality fruit that is intelligently displayed and carefully handled. Figs are never inexpensive or very dear; they are merely exotic and deserve their price and your purchase. Look for them at great fruit stands and Italian neighborhood stores, never in supermarkets.

Fresh figs are very high in mineral content. The dried version, by a mysterious act of God and science, contains four to five times the nutrition of the fresh fig and none of the succulence. I say eat the fresh figs in season for their downright sensuous fruit pleasure, and buy many packages of imported dried Greek string figs in October for delicious high nutrition during the winter months. Come spring, find Grandpa.

NUTRITIONAL DATA

Weight (g)	Calories	Protein (g)	Carbohydrates (g)	Crude Fiber (g)	Water (g)	Fat (g)	Cholesterol (mg)
50	37	0.4	9.6	0.6	39.6	0.2	—

Vitamin A (IU)	Vitamin C (mg)	Thiamine B-1 (mg)	Riboflavin B-2 (mg)	Niacin (mg)	Vitamin B-6 (mg)	Vitamin B-12 (mcg)	Folic Acid (mcg)	Pantothenic Acid (mg)
71	1	0.03	0.03	0.2	0.06	—	—	0.15

Sodium (mg)	Potassium (mg)	Calcium (mg)	Phosphorus (mg)	Magnesium (mg)	Iron (mg)	Zinc (mg)	Copper (mg)	Manganese (mg)
1	116	18	7	8	0.18	0.07	0.035	0.064

Raw—1 medium

PASSIONFRUIT

The passionfruit is tropical in its genealogy and is definitely exotic in its personality. It is globular, the size of a small tomato, a deep rich purple when ripe, with little fragrance. The flesh is also deep purple, with many small soft seeds that are inseparable from the fruit. Passionfruit can be halved and eaten out of its shell, juiced, or pressed for fruit punch, cocktails, purées, and sauces. Pressed and strained, it also makes unforgettable ice cream or passion mousse.

Passionfruit is only sweet, juicy, and exotic in its ripe, very dark-skinned state. Fruit that is eaten prematurely has a white milk just inside the skin that ruins your appetite for more passionfruit. Ripe, however, the passionfruit is refreshing and sweet, with a unique fruit texture.

I must confess, however, that the passionfruit's chief claim to fame is the flower. The passionflower could very well have created the term exotic. Legend and imagination have related its form to the Passion of Christ. The three styles represent the three nails used to crucify Christ; the ovary is the vinegar sponge that touched his lips; the five stamens represent the wounds in his hands, feet, and side; the crown, which is an impressive radiance of dark purple, white, and soft lavender, is the crown of thorns; and the Apostles are represented by the petals and sepals. It is one of the world's most impressive fruit buds, a remarkable piece of botanical handiwork its fruit has yet to live up to.

Passionfruit is sold only in tropical fruit markets and specialty stores. Look for it in all resort towns from Mexico to Peru. Passionfruit has very little nutritional value.

NUTRITIONAL DATA

Weight (g)	Calories	Protein (g)	Carbohydrates (g)	Crude Fiber (g)	Water (g)	Fat (g)	Cholesterol (mg)
18	18	0.4	4.2	2	13.1	0.1	—

Vitamin A (IU)	Vitamin C (mg)	Thiamine B-1 (mg)	Riboflavin B-2 (mg)	Niacin (mg)	Vitamin B-6 (mg)	Vitamin B-12 (mcg)	Folic Acid (mcg)	Pantothenic Acid (mg)
126	5	—	0.02	0.3	—	—	—	—

Sodium (mg)	Potassium (mg)	Calcium (mg)	Phosphorus (mg)	Magnesium (mg)	Iron (mg)	Zinc (mg)	Copper (mg)	Manganese (mg)
5	63	2	12	5	0.29	—	—	—

Purple—1 medium

KUMQUATS

Kumquats are usually served cold, after they've been simmered in sugar, as a dessert in Chinese restaurants. Kumquats are also a wonderfully refreshing, delightful tart-sweet fruit. I say tart-sweet because kumquats have a sweet rind and tart flesh. They are the opposite of oranges, which they look like in miniature. Kumquats are bright orange in color, with the same kind of peel texture as that of citrus fruits. They are technically not citrus fruits, but so far have remained classified with their larger look-alikes. Japanese-variety kumquats are round and tart. The oval Chinese variety is the only variety shipped in its fresh state.

Florida grows kumquats and ships them from early December until May. They are eaten whole; sometimes the pits are eaten, sometimes not. Kumquats are shipped in quart containers with their bright shiny green leaves. Fresh leaves are a sign of fresh fruit. Kumquats are the only fruit I know that arrives with its foliage. I think kumquats are an excellent winter fruit. They can be enjoyed fresh, simmered in sugar, brandied, pickled, made into a sauce, or used whole as excellent companions to roast fowl, especially duck, and pork. Kumquats are sold at Southern citrus stands and high-quality fruit markets. Some supermarkets carry kumquats around Christmastime.

Kumquats are nutritionally fair, with goodly amounts of vitamin C. It's the least you'd expect from a tiny orange.

NUTRITIONAL DATA

Weight (g)	Calories	Protein (g)	Carbohydrates (g)	Crude Fiber (g)	Water (g)	Fat (g)	Cholesterol (mg)
19	12	0.2	3.1	0.7	15.5	trace	—

Vitamin A (IU)	Vitamin C (mg)	Thiamine B-1 (mg)	Riboflavin B-2 (mg)	Niacin (mg)	Vitamin B-6 (mg)	Vitamin B-12 (mcg)	Folic Acid (mcg)	Pantothenic Acid (mg)
57	7	0.02	0.02		—	—	—	—

Sodium (mg)	Potassium (mg)	Calcium (mg)	Phosphorus (mg)	Magnesium (mg)	Iron (mg)	Zinc (mg)	Copper (mg)	Manganese (mg)
1	37	8	4	2	0.07	0.02	0.02	0.016

Raw—1 medium

PRICKLY PEARS

The prickly pear, or Indian fig, is the fruit of a cactus. It bears no resemblance to either a pear or a fig, but it is prickly, and Indians from Mexican desert lands cultivated it widely and made it one of their important food crops. The Spaniards took it back to Europe; it made a new home in Sicily, where it is cultivated and highly enjoyed. It is also a favorite fruit in Israel and is a successful commercial crop in California.

All domestic prickly pears are gently processed to remove the glasslike thorns that dot the surface of the fruit. They never get them all. The fruit itself is red, purple, deep orange, and rose-pink, with matching skin. The flesh contains many small hard seeds, which I find bothersome and inedible. In fact, I find prickly pears as a fruit bothersome and inedible. They have excellent flavor, something like watermelon? Or kiwi? Or perhaps pineapple? They also are pretty fair nutritionally. I have a good friend who knows fruit and food and is an excellent cook and baker who buys them by the box in season and gorges all season long. As for me, any fruit that grows thorns on the outside and produces hard, bony seeds inside and grows on a plant with eight- to ten-inch spikes doesn't really want to be eaten. Besides, prickly pears arrive in September, and I'd rather eat more-willing fruit like apples, figs, raspberries, kiwis, and grapefruit. That way nobody gets hurt. If you think prickly pears have something in store for you, look in really fine fruit and vegetable markets like Sunfrost and Italian neighborhood stores, but never in supermarkets.

When eating a prickly pear, it is best to use a knife and fork. The fruit should be cut lengthwise into wedges and the fruit sliced away from the skin. Seeds may be chewed or discarded, but the fruit itself is juicy, sweet, and pulpy. They are excellent with a splash of lime and make a wonderful addition to fruit or green-leaf salads. Picked green for shipping, most prickly pears require several days' ripening and will turn various shades of red and crimson.

Prickly pears have a high mineral content, especially calcium, magnesium, and phosphorus.

NUTRITIONAL DATA

Weight (g)	Calories	Protein (g)	Carbohydrates (g)	Crude Fiber (g)	Water (g)	Fat (g)	Cholesterol (mg)
103	42	0.8	9.9	1.9	90.2	0.5	—

Vitamin A (IU)	Vitamin C (mg)	Thiamine B-1 (mg)	Riboflavin B-2 (mg)	Niacin (mg)	Vitamin B-6 (mg)	Vitamin B-12 (mcg)	Folic Acid (mcg)	Pantothenic Acid (mg)
53	14	0.01	0.06	0.5	—	—	—	—

Sodium (mg)	Potassium (mg)	Calcium (mg)	Phosphorus (mg)	Magnesium (mg)	Iron (mg)	Zinc (mg)	Copper (mg)	Manganese (mg)
6	226	58	25	88	0.31	—	—	—

Raw—1 medium

KIWI FRUIT

Originally known as the Chinese gooseberry, this fruit of ancient Chinese origin is now cultivated heavily in New Zealand and to a lesser degree in California. The fruit is covered by a tough brown hairy skin. The flesh is lime green with a distinct bright center burst and a cluster of small black edible seeds. The kiwi flavor is tart-sweet and slightly crisp-textured, something like strawberry-flavored watermelon. Kiwis can be halved and squeezed out of their skins like Italian ices in a paper cup or peeled and sliced and served in salads, as a garnish, or puréed for topping. Kiwi makes an incredibly efficient meat tenderizer and has impressive medicinal qualities in addition to its delightful, refreshing taste. Before ripening, kiwis can last under refrigeration for three to four months and will then ripen at room temperature in three to four days. Because of their intense cultivation, easy handling, and storage, kiwis are available almost everywhere in the United States all year round. Look for them at almost every kind of fruit stand and supermarket produce department.

Kiwi fruit contains considerable amounts of protein, iron, calcium, and phosphorus salts, and as much vitamin C as is found in two lemons.

NUTRITIONAL DATA

Weight (g)	Calories	Protein (g)	Carbohydrates (g)	Crude Fiber (g)	Water (g)	Fat (g)	Cholesterol (mg)
76	46	0.8	11.3	0.8	63.1	0.3	—

Vitamin A (IU)	Vitamin C (mg)	Thiamine B-1 (mg)	Riboflavin B-2 (mg)	Niacin (mg)	Vitamin B-6 (mg)	Vitamin B-12 (mcg)	Folic Acid (mcg)	Pantothenic Acid (mg)
133	75	0.02	0.04	0.4	—	—	—	—

Sodium (mg)	Potassium (mg)	Calcium (mg)	Phosphorus (mg)	Magnesium (mg)	Iron (mg)	Zinc (mg)	Copper (mg)	Manganese (mg)
4	252	20	31	23	0.31	—	—	—

Raw—1 medium

JUICES AND SHAKES

*T*here is no place in the world of fresh fruit and vegetables where your imagination can have as much fun and create such healthy pleasures as in fresh juices and fruit shakes. Endless variations of vegetable-juice combinations extracted through a proper machine provide instant energy, delightful refreshment, and long-term health. The ordinary kitchen blender can turn a few pieces of fruit, some yogurt or orange juice, and ice into a thick, healthy, refreshing liquid meal. A handful of fresh spinach, parsley, three carrots, and an Acme juicer will produce a high-powered delicious vegetable juice that is *healthier than most meals.* It will lift your spirit and your energy. Your body and your taste buds will love it. With minimum investment in the proper and necessary appliances you can satisfy your desire for food pleasure and optimum health. Go for it!

VEGETABLE JUICES

*M*ost vegetable-juice drinks begin with carrot juice, since the carrot produces most liquid volume for raw weight. Carrot juice mixes well with all vegetable juices and neither dominates other vegetable flavors nor loses its own sweet taste.

All vegetables qualify for the juicer; however, it stands to reason some are better than others. Carrots, beets, celery, parsley, spinach, string beans, comfrey, and cucumbers are the most popular. Cabbage, potatoes, onions, and turnips can also be juiced, but most people find them too strong or too intestinally active for regular consumption. Tomatoes can also be juiced but produce a thin, watery fluid that requires additional strained seeded tomato purée for body and appetite appeal. Apples can also be added to any vegetable combination for sweetness and nutrition.

The two most popular vegetable juicers are the Champion and the Acme Juicerator. There is a giant vegetable extractor built for commercial use by the Ruby Manufacturing Company in California. It is expensive but highly efficient and can be cleaned by pouring in a glass of water. Panasonic, Braun, and some small companies manufacture vegetable-juice extractors that work in various ways: a grating plate and centrifugal extraction; a grating plate with gravity drain; or forcing the ground pulp through a strainer. Some machines eject the pulp; others retain the pulp in the extraction basket and require manual cleaning. The Acme Juicerator has a grinding plate, separates the juice by centrifugal force, and requires manual pulp removal after three pounds of vegetables. It is a highly reliable appliance, well-designed, with a powerful motor and a hard steel grinding plate that stays sharp. I wish it had a bigger basket so it would need less changing, but for clean, smooth juice, it is the best model for home use.

CARROT JUICE

California carrots always have the most juice and the sweetest, smoothest taste. Some carrot-juice–lovers prefer juice only from organically grown chemical-free carrots, which is understandable. But unless the carrots are grown in highly alkaline soft soil with easy irrigation and absorption, the carrots will not compare in sweetness or flavor with those grown commercially in California. It is surprising that even fresh-picked homegrown carrots fall short of the sweet California essence. It must be remembered that the carrot is a root and that it is the only edible part of the plant. Whatever is going on in the dynamics of the soil will present itself in the taste and texture of the root itself. California carrots are very low in any chemicals and are more easily digestible because of their softer fiber and higher juice content. I mean no offense to Michigan, Florida, Canada, or the millions of home gardeners, but I have juiced tons of carrots, and Californias are always the best.

Carrots should be soaked and washed, not scrubbed. Carrots too large for the extractor or juicer should be cut lengthwise. Once juiced, carrot juice had best be drunk immediately for the full nutritional benefit, natural energy, and best flavor.

VEGETABLE-JUICE
COMBINATIONS

*T*he following combinations indicate juice ratios in ounces and some specific benefits. A small piece of ginger or horseradish added to any of the vegetable combinations adds a nice spicy dash without the use of salt or pepper.

4 carrots, 3 string beans, 1 parsley	Quick energy/iron
4 carrots, 2 beets, 2 spinach	Iron/diuretic
4 carrots, 4 cantaloupes	Sweet energy
4 carrots, 2 celery, 2 beets	Kidney cleansing/ iron
4 carrots, 2 string beans, 2 Jerusalem artichokes	Adrenalin unlimited
4 carrots, 2 cabbage, 2 potatoes	Intestinal fortitude
4 carrots, 2 celery, 2 cucumbers	C-c-c-cool
2 celery, 2 spinach, 2 comfrey leaf, 2 pineapple	Body cleanser/ healer
2 carrots, 2 spinach, 2 celery, 2 cucumber/radish	Liquid salad/ digestive aid
4 carrots, 4 oranges	Proper balance

CITRUS JUICES

*O*ranges and grapefruits have their peak seasons when their juice is sweetest and most plentiful. They are also least expensive when they are at their best. Different orange and grapefruit varieties should be selected, and different combinations should be used at different times of the year for the most juice and flavor.

I recommend a good citrus juicer. Hand juicers are slow and clumsy to clean. There are large restaurant hand juicers that are designed with high leverage action, and I have seen them operated with incredible efficiency and speed. Most small-appliance manufacturers market a home electric citrus juicer. They are fast and efficient, but most of them have weak motors and with regular family use don't seem to last. Sunkist manufactures a great orange and grapefruit juicer that is not inexpensive but is a lifetime appliance. The Sunkist machine I use at home was built in 1928, has never been repaired, and has its original ceramic top globe. All used-restaurant-supply stores sell citrus juicers and almost always have a used Sunkist unit for around $150. New ones are $350. There are other units built like the Sunkist machine which cost around $200.

A good juicer removes the juice in seconds. Almost all have a vibrating strainer plate that removes the heavy pulp and seeds. I like the pulp and use a hand strainer that lets most of the citrus cells through and holds back only the seeds and webbing.

I say make the investment. It will pay off in better nutrition; more delicious drinks; and pure juice for fruit ades, cocktails, punches, and refrigerator drinks for the whole family to enjoy instead of sugary canned junk, soda, thawed juice, and artificially flavored water.

NUTRITIONAL DATA (Orange Juice)

Weight (g)	Calories	Protein (g)	Carbohydrates (g)	Crude Fiber (g)	Water (g)	Fat (g)	Cholesterol (mg)
248	111	1.7	25.8	0.3	219	0.5	—

Vitamin A (IU)	Vitamin C (mg)	Thiamine B-1 (mg)	Riboflavin B-2 (mg)	Niacin (mg)	Vitamin B-6 (mg)	Vitamin B-12 (mcg)	Folic Acid (mcg)	Pantothenic Acid (mg)
496	124	0.22	0.07	0.1	0.01	—	—	0.47

Sodium (mg)	Potassium (mg)	Calcium (mg)	Phosphorus (mg)	Magnesium (mg)	Iron (mg)	Zinc (mg)	Copper (mg)	Manganese (mg)
2	496	27	42	27	0.5	0.13	0.109	0.035

Fresh—1 cup

NUTRITIONAL DATA *(Grapefruit Juice)*

Weight (g)	Calories	Protein (g)	Carbohydrates (g)	Crude Fiber (g)	Water (g)	Fat (g)	Cholesterol (mg)
247	96	1.2	22.7	—	222.3	0.3	—

Vitamin A (IU)	Vitamin C (mg)	Thiamine B-1 (mg)	Riboflavin B-2 (mg)	Niacin (mg)	Vitamin B-6 (mg)	Vitamin B-12 (mcg)	Folic Acid (mcg)	Pantothenic Acid (mg)
*	94	0.1	0.05	0.5	—	—	—	—

Sodium (mg)	Potassium (mg)	Calcium (mg)	Phosphorus (mg)	Magnesium (mg)	Iron (mg)	Zinc (mg)	Copper (mg)	Manganese (mg)
2	400	22	37	30	0.49	0.13	0.082	0.049

Fresh—1 cup
*White grapefruit juice, 25 IU; pink or red grapefruit juice, 1,087 IU.

ORANGE-JUICE COMBINATIONS

*T*he following charts are keyed to seasonal citrus peaks. Oranges and grapefruits taste different and yield different amounts of juice at different times of the year. If you buy citrus varieties at their peak, you will always get the taste you like best and get the most juice for the least money.

March through May	⅔ Florida Valencias, ⅓ California Valencias
June and July	Florida Valencias
August	½ Florida Valencias, ½ California Valencias
September through November	California Valencias
December	⅔ California Valencias, ⅓ Florida Valencias
December through February	Any combination of non-navel oranges: California Valencias, Florida Valencias, Temples, tangelos, Honeybelles

GRAPEFRUIT-JUICE COMBINATIONS

*G*rapefruit combinations are best mixed to individual taste for tartness and sweetness.

March through May	Large or small Florida Goldens, sweet; small Florida Marsh Rubies, very sweet
June through August	Small Florida Goldens, sweet
September and October	Large California pink meats, semisweet; large or small Florida pink meats, tart;
November through February	Florida pink meats, sweet; large Florida Goldens, semisweet

FRUIT SHAKES

All combinations of fresh fruits make drinks, but more kinds of fruit in the mix is not necessarily better. Too many is usually too much. Except in special combinations, more than three different fruits tend to lose their distinct flavors and homogenize into a bland, sweet, pulpy taste. The three basic liquids for fruit shakes are dairy (yogurt, milk, ice cream), orange juice, and apple juice. Each liquid ingredient creates a different effect.

All fruit qualifies, even bruised or spotted fruit. Bananas in the flecked state are not only flavorful but act as a natural sweetener and give your drinks body and texture. At the Sunfrost Juice Bar we use tupelo honey or maple syrup to satisfy the desires of our customers whose taste buds need the sweet kick.

Use fruits in season and only in their ripe, juicy state. Some fruits are drier than others and require more liquid juice. Melons make their own juice. Berries need liquid; so do bananas, peaches, and mangos. Pineapples can be processed in a juice extractor for juice; pineapple chunks can be blended into fruit shakes, but they need additional liquid. There are only two rules: ripe fruit and your imagination. There are no bad fruit shakes. The fruit you shake is the fruit you drink.

Yogurt	Creates thicker texture, tart (rather than sweet) flavor
Milk	Creates medium-thick texture, medium-sweet flavor
Ice cream	Creates thick creamy texture, sweet flavor
Orange juice	Creates medium-thick texture, refreshing sweet flavor
Apple juice	Creates light texture, dark color, vey sweet flavor

WAKE UP, AMERICA

1 cup orange juice
1 egg
1 tablespoon protein powder (optional)

Splash maple syrup
2–3 ice cubes

Blend at medium speed for 30 seconds.

CREAMSICLE

½ cup orange juice
½ cup yogurt
Splash maple syrup
Dash vanilla
5–6 ice cubes

Blend at high speed for 10 seconds, then at low speed for 30 seconds.

ORANGE JULEP

6–8 mint leaves
1 tablespoon sugar
Ice cubes
½ lime
¾ cup orange juice
Mint sprig

Crush mint leaves in bottom of glass with sugar and droplets of water. Fill glass with ice. Squeeze lime over ice. Pour in orange juice. Serve with mint sprig and straw.

ORANGE SUNRISE

2–3 strawberries
2 tablespoons Simple Sugar Syrup (below)
Ice cubes
1 cup orange juice
1 slice lime

Mash strawberries and syrup in bottom of glass. Add ice cubes. Pour in orange juice. Serve with slice of lime.

Simple Sugar Syrup

Combine 2 cups sugar and 4 cups water in saucepan. Heat to boiling, stirring if necessary to help sugar dissolve. Cool.

Sunfrost Orangeade

2 cups orange juice
½ cup lime juice
½ cup Simple Sugar Syrup (above)
2 cups ice cubes
Mint sprigs

Place liquids and ice cubes in quart pitcher or container. Stir or shake. Serve with mint sprigs.

Golden Morning

Juice of 1 grapefruit
Juice of ½ lime
3–4 ice cubes

Blend until frothy. This is an excellent diet drink and delicious vitamin drink, good for low-cholesterol diets or as part of a cleansing or detoxification program.

Sunfrost Lemonade

1¼ cups lemon juice
¾ cup Simple Sugar Syrup (above)
1 lemon, sliced
Mint sprigs

Mix lemon juice, syrup, and 2 cups water. Place lemon slices in quart pitcher or container. Pour in lemonade. Chill. Serve over ice with mint sprigs.

OLD-FASHIONED STRAWBERRY LEMONADE

Make Sunfrost Lemonade (above), but purée 8 ripe strawberries with the water in blender before mixing.

LEMON COMFORT

Juice of 1 lemon
1 teaspoon honey
1 cup warm water

Mix all ingredients. Drink daily upon arising and before sleep for cleansing, better digestion, and liver maintenance.

LEMON-MINT ICED TEA

10 cups water
10 tea bags
4 tablespoons mint leaves, chopped
Juice of 8 lemons
1 recipe Simple Sugar Syrup (above)
1 lemon, sliced
Mint sprigs

Boil 10 cups water. Tie tea bags together. Remove water from heat. Add tea bags to water and steep for 20 to 30 minutes. Cool. Combine mint leaves with ¼ cup water in blender. Mix mint water with lemon juice, sugar syrup, and tea in pitcher or container. Add lemon slices. Chill. Pour into ice-filled glass with mint sprigs. Makes 16 8-ounce glasses.

STRAWBERRY-BANANA SMOOTHIE

6–8 strawberries
½ flecked ripe banana
½ cup apple cider, yogurt, or orange juice
Splash maple syrup
4–6 ice cubes

Blend at low speed for 20 seconds, then at high speed for 10 seconds.

TROPICAL STRAWBERRY COOLER

6–8 strawberries
½ mango, peeled
¼ flecked ripe banana
½ cup orange juice
Juice of ½ lime
5–6 ice cubes

Blend at low speed for 20 seconds, then at high speed for 10 seconds.

STRAWBERRY CREAM DREAM

8 strawberries
½ cup light cream, half-and-half, or vanilla ice cream
Splash maple syrup
5–6 ice cubes

Blend at low speed for 20 seconds, then at high speed for 10 seconds.

STRAWBERRY-PEACH SHAKE

6 strawberries
½ peach, peeled
½ cup orange juice or apple cider
1 teaspoon honey
5–6 ice cubes

Blend at low speed for 30 seconds, then at high speed for 10 seconds.

LIQUID BERRIES

6 strawberries
3 tablespoons blueberries
2 tablespoons raspberries, boysenberries, or blackberries
6 tablespoons apple cider or yogurt
1 teaspoon honey or maple syrup
5–6 ice cubes

Blend at low speed for 30 seconds, then at high speed for 10 seconds.

THE STRAWBERRY APPLE

> 10 strawberries
> ¾ cup apple cider
> 5–6 ice cubes

Blend at medium speed for 20 seconds.

STRAWBERRY-WATERMELON QUENCH

> 5–6 strawberries
> 1 cup seeded watermelon pieces
> Juice of ½ lime
> 6 mint leaves
> 3–4 ice cubes

Blend at low speed for 30 seconds, then at high speed for 10 seconds.

RASPBERRY SUPREME CREAM

> ½ cup raspberries
> ½ cup yogurt
> Splash maple syrup
> 5–6 ice cubes

Blend at low speed for 20 seconds, then at high speed for 10 seconds.

BLUEBERRY COW

> 6 tablespoons blueberries
> ½ cup yogurt
> ¼ cup half-and-half
> Splash maple syrup
> 5–6 ice cubes

Blend at low speed for 20 seconds, then at high speed for 10 seconds.

SANDIA

½ blender seeded watermelon pieces
Juice of 1 lime
¼ blender ice cubes

This recipe is the most refreshing of all summer-fruit blender drinks. Blend at high speed in short bursts several times, then at constant high speed for 20 seconds.

CANTALOUPE COOLER

1 cup cantaloupe pieces
¼ cup orange juice
Juice of ½ lime

Blend at medium speed for 30 seconds. Serve in ice-filled glass with straw.

BUDDHA'S MIST

¾ cup cantaloupe pieces
½ mango, peeled
⅓ Hawaiian papaya, peeled
Juice of 1 lime

Blend at low speed for 30 seconds. Fill glass with ice cubes. Pour in blended juice. Serve with straw and slice of lime.

MISTY GRAPE

1¼–1½ cups red or green seedless grapes, mashed
Juice of 1 lime
5–6 ice cubes
Lime slice

Blend at high speed until smooth. Pour over crushed ice or small ice cubes in old-fashioned glass. Serve with straw and slice of lime.

APRICOT EXOTICA

> 3–4 apricots
> ½ cup light cream, half-and-half, or yogurt
> ¼ cup orange juice
> 1 teaspoon honey or maple syrup
> 5–6 ice cubes

Blend at low speed for 30 seconds, then at high speed for 10 seconds.

PEACHES 'N' CREAM SMOOTHIE

> 1½ peaches, peeled
> ½ cup light cream or half-and-half
> 2 tablespoons orange juice
> 1 teaspoon honey
> 4–5 ice cubes

Blend at low speed for 30 seconds, then at high speed for 10 seconds.

PEACHES-AND-ANYTHING SMOOTHIE

> 1½ peaches, peeled
> ½ cup summer or tropical fruit such as banana, strawberries, pineapple, or blueberries (cut up if necessary)
> ½ cup orange juice, yogurt, or apple cider
> 1 teaspoon honey
> 5–6 ice cubes

Blend at low speed for 30 seconds, then at high speed for 10 seconds.

THICK BANANA SHAKE

> 1½ flecked ripe bananas
> ¾ cup yogurt, half-and-half, or light cream
> 1 teaspoon honey or maple syrup
> Dash vanilla
> 4–6 ice cubes

Blend at low speed for 20 seconds, then at high speed for 10 seconds.

Light Banana Shake

Make Thick Banana Shake (above), but substitute orange juice or apple cider for yogurt and eliminate vanilla.

Banana and Anything

> 1 flecked ripe banana
> ¾ cup fruit (anything goes), cut up if necessary
> ¾ cup orange juice or yogurt
> Juice of ½ lime
> 5–6 ice cubes

Blend at low speed for 30 seconds, then at high speed for 10 seconds.

Mango Crema

> ½ mango, peeled
> ½ flecked ripe banana
> 1 small piece pineapple
> ¼ cup light cream or half-and-half
> 2 tablespoons orange juice
> Splash maple syrup
> 4–5 ice cubes

Blend at low speed for 30 seconds, then at high speed for 10 seconds.

Papaya Maya

> ⅓ Hawaiian papaya, peeled, or ½ cup peeled tropical papaya pieces
> ½ flecked ripe banana
> 3–4 pieces pineapple
> ¼ cup orange juice
> Juice of ½ lime
> 1 teaspoon honey
> 5–6 ice cubes

Blend at low speed for 30 seconds, then at high speed for 10 seconds.

Papaya Pura

> 1 cup peeled papaya pieces
> Juice of 1 lime
> ½ cup water

Blend at medium speed for 20 seconds.

La Piña

> 1 cup pineapple pieces
> ½ cup orange juice (for light texture) or yogurt
>   (for creamy texture)
> ½ teaspoon maple syrup
> 5–6 ice cubes

Blend at medium speed for 20 seconds, then at high speed for 10 seconds.

Piña Colada Rica

> 6 pieces pineapple
> ½ flecked ripe banana
> 6 tablespoons light cream or half-and-half
> 3 tablespoons coconut cream
> 1 teaspoon shredded coconut
> ½ teaspoon maple syrup

Blend at low speed for 30 seconds, then at high speed for 10 seconds.

INDEX OF RECIPES AND MAJOR INGREDIENTS